J. M. Sanchez

CALCULATED ELECTRONIC PROPERTIES OF METALS

CALCULATED

ELECTRONIC PROPERTIES

OF METALS

by

V. L. Moruzzi
J. F. Janak
A. R. Williams

IBM Thomas J. Watson Research Center
Yorktown Heights, New York

PERGAMON PRESS INC.

New York/Toronto/Oxford/Sydney/Frankfurt/Paris

Pergamon Press Offices:

U.S.A. — Pergamon Press Inc., Maxwell House, Fairview Park, Elmsford, New York 10523, U.S.A.

U.K. — Pergamon Press Ltd., Headington Hill Hall, Oxford OX3OBW, England

CANADA — Pergamon of Canada, Ltd., 75 The East Mall, Toronto, Ontario M8Z5W3, Canada

AUSTRALIA — Pergamon Press (Aust) Pty. Ltd., 19a Boundary Street, Rushcutters Bay, N.S.W. 2011, Australia

FRANCE — Pergamon Press SARL, 24 rue des Ecoles, 75240 Paris, Cedex 05, France

WEST GERMANY — Pergamon Press GmbH, 6242 Kronberg/Taunus, Frankfurt-am-Main, West Germany

Library of Congress Cataloging in Publication Data

Moruzzi, Victor L., 1930-
Calculated electronic properties of metals.

Bibliography: p.
1. Atomic structure. 2. Metals--Electric properties. I. Janak, James F., joint author. II. Williams, Arthur R., 1941- joint author. III. Title. IV. Title: Electronic properties of metals.
QC173.M674 1978 546'.31 77-28496
ISBN 0-08-022705-8

This book was prepared using the interactive computer controlled text processing and formatting facilities and an experimental computer controlled photocomposer maintained at the IBM Thomas J. Watson Research Center.

Printed in the United States of America

PREFACE

In 1963 Herman and Skillman[1] published calculated electron densities, energy levels and wavefunctions for essentially all the atoms in the periodic table. This work has provided an extremely useful basis for theoretical calculations of many kinds and, perhaps more importantly, it provides a quantitative basis for thinking about the chemical trends in a vast array of chemical and physical problems. The Herman-Skillman calculations were approximate, but their self-consistent central-field local-exchange model of the atom was sufficiently realistic to generate a very useful picture of the interior of the atom. Furthermore, because their model is the same for all atoms, the variation of their results with atomic number is particularly realistic and useful.

The present work is similar to the Herman-Skillman work in many respects, certainly in spirit and to varying degrees in analytical technique. The aim of the present work is an understanding of condensed systems containing many atoms rather than individual isolated atoms. The calculations are somewhat more complicated and the resulting quantities of greater variety, but our objective is fundamentally the same, to provide an approximate but internally consistent picture of the electronic structure of (in our case) closely packed condensed systems and to show how this picture varies with the progressive filling of the atomic shells.

The calculations we present are based on three fundamental approximations, the local-density treatment of electronic exchange and correlation, the 'muffin-tin' approximation, which might be viewed as the analog of the atomic central field approximation, and the neglect of relativistic effects. These approximations limit the range of systems for which our calculations can be expected to be accurate to metals comprised of atoms possessing fewer than approximately 50 protons. There are however 32 such metallic elements and the number of systems comprised of mixtures of these 32 is, of course, infinite. Thus the relevance of our calculations, while not universal, is indeed broad.

CONTENTS

I. INTRODUCTION

The objective of this book is to provide detailed reference material describing the electronic structure of metals. By electronic structure we mean the distribution of electrons in space, momentum and energy. Many of the measurements performed on metals are interpretable in terms of these distributions. Furthermore, they form a basis for thinking about and interpreting the properties of liquid metals and combinations of metallic elements in both ordered and disordered phases. The relationship between these distributions and measured quantities is usually known only imperfectly and our ability to calculate them is also imperfect. Nonetheless, the large number of measurements which are understandable in terms of these distributions and the accuracy with which they can be calculated using modern theories and computational procedures together motivate their calculation and presentation. The theoretical framework we have used to calculate the distributions possesses two important attributes. First, it is sufficiently accurate to make the calculated quantities for each element quite realistic and second, the fact that this framework is completely independent of the system to which it is applied makes the trends in our calculated quantities with atomic number even more reliable. In addition to the distribution of electrons in space, momentum and energy, we also calculate and present several derivative quantities, including the total energy of the solid, the cohesive energy, the equilibrium nuclear separation, the bulk modulus, the difference between atomic and solid electron densities, and the exchange-correlation enhanced paramagnetic susceptibility.

Our calculations of these properties are based on three simplifying assumptions, the local-density theory of electronic exchange and correlation,[2,3] the muffin-tin approximation to the spatial form of the electron density,[4] and the neglect of relativistic effects. The muffin-tin approximation consists of assuming the electron density to be spherically symmetric inside each of a set of non-overlapping spheres centered on the nuclei, and constant in the region between the spheres. Numerous tests have shown this approximation to lead to negligible errors in the closely packed structures into which metals condense. Similarly, the importance of our neglect of relativistic effects is under control in the sense that we confine our attention to elements with atomic number less than 50, where relativistic effects are not large.

Unlike anisotropy and relativistic effects, we cannot avoid dealing with exchange and correlation by restricting the range of our study. To avoid exchange and correlation altogether would restrict us to atomic hydrogen. Fortunately there has been considerable progress over the past decade in the treatment of these complicated effects. The development of density-functional techniques[2,3,5,6] permits the study of inhomogeneous interacting electron systems such as atoms, molecules, and solids using the simple and computationally tractable techniques of independent particle dynamics. Density-functional theory puts the self-consistent-field techniques developed by Hartree,[7] Slater[8] and others on a rigorous theoretical footing. For the physics of metals perhaps the most important accomplishment of density-functional theory is that it permits exchange and correlation to be treated equivalently at various levels of approximation. In the calculations we present here, an approximate version of density-functional theory, the local-density approximation introduced by Kohn and Sham,[2] was used throughout.

As the discussion above indicates, our treatment of exchange and correlation is the only approximation we make which is not controllable. We therefore devote the next section to a description of density-functional theory, the local-density approximation our calculations employ and a review of the existing evidence indicating its accuracy. In the following Chapter, Chapter II, we describe the numerical techniques we use to carry out the self-consistent-field calculations.

1. TREATMENT OF EXCHANGE AND CORRELATION

The motion of different electrons in condensed systems is not independent; it is highly correlated. This fact would seem to make the *ab-initio* theoretical description of condensed systems completely intractable. The motion is ironically so highly correlated that the concept of a quasi-particle makes possible the description of an interacting electron gas in terms of independent particle techniques. The net result of these correlations is that following each electron as it moves through the system is a depression in the distribution of other electrons; the depression is called the exchange-correlation hole and the electron together with its hole constitute what we call a quasi-particle. It is the motion of such quasi-particles which is approximately independent and it is this approximate independence which is crucial in rendering the description of many-electron problems, such as metals, tractable. Density-functional theory provides a rigorous theoretical framework in which the quasi-particle concept can be exploited using self-consistent-field, independent-particle analysis. The calculations we present in this book employ an approximation within the otherwise exact density-functional framework. To make the content of our only important approximation clear we now review the formulation of density-functional theory, the local-density approximation we use within it and the existing evidence we have bearing on the accuracy of this approximation.

2. DENSITY FUNCTIONAL THEORY

The formal development of density-functional theory rests on two facts proven by Hohenberg and Kohn.[5] First, properties of the ground state (assumed nondegenerate) of an inhomogeneous interacting electron gas, such as that comprising an atom, molecule, or solid, are (with the exception of the trivial dependence on the external potential) system-independent functionals of the electron density. The second fact involves one of these functionals, the total energy $E\{\rho\}$. They showed that $E\{\rho\}$ takes on its minimum value for the true electron density. Kohn and Sham[2] exploited this minimum property, using it to find the electron density. Knowing that the Hartree self-consistent-field method is a procedure for minimizing a particular approximation to the total-energy functional, they wrote the exact total energy as the sum of the Hartree energy $E_H\{\rho\}$ plus whatever is left:

$$E\{\rho\} = E_H\{\rho\} + E_{xc}\{\rho\} \qquad 1.1$$

where "xc" stands for exchange and correlation, the effects ignored in the Hartree approximation. They then showed that minimization of $E\{\rho\}$ leads once again to the Hartree self-consistent-field equations

$$(-\nabla^2 + V(\vec{r}) - \varepsilon_i)\Psi_i(\vec{r}) = 0 \qquad 1.2$$

$$\rho(r) = \sum_i |\Psi_i(\vec{r})|^2\, \Theta(E_f - \varepsilon_i)\,. \qquad 1.3$$

As in the Hartree method, the orbitals used to construct the electron density are determined by single-particle equations containing a potential energy term which in turn depends on the electron density, so that finding the implicit solution requires the iteration of these equations until the electron density generated by the potential is identical to that used to construct the potential. The only difference in the single-particle equations resulting from the inclusion of $E_{xc}\{\rho\}$ in the total energy is an additional term in the effective single-particle potential $V(\vec{r})$:

$$V(r) = 2\int dr' \frac{\rho(\vec{r}\,') - \rho_+(\vec{r}\,')}{|\vec{r} - \vec{r}\,'|} + \frac{\delta E_{xc}\{\rho\}}{\delta\rho(\vec{r})}\,. \qquad 1.4$$

In these equations ρ is the electron density, $\rho_+(r)$ is the distribution of positive charge, $\Theta(x)$ is unity for $x>0$, and zero for $x<0$, E_f is the Fermi energy, and the ε_i are Lagrange multipliers introduced to fix the number of electrons as the orbitals constituting the electron density are varied. The ε_i play the role of effective single particle energies, but their identification as such requires additional arguments given later in this section.

Although this development is rigorously justified only for the ground state and although only rather crude approximations to the exchange-correlation energy functional are known at present, the significance of the Hohenberg-Kohn-Sham development is enormous, because it reduces the many-body problem to a set of one-body problems coupled only through the effective one-body potential V(r).

Approximations enter this development only because we do not know, and must therefore approximate, $E_{xc}\{\rho\}$. We have used in all the calculations the particularly intuitive approximation introduced by Kohn and Sham

$$E_{xc}\{\rho\} \approx \int d^3r\rho(\vec{r})\varepsilon_{xc}^{(h)}(\rho(\vec{r})) \tag{1.5}$$

where $\varepsilon_{xc}^{(h)}(x)$ is the contribution of exchange and correlation to the total energy (per electron) in a homogeneous, but interacting, electron gas of density x. The function $\varepsilon_{xc}^{(h)}(x)$ is itself known only approximately,[9] but it is known numerically to an accuracy which makes equation 1.5 the principal approximation of our work.[3]

Equations 1.2, 1.3, and 1.5 embody the notion that the energy of a given electron is lowered by the tendency of the other electrons to move out of its path. The quantitative implications of this tendency are very subtle and an attractive aspect of the approximation (1.5) is that it exploits our considerable understanding of these effects in the homogeneous interacting electron gas.[3] The local-density approximation (eq. 1.5) assumes that the contribution of exchange and correlation to the total energy can be accumulated additively from each portion of a non-uniform gas as if it were locally uniform. The approximation clearly improves as the system becomes more nearly uniform and cannot be formally justified for systems as inhomogeneous as atoms. The principal justification for its use in systems composed of real atoms is empirical. Before going on to a review of the empirical evidence that exists, we can offer the reader some insight into the qualitative types of situations in which this approximation might be expected to break down.

The fundamental assumptions of the local-density approximation are 1) that the exchange-correlation hole is centered on the electron, 2) that the electron density supporting the hole is electron-gas-like and approximately uniform on the scale of the hole itself. By electron-gas-like we mean possessing a continuous energy spectrum which permits many low energy excitations to form in order to screen the electronic charge.[10] This is clearly not the case in the atomic core and it is not therefore surprising that the local-density approximation systematically overestimates correlation and underestimates exchange in systems of this type. Fortunately atomic cores play a relatively passive role as atoms condense to form metals. As is discussed below and indicated by our numerical results, our calculations of cohesive energies (total energy per atom of the condensed phase minus the atomic total energy) systematically exploit cancellation of errors made in the atomic cores which are considerably larger than the small total-energy differences of interest.

Consider the assumption of the local-density theory that the electron density is approximately uniform over the range of the exchange-correlation hole. Since the latter is roughly the size of the atom, this assumption is seldom formally justifiable in systems comprised of atoms. Gunnarsson and Lundqvist[6] have made two important observations in this context. First, the quasi-particle implied by the local-density approximation is perfectly neutral. This rigid neutrality is particularly important, because the energies associated with even small charge imbalances are quite large. Second, the total energy depends on only the spherical average of the exchange-correlation hole and is therefore insensitive to many of the distortions of the hole caused by inhomogenieties in the system. The total energy is, however, sensitive to inhomogeneities which cause the hole to be off-center. The extreme example of off-centeredness is provided by an electron passing through a metal surface. The hole, which was centered on the electron as it approached the surface from inside the metal, separates from the electron altogether, becoming the electron's image, as the electron leaves the metal. This phenomenon is not confined to macroscopic surfaces; it occurs when an electron moves to the outer parts of an atom. The hole remains in the interior of the atom where it pulls on the electron with a simple $1/r$ Coulomb potential. Local-density potentials, not reflecting this phenomenon, fall to zero exponentialy rather than as $1/r$, as they should. A manifestation of this phenomenon which appears to affect our results occurs when electrons in the transition and noble metals pass outside the d-shell. The theory describes the hole as if it were supported entirely by the s and p electrons outside the d-shell, whereas it seems likely that the nearby d-electrons contribute to the hole. (We are not suggesting here that the s and d electrons do not overlap, only that some portions of the electron density can be thought of as derived primarily from s and d electrons.)

We consider now the empirical evidence bearing on the accuracy of the local-density approximation. While the approximation has been used to study small molecules[11] and simple metal surfaces,[12] the most relevant accuracy test is a comparison of our computed properties with measurements where this is possible. Such a comparison appears in Figure 1.1, where we present measured and computed binding properties for the twenty-six metals in the third and

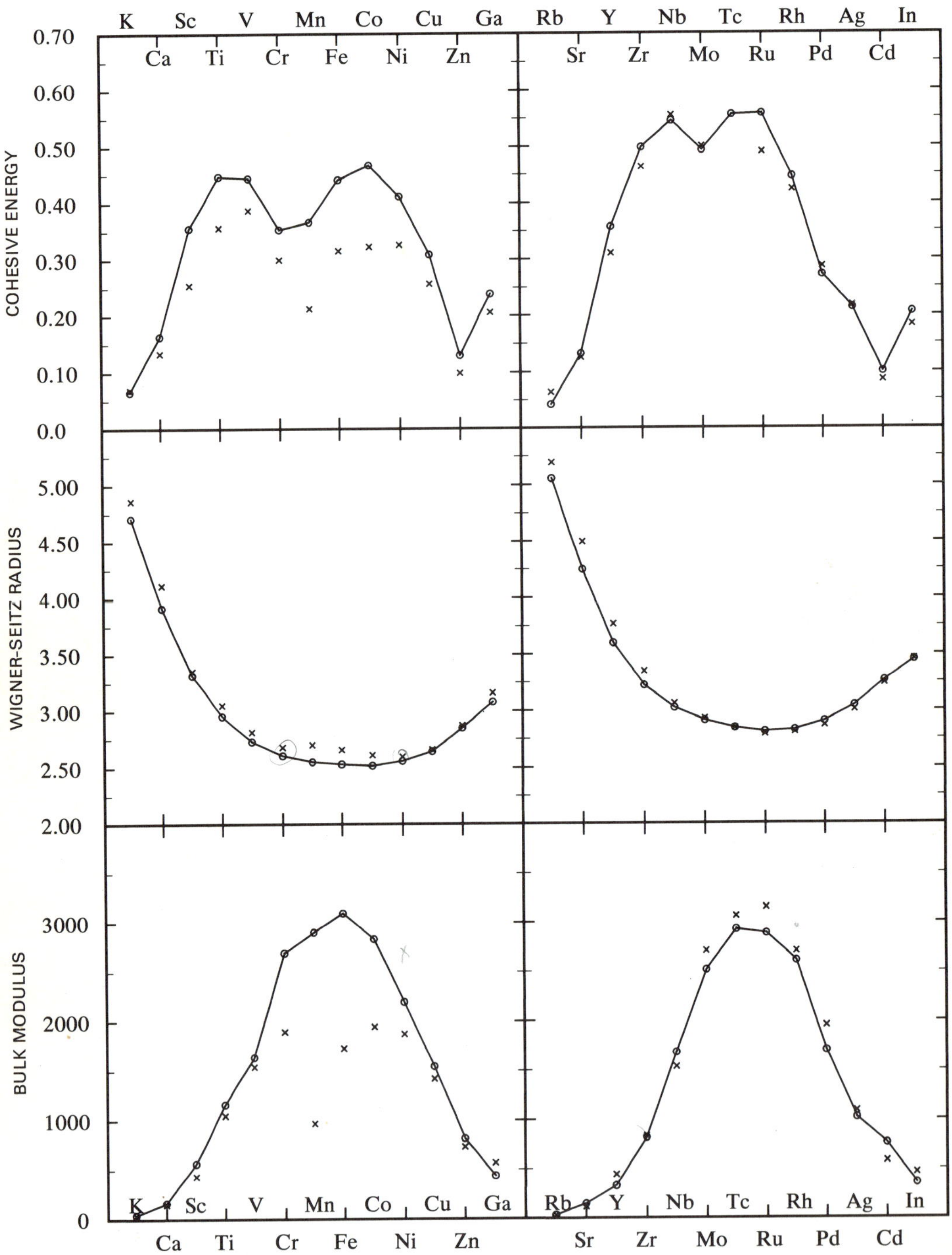

Figure 1.1 Cohesive properties. Top row- cohesive energy (Ry/atom). Middle row- Wigner-Seitz radius (a.u.). Bottom row- bulk modulus (Kbar). Measured values are indicated by crosses.

the fourth rows of the periodic table.[13,14] Three binding properties are presented: the cohesive energy, the atomic density and the bulk modulus. In terms of the binding curve (the total energy of the solid relative to that of the atom as a function of lattice constant) these three properties are the depth, position and curvature of the equilibrium-producing minimum. The calculated values are connected by a line; the measured values are indicated by crosses.[15]

In assessing the degree of agreement with measurement shown in Figure 1.1, the reader should bear in mind that, of the three properties considered, only the cohesive energy involves atomic calculations. The most serious deviations from measurement in the purely solid properties, the atomic density and bulk modulus, occur for the magnetic 3d transition metals Cr, Mn, Fe, Co and Ni. The analysis of Janak and Williams[16] indicates that these deviations stem largely from the assumed absence of magnetic order in the calculations which produced Figure 1.1. The small remaining discrepancies in the solid properties suggest that the calculations provide a good description of the metallic ground state.

With regard to the differences between the computed and measured cohesive energies, note that accounting for magnetic order in the solid can only (and does) lower the total energy of the solid, thereby increasing the discrepancy. The source of these discrepancies is probably, therefore, the atomic calculations. We can infer an additional fact about these discrepancies. While our atomic calculations allow for spin polarization, they yield only an approximate average over the different multiplets which can be formed from the single configuration for which the calculation is done. In this regard atomic manganese is unique among the 3-d transition metals, because one member of the ground state ^6S multiplet consists of a single Slater determinant. Furthermore, the single determinant describing the half-filled 3-d shell corresponds to a spherically symmetric electron density, which is assumed by the calculations. The fact that the cohesive energy predicted for Mn is not closer to the measured value than for the other 3-d transition metals suggests that the non-trivial multiplet structure of the open shell atoms is not the source of the discrepancies.

As mentioned above, the accuracy of our computed cohesive energies greatly exceeds the absolute accuracy of either the atomic or solid calculations. The basis for this statement is contained in Table 1.1, where we compare our local-density atomic total energies with those given by the Hartree-Fock approximation.[17] The Hartree-Fock approximation, it should be remembered, provides a strict upper bound to the total energy. Therefore, in those cases where the local-density total energy is higher (less negative) than the Hartree-Fock total energy, the difference between the two is a strict lower bound on the error in the local-density result. Table 1.1 therefore indicates that the absolute total-energy error is a rapidly increasing function of the atomic number and is larger than a Rydberg for most of the metals we consider. (Note that the fractional difference between local-density and Hartree-Fock total energies decreases with atomic number; for indium it is only 0.05 percent.) In light of these large absolute errors, the accuracy of our predicted cohesive energies, particularly those for the

atom	LSD total energy (ry)	HF total energy (ry)	LSD ground config.	exp. config.
H	-0.976	-1.000	1s	1s
Li	-14.709	-14.865	$1s^2 2s$	$1s^2 2s$
Be	-28.909	-29.146	$1s^2 2s^2$	$1s^2 2s^2$
Na	-322.902	-323.718	3s	3s
Mg	-398.274	-399.229	$3s^2$	$3s^2$
Al	-482.637	-483.754	$3s^2 3p$	$3s^2 3p$
K	-1196.382	-1198.330	4s	4s
Ca	-1351.442	-1353.517	$4s^2$	$4s^2$
Sc	-1517.322	-1519.472	$3d4s^2$	$3d4s^2$
Ti	-1694.591	-1696.740	$3d^3 4s$	$3d^2 4s^2$
V	-1883.529	-1885.607	$3d^4 4s$	$3d^3 4s^2$
Cr	-2084.386	-2086.284	$3d^5 4s$	$3d^5 4s$
Mn	-2297.211	-2299.252	$3d^5 4s^2$	$3d^5 4s^2$
Fe	-2522.369	-2524.582	$3d^{6.4} 4s^{1.6}$	$3d^6 4s^2$
Co	-2760.288	-2762.617	$3d^{7.9} 4s^{1.1}$	$3d^7 4s^2$
Ni	-3011.233	-3013.632	$3d^9 4s$	$3d^8 4s^2$
Cu	-3275.464	-3277.928	$3d^{10} 4s$	$3d^{10} 4s$
Zn	-3553.007	-3555.697	$3d^{10} 4s^2$	$3d^{10} 4s^2$
Ga	-3843.556	-3846.522	$3d^{10} 4s^2 4p$	$3d^{10} 4s^2 4p$
Rb	-5872.477	-5876.716	5s	5s
Sr	-6258.683	-6263.092	$5s^2$	$5s^2$
Y	-6658.819	-6663.369	$4d5s^2$	$4d5s^2$
Zr	-7073.319	-7077.938	$4d^3 5s$	$4d^2 5s^2$
Nb	-7502.361	-7506.984	$4d^4 5s$	$4d^4 5s$
Mo	-7946.087	-7950.739	$4d^5 5s$	$4d^5 5s$
Tc	-8404.434	-8409.215	$4d^6 5s$	$4d^5 5s^2$
Ru	-8877.807	-8882.913	$4d^7 5s$	$4d^7 5s$
Rh	-9366.418	-9371.675	$4d^9$	$4d^8 5s$
Pd	-9870.419	-9875.844	$4d^{10}$	$4d^{10}$
Ag	-10389.75	-10395.40	$4d^{10} 5s$	$4d^{10} 5s$
Cd	-10924.44	-10930.27	$4d^{10} 5s^2$	$4d^{10} 5s^2$
In	-11474.28	-11480.34	$4d^{10} 5s^2 5p$	$4d^{10} 5s^2 5p$

Table 1.1. Atomic total energies and ground-state configurations obtained in the spin-polarized local-density (LSD) approximation, compared to configuration-averaged Hartree-Fock total energies,[17] and experimental ground-state configurations.

4-d transition metals, indicates that the aspect of the problem responsible for the large absolute error, i.e. the deeply bound core states, is unimportant for valence-electron-derived properties.

We close this section with a discussion of excitation spectra, such as kinetic-energy distributions of photoemitted electrons, and the relevance of our calculations to such spectra. The ε_i's appearing in equation 1.2 are related to excitation energies by the following expression:

$$\varepsilon_i = \partial E/\partial n_i \qquad 1.6$$

i.e. if the occupation of the ith orbital is changed by an infinitesimal amount δn_i, then the total energy of the system changes by $\varepsilon_i \delta n_i$. The connection between the ε_i's and excitation energies is complicated by two factors. First, real excitations involve whole electrons; the excitation energy corresponds, therefore, not simply to the first derivative of the total energy, but to an entire Taylor series,

$$E(n_i+\delta n_i) = E(n_i)+(\partial E/\partial n_i)\delta n_i+(1/2)(\partial^2 E/\partial n_i^2)(\delta n_i)^2+... \qquad 1.7$$

Numerical examples of the convergence of such series are given by Slater.[8] The importance of second and higher order terms, which correspond to relaxation, depends primarily on the localization of the electron density associated with the ith state from which the electron was excited. These terms should not contribute significantly to the energy required to excite delocalized continuum electrons.

The second complication obscuring the relationship between the ε_i's and excitation energies is the fact that the system in which the ith state is unoccupied is not, in general, a ground state and is not therefore amenable to description by density-functional theory. In some cases, such as ionization potentials, both of the states involved are ground states. In others, such as systems containing a core hole, the ionized system is in a quasi-stationary state; density-functional theory appears to yield reasonable results in such cases. These arguments indicate that at least an approximate relationship between the ε_i's of equation 1.2 and excitation energies exists. Some of the empirical evidence bearing on this relationship is the following. First, the occupied band width of metallic aluminum is given essentially exactly by the ε_i's of the theory. Second, the d-bands of copper, which experimentally lie 2.1 eV below E_f, appear only 1.5 eV below E_f in the theory.[18,19] This discrepancy seems too large to be accounted for by the difference between the ground-state theory and the approximate theory of excitations discussed by Sham and Kohn.[20] It appears more likely that this discrepancy stems from the use of the local-density approximation in the ground state theory. The discrepancy increases with the localization of the states in question. Thus, the d-bands of zinc, which experimentally lie 10 eV below E_f, are placed about a volt too high by the theory. The energy distribution of the ε_i's which we provide for each metal in the tabular portion of the text can therefore be used to interpret excitation spectra, but the lack of formal justification and the empirical indication that localized states are somewhat too high must be kept in mind.

II. CALCULATIONS

In this chapter, we describe in some detail how the calculations leading to the tables which appear later in this book were performed. Some aspects of this discussion, concerned with definitions and the overall structure of the calculations, will be of interest to the general reader. Other aspects, concerned with computational detail, will be of interest mainly to the specialist. In general terms, our object has been to perform all the calculations required to go from a given approximation for the exchange-correlation functional $E_{xc}\{\rho\}$ to experimentally measurable quantities, maintaining a degree of computational accuracy that does not obscure the dependence of the results on the exchange-correlation functional. Since the only input to these calculations is the atomic number Z, we have confidence that the calculations correctly display all the effects of the chosen approximation for exchange and correlation on trends with atomic number.

1. SELF-CONSISTENT SOLID CALCULATIONS

The calculations for solids which appear in this book were all performed using the KKR method[21] to solve the self-consistent single-particle equations. The electron density was assumed to be of muffin-tin form, i.e., spherically symmetric inside touching, but nonoverlapping, spheres, and constant in the region between the spheres.

The appropriate one-electron potential corresponding to a charge density of this form is[4]

$$V(r) = \frac{-2Z}{r} + \frac{8\pi}{r}\int_0^r r'^2\rho(r')dr' + 8\pi\int_r^R r'\rho(r')dr' + \frac{CZ_{out}}{a} + \mu_{xc}[\rho(r)] - \mu_{xc}(\rho_{out}) \qquad 2.1$$

for r<R, and zero for r>R. In this equation the potential is in rydbergs if the electronic charge density $\rho(r)$ is in atomic units (electrons/Bohr^3), and R is the radius of the muffin tin (inscribed sphere radius). The quantity $\mu_{xc} = \delta E_{xc}\{\rho\}/\delta\rho(r)$ is the exchange-correlation potential discussed in Chapter I; it is a local function of the charge density in the local approximation. Z_{out} is the interstitial charge per unit cell, and

$$\rho_{out} = Z_{out}/\Omega_{out} \qquad 2.2$$

is the interstitial charge density if Ω_{out} is the difference between the volumes of the unit cell and the muffin tin. The reference energy has been chosen so that the potential is zero in the interstitial region; this is the source of the $\mu_{xc}(\rho_{out})$ term. With respect to this choice of energy reference, the entire contribution of the interstitial charge density to the electrostatic part of the muffin-tin potential is represented by the C Z_{out}/a term in eq. (2.2). Values of C for the three cubic lattices are 3.116686 (simple cubic), 4.085521 (body-centered cubic), and 4.832066 (face-centered cubic), all in rydberg-bohrs.

The 'muffin-tin zero', or the relation between the chosen reference energy (zero potential in the interstitial region) and the vacuum level, is determined by the electrical double layer at the crystal surface. Since no model for the surface dipole layer was included in these calculations, they provide no relationship between the muffin-tin zero and the vacuum level. This relationship might be established by using the measured work functions (Fermi energy relative to vacuum).

Given the potential V(r) obtained from eq. (2.1), the energy-band problem is to solve Schrödinger's equation for the eigenvalues $\varepsilon_n(\vec{k})$ and Bloch functions $\Psi_{nk}(\vec{r})$, in terms of which a new charge density is obtained from

$$\rho'(r) = \sum \Theta(E_f - \varepsilon_{nk})\, |\Psi_{nk}(r)|^2, \qquad r \leq R \qquad 2.3$$

(here $\Theta(x)$ is unity for $x>0$ and zero for $x<0$; it represents the Fermi function at the absolute zero of temperature). The charge density $\rho'(r)$ obtained from eq. (2.3) will, in general, be different from the charge density $\rho(r)$ used to construct the potential from eq. (2.1). The entire procedure must be iterated until the two charge densities agree; the result is the self-consistent ground-state charge density. A first guess at the charge density is needed in order to start off this iteration procedure (we used a superposition of Herman-Skillman atomic charge densities) but the final self-consistent charge density is totally independent of this first guess (although a large number of iterations will be required if the guess is a poor one); such questions as differences in atomic configurations in the initial charge density are irrelevant, and our experience is that the self-consistent solution (at least for nonmagnetic solids) is unique.

Because of the strength of the Coulomb forces, the procedure used to form the input charge density for each iteration may lead to instabilities. Only when the kinetic, Coulomb (Hartree) and approximate exchange-correlation forces achieve a delicate balance will the charge density become self-consistent. If, for example, for one particular iteration the electronic charge happens to be too far from the nucleus, the potential from eq. (2.1) will be overly attractive, and will pull charge in closer to the nucleus during the next iteration. The straightforward procedure of using the output electron density of a given iteration as input to the next does not usually converge: the inward rush of charge to screen a too-attractive potential overcompensates, producing a too-repulsive potential which pushes electrons even farther from the nucleus during the next iteration. The standard procedure for damping these growing oscillations is to use a linear combination of input and output charge densities from one iteration as the input charge density for the next iteration. The appropriate linear combination depends on the system, and is chosen by trial and error to obtain a good rate of convergence, but the final self-consistent result is once again independent of the choice.

We have used a different mixing scheme, taking the input charge density for each iteration to be a linear combination of the *output* charge densities of the two previous iterations. If Δ_i is the integrated rms difference between input and output charge densities for the ith iteration, the linear combination is formally

$$\rho_{i+1}^{(in)} = \frac{\Delta_i \rho_{i-1}^{(out)} + \Delta_{i-1} \rho_i^{(out)}}{\Delta_i + \Delta_{i-1}} . \qquad 2.4$$

Since this linear combination involves only output charge densities, the effects of a poor input charge density disappear rapidly, and the weights are such that the linear combination is always biased toward that charge density which is closest to self-consistency. In practice, this mixing scheme works at least as well as the conventional scheme, and the fact that convergence can be achieved using only output charge densities is interesting.

We turn now to the details of the calculations required for one iteration. All states were treated nonrelativistically, and core states (which were not frozen, but recomputed for each iteration) were treated differently from valence states. A core state was arbitrarily defined as any state for which the difference in eigenvalues of the radial Schrödinger equation using zero-slope and zero-value boundary conditions at the Wigner-Seitz radius (~ the bandwidth) was less than 0.025 ry. The actual wavefunction for each core state was obtained by matching to a constant-potential solution outside the muffin-tin, and the core charge density was taken to be the sum of the squares of these wavefunctions, each fully occupied. Any core charge outside the muffin tin, due to the tails of these orbitals, was treated as part of the interstitial charge. In practice, very little change in the total energy, pressure, charge density, etc. was found if the uppermost core states were treated as bands instead of as described above (the worst cases occur for the outermost p-orbitals at the beginning of each transition series, but in no case was any core orbital higher than 1 Ry below the muffin-tin zero).

The valence states and energies were found using the "fast" KKR method we described in reference 22; the reader is referred to this paper for details. Basically, this procedure consists of an exact transformation of the KKR equations to a mixed plane wave-atomic orbital representation in which the $\vec{k}$ and energy dependence of the secular matrix are amenable to polynomial description. This transformation is performed on a relatively coarse $\vec{k}$ and ε mesh and the result is interpolated to a fine mesh. The accuracy of the $\vec{k}$-expansion is controlled by restricting the use of the procedure to $\vec{k}$-points in the neighborhood of points on the coarse mesh of $\vec{k}$-values, and the accuracy of the energy expansion is controlled by dividing the energy range to be scanned for solutions into "panels", each of sufficiently small width.

The expansion of the valence wavefunctions in spherical harmonics was truncated at $l_{max}=4$, and the valence energies were found to a tolerance of 10^{-4} ry (the energy and wavevector meshes in the "fast" procedure were chosen to give an overall accuracy of 10^{-3} ry). While the valence-state calculation was performed mostly in single-precision arithmetic (approximately 7 significant figures), the core calculation was performed entirely in double precision (with an eigenvalue tolerance of 10^{-6} ry); it was also found necessary to perform the core calculation on a finer radial mesh in order to find the deep-lying core states to the required precision. Finally, the iteration to self-consistency was continued until the total energy (calculated as described below) was stable to 10^{-3} ry; at this level of self-consistency, Δ (the integrated rms difference between input and output charge densities) was of the order of 10^{-4} – 10^{-5} electrons.

2. FREE ATOM CALCULATION

Calculations for isolated atoms (assumed spherically symmetric) are needed in order to obtain cohesive energies, and are useful for examining changes in the electron density associated with metallic bonding. Our self-consistent atomic programs integrate the radial Schrödinger equation both outward from the nucleus and inward from a large radius (~30 Bohr radii), matching value and slope at the classical turning point.

It was first pointed out by Hartree[23] in 1937 that the total energy of the atom in the Hartree-Fock model can be lowered by introducing spin polarization. Later, Hattox and coworkers[24] used spin-polarized Xα atomic calculations to get multiplet splittings and lower ground-state energies in atoms, in order to get a better cohesive energy for V. More recently, Gunnarson *et. al.*[25] have suggested that the energy lowering due to spin polarization can be as large as 20% of the cohesive energy; we have now verified this for a number of elements.[13,14]

In comparing the total energies of atoms and solids, we exploit a large cancellation of errors (originating in the atomic cores) by using the same exchange-correlation functional for both. Thus, the spin-polarized exchange-correlation functional used for the atomic calculation must reduce to the one used for the solid calculation when spin polarization vanishes. We used the form suggested by von Barth and Hedin,[26]

$$\varepsilon_{xc}(\rho\uparrow,\rho\downarrow) = \varepsilon_{xc}^{P}(r_s) + [\varepsilon_{xc}^{f}(r_s) - \varepsilon_{xc}^{P}(r_s)]f(\rho\uparrow,\rho\downarrow) \qquad 2.5$$

where

$$f(\rho\uparrow,\rho\downarrow) = [(2\rho\uparrow/\rho)^{4/3} + (2\rho\downarrow/\rho)^{4/3} - 2]/(2^{4/3} - 2), \qquad 2.6$$

$\rho\uparrow$ and $\rho\downarrow$ being the spin-up and spin-down components of the total charge ρ ($\rho = \rho\uparrow + \rho\downarrow$), and r_s being defined by

$$(4/3)\pi r_s^{\,3} = 1/\rho. \qquad 2.7$$

The paramagnetic and ferromagnetic exchange-correlation energies in eq. (2.5) are given by

$$\varepsilon_{xc}^{\,i} = \varepsilon_x^{\,i}(r_s) - c_i G(r_s/r_i), \quad i=p,f \qquad 2.8$$

where $\varepsilon_x^{\,P} = -0.91633/r_s$, $\varepsilon_x^{\,f} = 2^{1/3}\varepsilon_x^{\,P}$,

$$G(x) = (1 + x^3)\ln(1 + 1/x) - x^2 + x/2 - 1/3\,, \qquad 2.9$$

and c_p, c_f, r_p, r_f are constants. These constants were chosen to fit eq. (2.5) to ε_{xc} for the homogeneous electron gas. Since the effects of exchange and correlation in the electron gas are less well known in the ferromagnetic than in the paramagnetic limit, we chose $c_p = 0.045$, $r_p = 21$, $c_f = c_p/2$, $r_f = 2^{4/3} r_p$ (RPA scaling of the spin dependence) so that the version of (2.5) used for the atomic calculations reduces to the Hedin-Lundqvist[3] form for $\varepsilon_{xc}{}^p$ (which we used for the crystal calculations) when $\rho\uparrow = \rho\downarrow$.

For open-shell atoms, the charge density was kept spherically symmetric by the simple expedient of averaging the charge density of all m_l's (e.g., for one p-electron, the charge density is $(\varphi^2{}_{m_l=1} + \varphi^2{}_{m_l=0} + \varphi^2{}_{m_l=-1})/3$); this is the same as summing the squares of radial wavefunctions with integral coefficients. The electron configuration giving the lowest total energy is not always the same as the experimental ground-state configuration, and in two cases (Fe and Co) it was found necessary to introduce nonintegral configurations (i.e., $(3d\uparrow)^5(3d\downarrow)^{1.404}(4s\uparrow)^1(4s\downarrow)^{0.596}$ for Fe, and $(3d\uparrow)^5(3d\downarrow)^{2.897}(4s\uparrow)^1(4s\downarrow)^{0.103}$ for Co). The use of nonintegral occupation numbers has been discussed by Slater and coworkers;[27] the problem is that any configuration with integral occupations for these two atoms will also have unoccupied levels lying lower in energy than the highest occupied level. The nonintegral occupation numbers are chosen to produce a self-consistent potential which makes the $3d\downarrow$ and $4s\downarrow$ levels degenerate, so that both can be partially occupied without violating Fermi statistics; since $\partial E/\partial n_i = \varepsilon_i$, this configuration also minimizes the total energy.

If one examines a plot of the positions of the 3d and 4s atomic levels, calculated using the self-consistency procedure outlined above, as functions of atomic number through the transition series, it becomes apparent that the incorrect and nonintegral configurations are occurring because the 3d levels are dropping down through the 4s levels too early in the transition series. It is difficult to say whether this is due to the approximations made in treating exchange and correlation effects, or whether it is due to our method of enforcing spherical symmetry on the atom.

The spin-polarized total energies and ground-state configurations obtained for the isolated atoms are compared to experimental configurations and configuration-averaged Hartree-Fock total energies[17] in Table 1.1. The relatively large differences in total energies are probably due mainly to nonlocal exchange effects, which are included in the Hartree-Fock calculations, but neglected in our local-density calculations. The reader is reminded that, since the Hartree-Fock energies represent a minimization of the exact many-body Hamiltonian within a set of approximate many-body wavefunctions, these energies are upper bounds to the true energies. Since the local-density total energies are systematically higher than their Hartree-Fock counterparts, they are more in error.

3. TOTAL ENERGY

The total energy

$$E = T + U + E_{xc}[\rho] \qquad 2.10$$

was evaluated for both solids and atoms using expressions given in reference 4. The pressure P was evaluated using the virial theorem,[4] and the bulk modulus B from

$$B = -\Omega \frac{\partial P}{\partial \Omega} = -\frac{1}{3} a \frac{\partial P}{\partial a} \qquad 2.11$$

where a is the lattice constant and Ω is the unit cell volume. In this last expression, the derivative was obtained numerically from calculated pressures at several lattice constants.

To the electronic energy and its volume derivatives was added an estimate of the zero-point lattice energy using the Debye model. Given experimental values[28,29] of the Debye temperature $\Theta_D{}^o$ at some atomic volume Ω_o, and the Grüneisen constant γ, we assumed that the zero-point energy per atom was

$$E_o = (9/8)k\Theta_D \qquad 2.12$$

where k is Boltzmann's constant, and we assumed the volume dependence of the zero-point energy was given by

$$\Theta_D = \Theta_D{}^o(\Omega_o/\Omega)^{\gamma}, \qquad 2.13$$

where it was also assumed that γ is volume-independent. The appropriate volume derivatives of E_o were then added to the electronic pressure and bulk modulus; these lattice contributions are small fractions of the electronic contributions for all the materials considered here. The values of $\Theta_D{}^o$ and γ used in these calculations are given in Table 2.1.

Material	Θ_D (K)	Grüneisen γ	Exp. Atomic Volume (Bohr^3)	Calc. Equil. Volume (Bohr^3)
H	0	—	—	20.0
Li	344	1.18	143.7	131.3
Be	1440	1.18	54.54	52.02
Na	158	1.31	254.5	228.0
Mg	400	1.48	151.4	149.4
Al	428	2.19	109.9	109.2
K	91	1.37	481.3	437.7
Ca	230	1.16	291.1	251.4
Sc	360	1.17	168.7	153.2
Ti	420	1.18	120.3	108.4
V	380	1.05	93.48	85.56
Cr	630	1.30	80.63	74.44
Mn	410	2.07	82.84	69.83
Fe	467	1.66	78.95	68.11
Co	445	1.93	74.72	67.13
Ni	450	1.88	73.42	70.36
Cu	343	2.00	78.92	77.48
Zn	327	2.01	99.35	97.16
Ga	320	2.00	132.4	122.3
Rb	56	1.67	598.9	540.7
Sr	147	1.00	373.6	321.1
Y	280	0.89	194.7	195.2
Zr	291	0.83	139.9	140.0
Nb	275	1.58	119.2	115.7
Mo	450	1.60	102.7	103.4
Tc	350	2.60	97.25	96.54
Ru	600	3.20	92.54	93.18
Rh	480	2.23	93.70	94.73
Pd	274	2.28	99.67	102.7
Ag	225	2.36	111.9	118.2
Cd	209	2.23	142.9	145.7
In	108	2.37	179.2	172.1

Table 2.1. Values used for the Debye temperature[28] Θ_D and Grüneisen constant[29] γ, and a comparison of experimental[28] and calculated equilibrium atomic volumes.

The energy-band calculations tabulated later in this book are given for each material for a lattice constant a_o near the one for which the computed pressure vanishes. Values are also given for the pressure P_o at the lattice constant a_o, and the bulk modulus B. The calculated equilibrium lattice constant can then be found from eq. (2.11) as

$$a_{eq} = a_o + P_o a_o / 3B, \tag{2.14}$$

and the values of the corresponding computed equilibrium atomic volumes are also given in Table 2.1.

The calculated cohesive energy is defined as the total energy of the atom minus the total energy of the solid (including the zero-point energy); thus positive values of the cohesive energy represent stability of the solid relative to the free atom. The importance of using the same exchange-correlation approximation in both atomic and solid-state calculations is apparent in Table 1.1, where it is seen that the effect of the exchange-correlation approximation on the atomic total energy can be quite large compared to the cohesive energy.

4. DENSITY OF STATES

To calculate the density of states (number of states with energies between E and E+dE), defined by

$$N(E) = \int \frac{dS}{|\nabla_k \varepsilon|} \tag{2.15}$$

where the integral goes over the constant-energy surface S defined by $\varepsilon(k)=E$, and the integrated density of states (total number of states below E), defined by

$$I(E) = \int_{-\infty}^{E} N(E)dE \tag{2.16}$$

we used the Gilat-Raubenheimer method[30] with 1440 k-points in the irreducible wedge of the

Brillouin zone. The charge density $\rho(r)$ was constructed by a variant of this method, in which the contribution of each k-mesh cube was taken to be $|\Psi_k(r)|^2$ at the cube center, times the contribution of the cube to I(E).

The Fermi energy was obtained by solving

$$Z = I(E_f) \ . \qquad 2.17$$

If $\rho(r,E)$ is the charge density calculated as though the Fermi energy were at E, we calculate $\rho(r,E)$ at the three energies E_f^o, and $E_f^o \pm \Delta/2$, where E_f^o is an initial guess at the Fermi energy, and Δ is a small energy tolerance. The charge density and its energy derivative are then evaluated at the correct Fermi energy E_f (the solution of eq. 2.17) by an interpolation procedure in which it is assumed that $\partial\rho(r,E)/\partial E$ in the neighborhood of E_f is equal to N(E) times a function depending on r, but only weakly on E. In this way, the strong dependence of $\rho(r,E)$ on E through its dependence on the density of states is handled without approximation, and only the relatively weak remaining dependence on the energy is actually interpolated.

Because E_f may be greater than E_f^o, and because the Gilat-Raubenheimer method finds the contribution of each k-mesh cube to the density of states for a range of energies about the energy at the cube center, it is necessary to scan k-space for roots of the KKR matrix up to some energy E_{max} which is somewhat larger than $E_f^o + \Delta/2$. In practice, allowing an extra 0.2 ry seems to guarantee that all states near E_f will have been properly counted.

In the tabulations appearing later in this book, plots are given of $r^2\rho(r)$ and of $r^2(\partial\rho(r,E)/\partial E)_{E = E_f}$ vs. r. Multiplying by r^2 brings out structure in these plots which would not otherwise be visible, but makes it difficult to obtain ρ and $(\partial\rho(r,E)/\partial E)_{E_f}$ at r=0. Accordingly, these two numbers are given separately in the tabulations.

Plots are also given of the difference in the charge densities of the solid and the isolated atom, and also of r^2 times this difference (the "radial charge density difference"). These plots give some indication of the charge density displacement in forming the metallic bonds, and the radial charge density difference shows where the important contributions to the potential difference arise. $\rho_{solid} - \rho_{atom}$ is negative at the nucleus for the simple metals, showing that the valence s electrons have been converted to p electrons in order to form the metallic bonds. However, in the transition series Ca-Ga and Sr-In, this difference is positive for those materials for which the atomic calculations give less than about 1.5 s electrons in the ground-state configuration (see Table 1.1). While the atomic calculations sometimes give the wrong ground-state configuration, the increase in the charge density at the nucleus in forming the solid is nevertheless a real effect (see, e.g., Cr, Pd, Cu, or Ag). This is because, in the cell corresponding to a given atom, the tails of the d-orbitals of the neighboring atoms actually have s character. This increase in the s charge density, due to the tails of the d-orbitals on neighboring atoms, leads to a charge density at the nucleus which can be larger than that of the free atom.

5. SUSCEPTIBILITY ENHANCEMENT

If

$$\gamma(r) = \frac{1}{N(E_f)}\left(\frac{\partial\rho(r,E)}{\partial E}\right)_{E = E_f} = \sum |\Psi_i|^2 \delta(E_f - \varepsilon_i)/N(E_f) \qquad 2.18$$

where the derivative is obtained as described below eq. 2.17, it was shown by Vosko and Perdew,[31] and also by Gunnarsson,[32] that the exchange-correlation integral J in the susceptibility enhancement

$$\frac{\chi}{\chi_o} = \frac{1}{1 - J\,N(E_f)} \qquad 2.19$$

is given approximately by

$$J = \int d^3r\, \gamma^2(r) K(r)\ . \qquad 2.20$$

Here K(r) is a functional derivative of the exchange-correlation functional E_{xc} with respect to magnetization:

$$K(r) = -\frac{1}{2}\left(\frac{\partial^2 E_{xc}}{\partial m^2}\right)_{m(r)=0} \qquad 2.21$$

where E_{xc} is defined in eqs. 1.5 and 2.5-2.9 ($m=\rho\uparrow-\rho\downarrow$; $\rho\uparrow+\rho\downarrow=\rho(r)$). The integral J and the susceptibility enhancement have been evaluated for each element,[33] and the results are given in the tables. The negative values of χ/χ_o for Fe and Ni show that these metals have ferromagnetic instabilities. This is borne out by spin-polarized energy-band calculations,[16,34] which, in addition to showing that the ferromagnetic state has the lower energy, can also be used to find a priori values of the magnetic moment, changes in the mechanical properties due to the magnetization, etc., as is discussed in Chapter V. With regard to Co, we note that equation 2.19 is a lower bound to the susceptibility,[31] and is accurate to about 10%.[33] Thus fcc Co, which is almost ferromagnetic according to this approximate calculation of its susceptibility, is in fact found to be ferromagnetic when spin-polarized calculations are performed (see Chapter V).

The conduction-electron contribution to the Knight shift is, apart from constants, $(\chi/\chi_o)(\partial\rho(r,E)/\partial E)_{E=E_f;r=0}$. Both quantities in this expression appear in the tabulation, but the product should not be compared to experimental Knight shifts, because core-polarization and orbital-moment contributions, both of which can be significant, have been ignored.

III. TRENDS

In this Chapter, we examine the trends in several of the calculated quantities with atomic number Z. For those quantities which have been measured, a comparison to experimental trends provides a test of the adequacy of the theory. For those quantities which have not yet been measured, information about the underlying physics can still be obtained by examining the theoretical trends.

A comparison of the calculated and experimental values of the cohesive energy, Wigner-Seitz radius, and bulk modulus for the 3-d and 4-d transition series has already been presented in Figure 1.1. While the sources of disagreement with experiment were discussed in Chapter I, no mention was made there of the basic reason for the parabolic shape of these curves. This has been shown[35] to be a consequence of the change from bonding to antibonding character of the d-orbitals as the Fermi energy increases through the d band. For materials at the beginning of a transition series, the bonding orbitals are being filled; this leads to increased cohesion, to a decrease in the atomic volume, because of the increased mechanical attraction gained by filling more and more bonding orbitals, and to an increase in the bulk modulus, because of the increased compression of the s electrons[36,37] (due simply to the decrease in atomic volume). These effects are all maximized near the middle of the transition series, when the bonding orbitals are filled, and then the trend is reversed when the antibonding orbitals begin to be filled.

Perhaps the simplest way to measure the compression of the s-electron gas is through the interstitial charge density ρ_{out}, which is defined in eq. 2.2. This is plotted against atomic number Z in Figure 3.1. The parabolic behavior of this quantity in going through each transition series directly reflects the parabolic shape of the atomic volume vs. Z, as shown in Figure 1.1. This is not to say, however, that the interstitial charge density corresponds to a fixed number of electrons; both Z_{out} (the total interstitial charge) and the effective charge corresponding to a charge density ρ_{out} spread throughout the entire atomic volume also show peaks near the middle of each transition series.

The interstitial charge density plotted in Figure 3.1 is of use in the theory of alloys; Miedema et. al.[38] have shown that one contribution to the heat of formation of a binary alloy is proportional to the square of the difference in ρ_{out} values for the two constituent metals. The interested reader is referred to the above paper for details, but it should be clear that the shape of ρ_{out} vs. Z, and the large differences in ρ_{out} values between, say, transition metals and alkali metals, have important implications for the formation of alloys.

That the increase in bulk modulus in going through a transition series is due to compression of the interstitial electrons is demonstrated by Figure 3.2, where the bulk modulus of each element is plotted as a function of the value of r_s obtained from eq. 2.7 using the interstitial charge density ρ_{out}. The curve plotted in this figure is the bulk modulus of the homogeneous electron gas, using the Hedin-Lundqvist[3] expression for the correlation energy per particle. The good agreement with the homogeneous-gas result, when r_s is defined for each element in terms of ρ_{out}, suggests that most of the contributions to the bulk modulus are arising from electrons in the interstitial region of the unit cell. Thus the dependence of the bulk modulus on atomic number, as shown in Figure 1.1, is a combination of the dependence of ρ_{out} on Z, as shown in Figure 3.1, and the rapid increase of the bulk modulus of the homogeneous electron gas with decreasing r_s, as shown in Figure 3.2. Interestingly, if the agreement shown in Figure 3.2 were perfect, it would imply that what happens inside the muffin tin is important, in terms of the bulk modulus, only insofar as it determines the value of ρ_{out}. While the correlation between the calculated bulk moduli and those of the homogeneous electron gas is less than perfect, it is nevertheless not too bad an approximation to think of the electronic structure inside the muffin tin as "rigid" here: changes of quantities inside the muffin tin with changes in volume do not seem to be of primary importance in determining the bulk modulus. The degree to which this is true is one measure of the adequacy of the rigid muffin-tin approximation[39] for calculating the frequencies of long-wavelength phonons.

Figure 3.3 shows the exchange-correlation integral J defined in eq. 2.20, and the density of states at the Fermi energy, $N(E_f)$, plotted as functions of atomic number. The density of states is largest toward the end of a transition series because the antibonding states at the top of the d-band are the most-localized d states; they have the smallest interaction with the neighboring atoms, and thus the smallest energy spread due to these interactions. The height of the peak is lower in the 4d transition series than in the 3d series because the 4d orbitals extend farther from the nucleus than the 3d orbitals; since there is no sizable increase in the atomic volumes of the 4d transition elements compared to the corresponding 3d elements, the 4d orbitals on neighboring atoms must overlap more than do the 3d orbitals. This increased overlap leads to a larger interaction between neighbors, and thus a larger bandwidth and a smaller density of states.

The overall gradual decrease of the exchange-correlation integral J with increasing Z is a consequence of the dependence of K(r) in eq. 2.21 on the total charge density $\rho(r)$. K varies roughly as an inverse power of $\rho(r)$ and, as $\rho(r)$ gradually increases with increasing Z, K(r) and J gradually decrease[33]. The parabolic behavior of J within a transition series is caused by

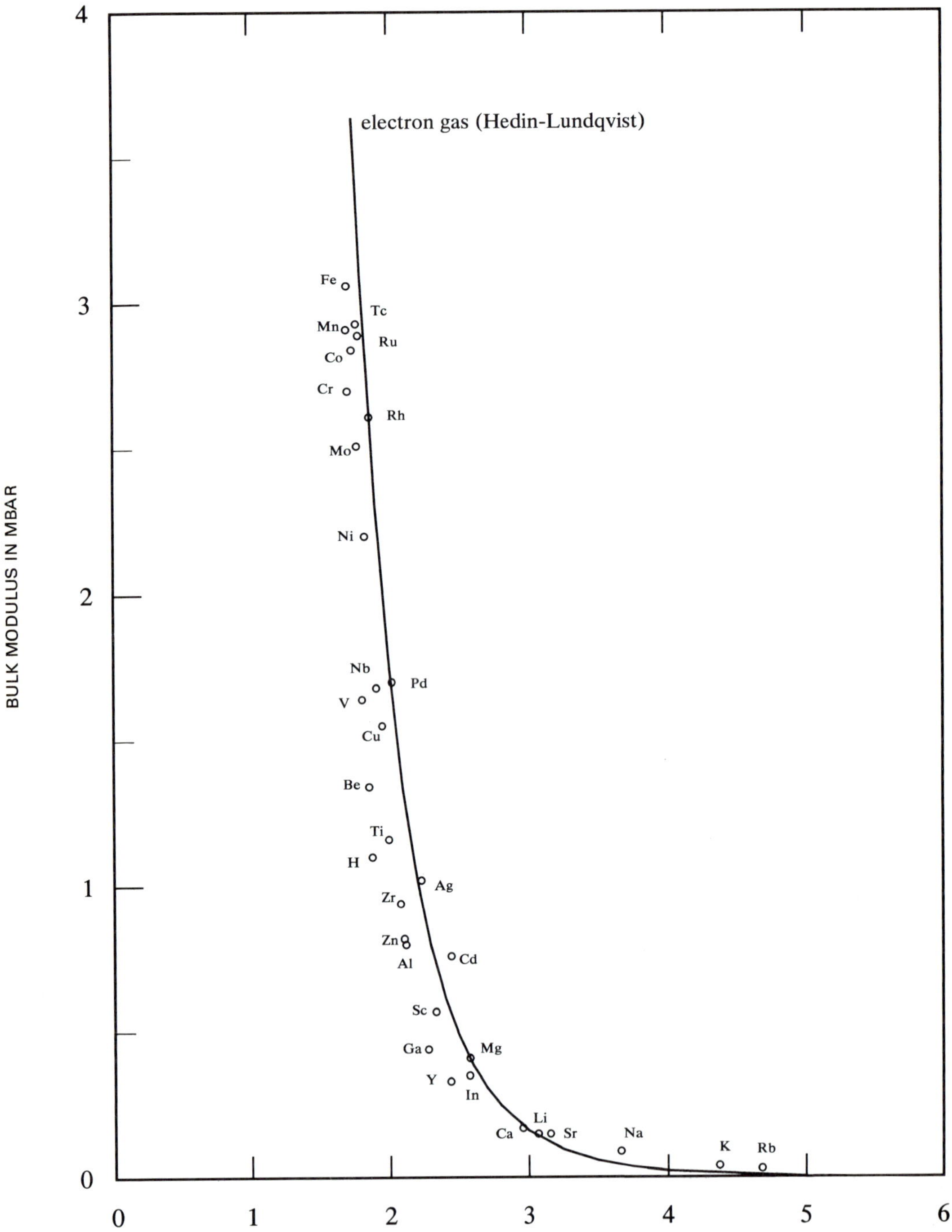

Figure 3.2 Bulk modulus vs r_s. The solid curve is the electron gas result. The points are the calculated bulk moduli (Mbar) at the r_s values derived from the interstitial charge density.

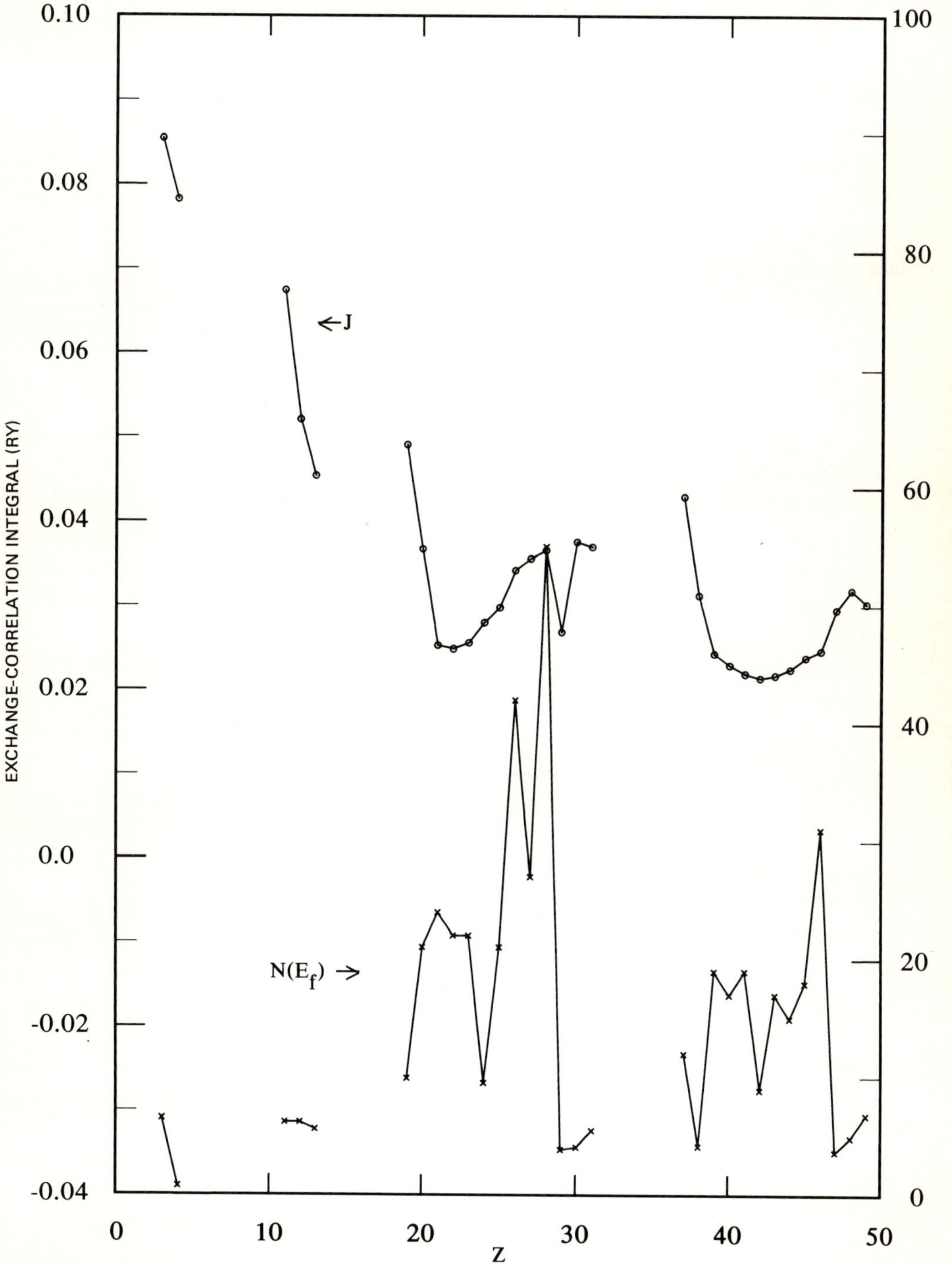

Figure 3.3 Exchange-correlation integral (upper curve; left scale) and density of states at the Fermi energy (lower curve; right scale) as functions of atomic number Z.

the change in character of the Fermi-energy wavefunctions as the Fermi energy increases through the d-band; since K(r) depends inversely on the charge density, it is largest far from the nucleus, where the charge density is smallest. At the beginning of a transition series, the Fermi-energy wavefunctions change from plane-wave-like to a more-localized d-character, resulting in a decrease in the value of $\gamma(r)$ (defined in eq. 2.18) in the region where K is largest, and thus to a decrease in J. Toward the end of the transition series, on the other hand, the Fermi-energy wavefunctions consist primarily of the highly-localized antibonding d-orbitals at the top of the d band. Since the integral of $\gamma(r)$ over a unit cell is unity, this increasing localization must lead to an increase in the amplitude of $\gamma(r)$, and this increase is enough to compensate for the smaller values of K(r) in the region where $\gamma(r)$ is largest.

It follows that ferromagnetism is a consequence of the localization of the orbitals near the top of the d-band; this localization produces both a maximum in the value of $N(E_f)$ and in the integral J. Both maxima are necessary: if J did not increase toward the end of the 3d transition series, and remained at its value for Ti, then only Ni would be predicted to have a ferromagnetic instability. On the other hand, if J did not gradually decrease with increasing Z, and were as large in the 4d series as it is in the 3d series, then ferromagnetism would be predicted to occur in Pd, even though the densities of states are smaller in the 4d series than in the 3d series. Thus the occurrence of ferromagnetism in the periodic table depends in a fundamental way on J having the dependence on Z shown in Figure 3.3.

In Figure 3.4, we show the limit as $r \to 0$ of $V(r)+2Z/r$, that is, the constant term in the potential at the origin, as a function of atomic number Z. This quantity is very nearly proportional to $Z^{3/2}$, and we have accordingly plotted the limit as $r \to 0$ of $Z^{-3/2}(V(r)+2Z/r)$ as a function of Z. An interesting feature of this plot is that, in Thomas-Fermi theory,[40] $V(r)+2Z/r$ should tend to a constant times $Z^{4/3}$ as $r \to 0$. The better treatment of the kinetic energy used in the present calculations apparently leads to a similar rule, but with the exponent closer to 3/2 than 4/3.

IV. ELECTRON AND STATE DENSITIES

In this Chapter, we present the results of calculations on the 32 metallic elements at lattice constants a_o close to the theoretical equilibrium lattice constants (corresponding to zero pressure). Calculations were usually performed for the observed crystal structure (fcc or bcc), but elements with more complicated structures were treated as fcc. The energy difference between structures is much smaller than the cohesive energy, and most of the properties given in the tables and graphs below depend only weakly on the structure. Exceptions are the density of states, and, through it, the spin susceptibility enhancement.

Four pages are devoted to each element. On the first page, we present a table of constants associated with the material. The atomic number and the lattice are input parameters, but all other quantities in this table are computed results. Included are:

1) The atomic number Z.
2) The assumed lattice (fcc or bcc).
3) The lattice constant a_o in atomic units (Bohrs). Except for the bulk modulus, which is evaluated at the equilibrium lattice constant, all subsequent quantities, graphs, and tables correspond to a_o, which differs slightly from the equilibrium lattice constant implied by eq. 2.14.
4) The pressure at a_o, which can be used in conjunction with the bulk modulus to find the calculated equilibrium lattice constant according to eq. 2.14 (pressure in kilobars is 1.471×10^5 times pressure in Rydbergs/Bohr^3).
5) The total energy of the solid, E_{solid}, in Rydbergs/atom.
6) The total energy of the free atom, E_{atom}, in Rydbergs.
7) The zero-point lattice energy, E_{zero} (see eq. 2.12), in Rydbergs/atom.
8) The cohesive energy, defined as $E_{atom}-(E_{solid}+E_{zero})$, also in Rydbergs/atom.
9) The bulk modulus in Mbars (1 Mbar=10^{12} dyne/cm^2), which measures the slope of the pressure curve at the calculated equilibrium lattice constant (see eq. 2.11).
10) The Fermi energy in Rydbergs relative to the muffin-tin zero.
11) The number of interstitial electrons per atom (Z_{out} in eq. 2.1).
12) The paramagnetic spin susceptibility enhancement, χ/χ_o, as defined in eq. 2.19.
13) The density of states (both spins) at the Fermi energy, $N(E_f)$, in states per eV-atom.
14) The charge density at the nucleus, ρ_o, in electrons per Bohr^3.
15) The energy derivative of the charge density at the Fermi energy and at the nucleus

(r=0), in electrons/Ry-Bohr^3. Items 14 and 15 are presented in this table because they would be difficult to obtain from the graphs to be described below.

Also on the first page is a table giving the band energies in Rydbergs (relative to the muffin-tin zero) at high symmetry points in the Brillouin zone, and a plot of the calculated energy bands along symmetry lines, in an energy range from the bottom of the valence bands to just above the Fermi energy.

On the second page appear graphs of the calculated density of states and charge densities. In the density of states graph, we show the density of states (both spins, in states/eV-atom) and the integrated density of states (electrons/atom) as functions of energy (eV) relative to the Fermi energy. The Fermi energy is defined as the energy where the integrated density of states equals $Z_{valence}$, the number of valence electrons per atom. The charge density graph shows r^2 times the total and the valence charge densities inside the muffin tin (the total charge density is always greater than the valence charge density) as functions of $\sqrt{r}$. The radial charge density is in units of electrons/Bohr, and the radius is in Bohrs. Multiplying the charge density by r^2 brings out structure which would not otherwise be visible.

On the third page, we plot the function $r^2 \partial\rho/\partial E$ evaluated at $E=E_f$, used in eq. 2.18, vs. $\sqrt{r}$. The second graph shows the difference between the charge densities of the solid and the free atom, and gives information about the charge redistribution upon the formation of the solid. The two functions plotted vs. $\sqrt{r}$ emphasize different regions of space. The function $\rho_{solid} - \rho_{atom}$ (the solid curve) reflects configurational differences between the solid and the atom, while the radial charge density difference (which is the same function multiplied by r^2, and is shown dotted) exhibits the charge redistribution at larger values of r.

On the fourth page we present a table giving the muffin-tin potential and charge density. Here we list the radius (Bohr), the effective one-electron potential in terms of 2Z(r) (=$-$rV(r)) in Rydberg-Bohrs, and $r^2\rho(r)$, the total radial charge density, in electrons/Bohr. The radial points are defined by a quadratic mesh: if R_{mt} is the muffin-tin radius, then $r=R_{mt}(i/128)^2$, i=1,...,128.

Hydrogen

Atomic Number	1
Lattice	fcc
Lattice constant a_o	4.30 a u.
Pressure at a_o	6.9 Kbar
Total energy (solid)	-1.076 Ry
Total energy (atom)	-0.976 Ry
Zero point energy	0.000 Ry
Cohesive energy	0.100 Ry
Bulk modulus	1.10 Mbar
Fermi energy	0.724 Ry
Interstitial electrons (per atom)	0.187
χ/χ_o	1.40
$N(E_f)$	0.10 states/eV-atom
ρ_o	0.30 electrons/Bohr^3
$(d\rho(r,E)/dE)_{E=E_f,r=0}$	0.5120 electrons/Bohr^3-Ry

Symmetry point	Band energies in Rydbergs (relative to muffin tin zero)
Γ	-0.496
X	
L	0.809
W	
K	

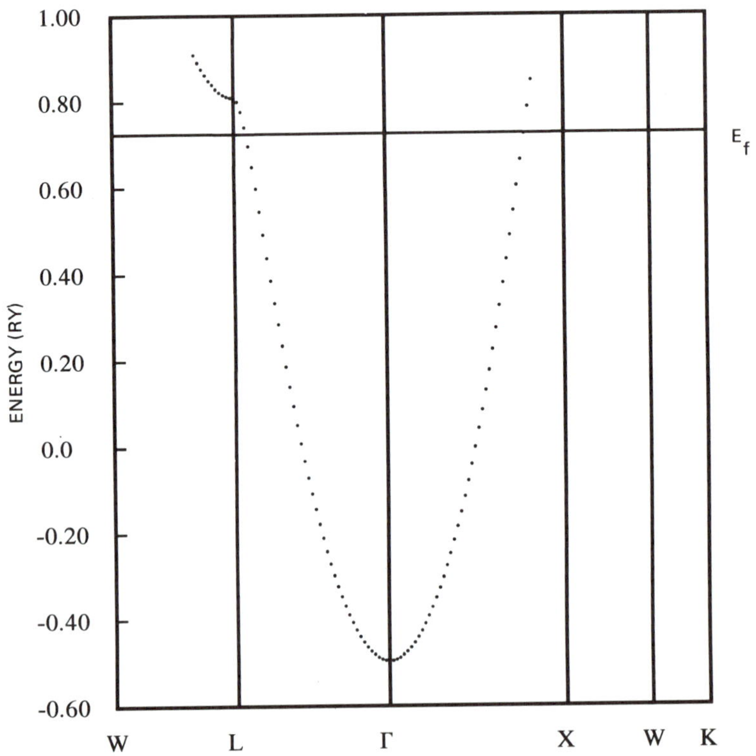

Z=1

Moruzzi
Janak
Williams

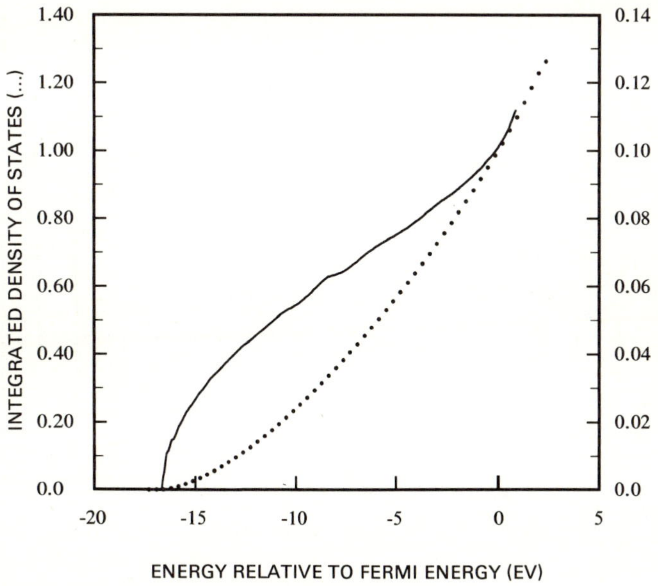

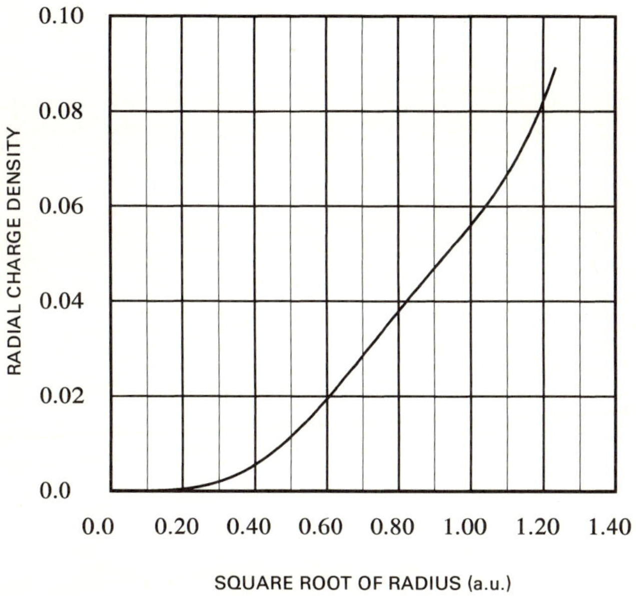

Moruzzi
Janak
Williams

Z=1

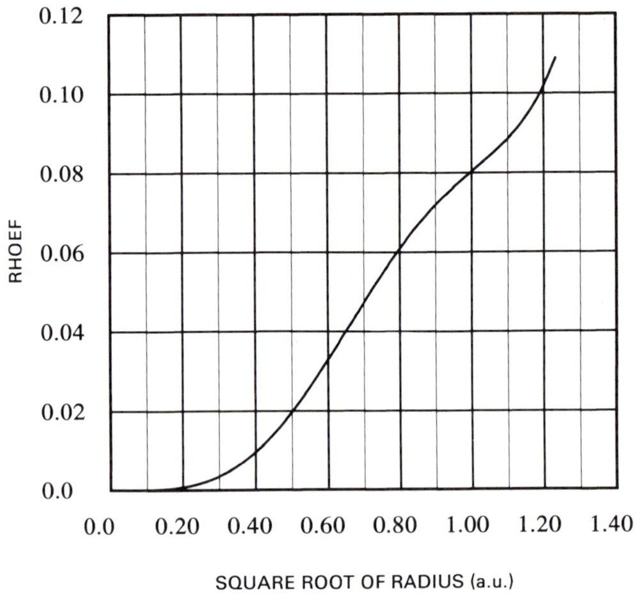

RHOEF=$r^2 \partial\rho(E,r)/\partial E$ (at $E=E_f$)

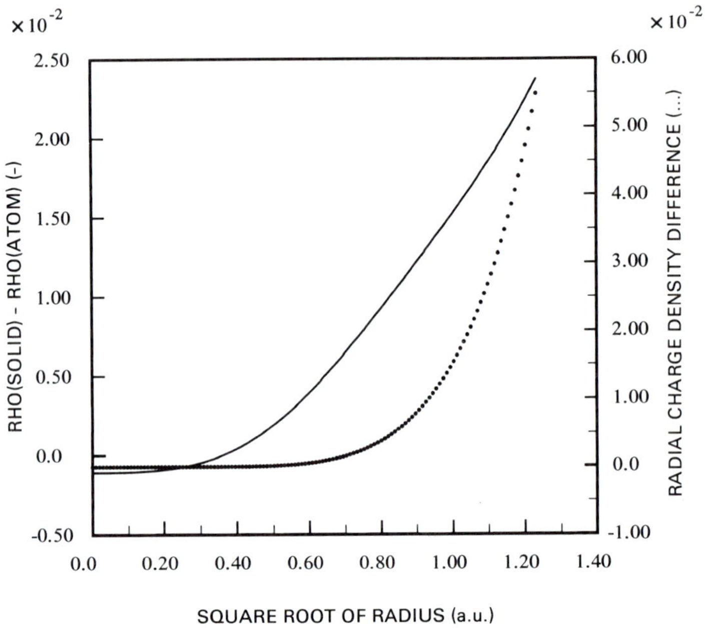

Z=1

Hydrogen

r	2z(r)	$r^2\rho$	r	2z(r)	$r^2\rho$
9.279046E - 05	1.999861E+00	2.595973E - 09	3.920397E - 01	1.350958E+00	2.180489E - 02
3.711618E - 04	1.999446E+00	4.151243E - 08	4.041952E - 01	1.330469E+00	2.267286E - 02
8.351142E - 04	1.998758E+00	2.099617E - 07	4.165363E - 01	1.309725E+00	2.354833E - 02
1.484647E - 03	1.997791E+00	6.627219E - 07	4.290631E - 01	1.288733E+00	2.443039E - 02
2.319762E - 03	1.996548E+00	1.615272E - 06	4.417754E - 01	1.267501E+00	2.531813E - 02
3.340457E - 03	1.995026E+00	3.342601E - 06	4.546732E - 01	1.246037E+00	2.621063E - 02
4.546732E - 03	1.993224E+00	6.177661E - 06	4.677567E - 01	1.224352E+00	2.710704E - 02
5.938590E - 03	1.991143E+00	1.050954E - 05	4.810258E - 01	1.202454E+00	2.800652E - 02
7.516026E - 03	1.988779E+00	1.678125E - 05	4.944804E - 01	1.180349E+00	2.890827E - 02
9.279046E - 03	1.986139E+00	2.548734E - 05	5.081205E - 01	1.158051E+00	2.981147E - 02
1.122764E - 02	1.983210E+00	3.717106E - 05	5.219463E - 01	1.135570E+00	3.071541E - 02
1.336183E - 02	1.979994E+00	5.242138E - 05	5.359577E - 01	1.112911E+00	3.161938E - 02
1.568159E - 02	1.976489E+00	7.186962E - 05	5.501546E - 01	1.090090E+00	3.252274E - 02
1.818693E - 02	1.972693E+00	9.618612E - 05	5.645372E - 01	1.067115E+00	3.342488E - 02
2.087785E - 02	1.968606E+00	1.260764E - 04	5.791053E - 01	1.043996E+00	3.432526E - 02
2.375436E - 02	1.964224E+00	1.622770E - 04	5.938590E - 01	1.020748E+00	3.522339E - 02
2.681644E - 02	1.959543E+00	2.055519E - 04	6.087982E - 01	9.973801E - 01	3.611886E - 02
3.006411E - 02	1.954563E+00	2.566874E - 04	6.239231E - 01	9.739051E - 01	3.701131E - 02
3.349736E - 02	1.949282E+00	3.164888E - 04	6.392335E - 01	9.503350E - 01	3.790044E - 02
3.711618E - 02	1.943695E+00	3.857757E - 04	6.547295E - 01	9.266832E - 01	3.878604E - 02
4.092059E - 02	1.937800E+00	4.653761E - 04	6.704111E - 01	9.029613E - 01	3.966797E - 02
4.491058E - 02	1.931597E+00	5.561241E - 04	6.862782E - 01	8.791847E - 01	4.054619E - 02
4.908615E - 02	1.925080E+00	6.588516E - 04	7.023310E - 01	8.553644E - 01	4.142070E - 02
5.344731E - 02	1.918247E+00	7.743882E - 04	7.185693E - 01	8.315167E - 01	4.229163E - 02
5.799404E - 02	1.911098E+00	9.035538E - 04	7.349932E - 01	8.076559E - 01	4.315917E - 02
6.272632E - 02	1.903625E+00	1.047153E - 03	7.516027E - 01	7.837963E - 01	4.402361E - 02
6.764424E - 02	1.895833E+00	1.205973E - 03	7.683978E - 01	7.599526E - 01	4.488533E - 02
7.274771E - 02	1.887714E+00	1.380783E - 03	7.853785E - 01	7.361413E - 01	4.574481E - 02
7.803673E - 02	1.879266E+00	1.572313E - 03	8.025447E - 01	7.123780E - 01	4.660261E - 02
8.351141E - 02	1.870489E+00	1.781278E - 03	8.198965E - 01	6.886784E - 01	4.745944E - 02
8.917159E - 02	1.861381E+00	2.008343E - 03	8.374339E - 01	6.650615E - 01	4.831604E - 02
9.501743E - 02	1.851933E+00	2.254142E - 03	8.551569E - 01	6.415411E - 01	4.917338E - 02
1.010488E - 01	1.842152E+00	2.519264E - 03	8.730654E - 01	6.181393E - 01	5.003237E - 02
1.072658E - 01	1.832033E+00	2.804257E - 03	8.911596E - 01	5.948713E - 01	5.089418E - 02
1.136683E - 01	1.821574E+00	3.109606E - 03	9.094393E - 01	5.717574E - 01	5.176002E - 02
1.202564E - 01	1.810778E+00	3.435763E - 03	9.279046E - 01	5.488166E - 01	5.263124E - 02
1.270301E - 01	1.799634E+00	3.783114E - 03	9.465555E - 01	5.260677E - 01	5.350929E - 02
1.339894E - 01	1.788146E+00	4.151978E - 03	9.653919E - 01	5.035341E - 01	5.439575E - 02
1.411343E - 01	1.776314E+00	4.542649E - 03	9.844140E - 01	4.812338E - 01	5.529238E - 02
1.484647E - 01	1.764136E+00	4.955325E - 03	1.003621E+00	4.591913E - 01	5.620091E - 02
1.559808E - 01	1.751612E+00	5.390167E - 03	1.023014E+00	4.374267E - 01	5.712347E - 02
1.636823E - 01	1.738743E+00	5.847242E - 03	1.042593E+00	4.159636E - 01	5.806215E - 02
1.715695E - 01	1.725527E+00	6.326597E - 03	1.062358E+00	3.948247E - 01	5.901921E - 02
1.796423E - 01	1.711965E+00	6.828181E - 03	1.082308E+00	3.740361E - 01	5.999694E - 02
1.879007E - 01	1.698058E+00	7.351901E - 03	1.102443E+00	3.536242E - 01	6.099788E - 02
1.963446E - 01	1.683807E+00	7.897578E - 03	1.122765E+00	3.336116E - 01	6.202500E - 02
2.049741E - 01	1.669213E+00	8.464985E - 03	1.143270E+00	3.140278E - 01	6.308067E - 02
2.137892E - 01	1.654277E+00	9.053834E - 03	1.163963E+00	2.948967E - 01	6.416851E - 02
2.227899E - 01	1.639001E+00	9.663772E - 03	1.184841E+00	2.762482E - 01	6.529129E - 02
2.319762E - 01	1.623387E+00	1.029439E - 02	1.205904E+00	2.581145E - 01	6.645250E - 02
2.413480E - 01	1.607438E+00	1.094521E - 02	1.227154E+00	2.405195E - 01	6.765574E - 02
2.509054E - 01	1.591156E+00	1.161572E - 02	1.248588E+00	2.234991E - 01	6.890452E - 02
2.606484E - 01	1.574543E+00	1.230535E - 02	1.270208E+00	2.070800E - 01	7.020313E - 02
2.705770E - 01	1.557604E+00	1.301346E - 02	1.292014E+00	1.912987E - 01	7.155538E - 02
2.806911E - 01	1.540341E+00	1.373939E - 02	1.314005E+00	1.761875E - 01	7.296556E - 02
2.909909E - 01	1.522759E+00	1.448243E - 02	1.336183E+00	1.617774E - 01	7.443851E - 02
3.014762E - 01	1.504863E+00	1.524181E - 02	1.358544E+00	1.481084E - 01	7.597858E - 02
3.121471E - 01	1.486655E+00	1.601675E - 02	1.381093E+00	1.352118E - 01	7.759112E - 02
3.230036E - 01	1.468143E+00	1.680643E - 02	1.403827E+00	1.231270E - 01	7.928103E - 02
3.340456E - 01	1.449328E+00	1.761000E - 02	1.426745E+00	1.118926E - 01	8.105379E - 02
3.452733E - 01	1.430221E+00	1.842657E - 02	1.449850E+00	1.015453E - 01	8.291525E - 02
3.566865E - 01	1.410823E+00	1.925527E - 02	1.473141E+00	9.212422E - 02	8.487141E - 02
3.682853E - 01	1.391143E+00	2.009516E - 02	1.496617E+00	8.367127E - 02	8.692831E - 02
3.800697E - 01	1.371185E+00	2.094536E - 02	1.520279E+00	7.622689E - 02	8.909237E - 02

Z=1

Lithium

Atomic Number	3
Lattice	bcc
Lattice constant a_o	6.42 a.u.
Pressure at a_o	-1.06 Kbar
Total energy (solid)	-14.832 Ry
Total energy (atom)	-14.709 Ry
Zero point energy	0.003 Ry
Cohesive energy	0.121 Ry
Bulk modulus	0.15 Mbar
Fermi energy	0.313 Ry
Interstitial electrons (per atom)	0.350
χ/χ_o	2.25
$N(E_f)$	0.48 states/eV-atom
ρ_o	13.52 electrons/Bohr^3
$(d\rho(r,E)/dE)_{E=E_f,r=0}$	0.8638 electrons/Bohr^3-Ry

Symmetry point	Band energies in Rydbergs (relative to muffin tin zero)
Γ	0.052
H	
N	0.329
P	

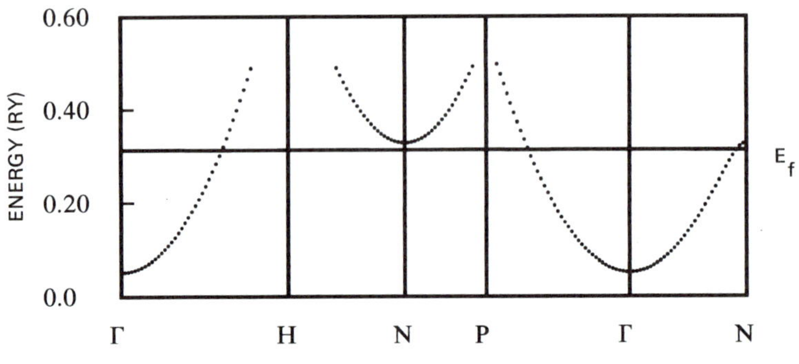

Z=3

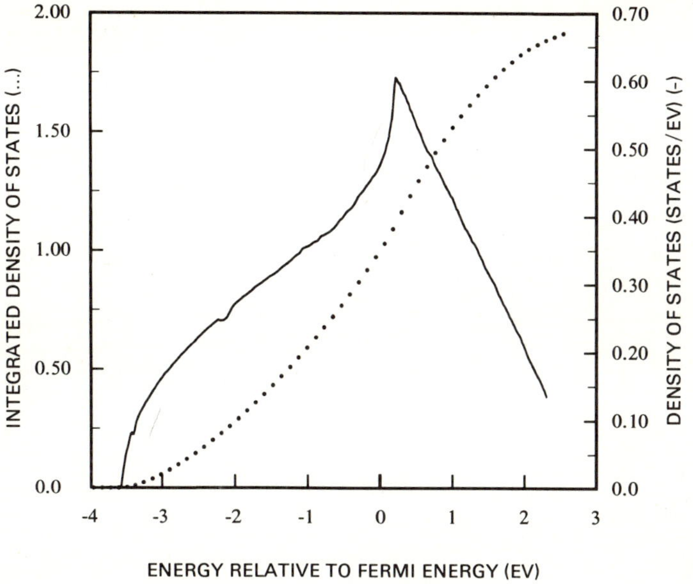

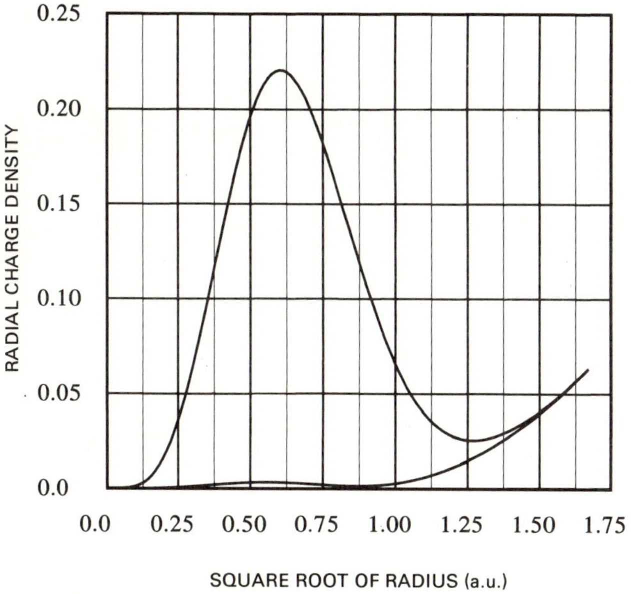

Moruzzi
Janak
Williams

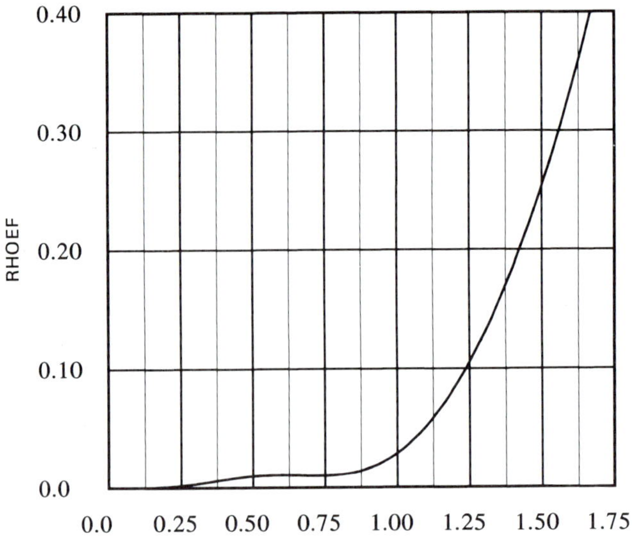

RHOEF=$r^2\partial\rho(E,r)/\partial E$ (at $E=E_f$)

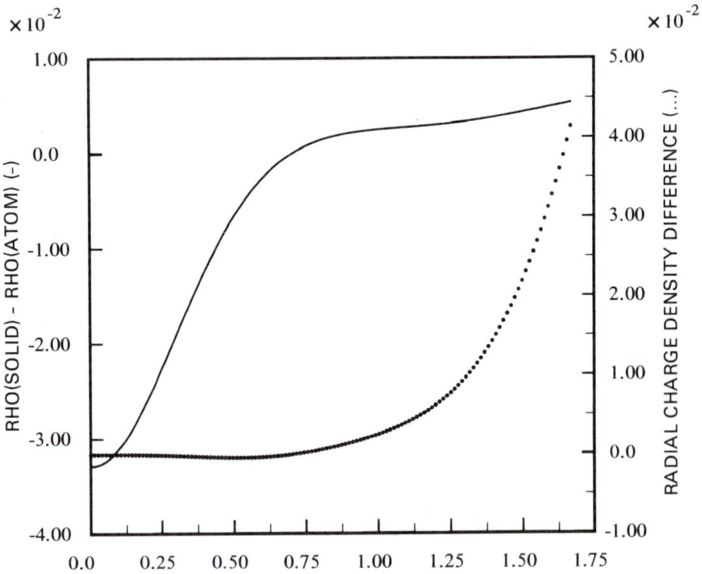

Moruzzi
Janak
Williams

Z=3

Lithium

r	2z(r)	$r^2\rho$	r	2z(r)	$r^2\rho$
1.696741E - 04	5.998793E+00	3.889143E - 07	7.168731E - 01	2.328291E+00	1.332994E - 01
6.786964E - 04	5.995169E+00	6.203654E - 06	7.391004E - 01	2.269030E+00	1.266540E - 01
1.527067E - 03	5.989122E+00	3.124654E - 05	7.616671E - 01	2.210379E+00	1.200981E - 01
2.714786E - 03	5.980632E+00	9.805351E - 05	7.845731E - 01	2.152307E+00	1.136619E - 01
4.241850E - 03	5.969680E+00	2.372058E - 04	8.078184E - 01	2.094773E+00	1.073725E - 01
6.108265E - 03	5.956239E+00	4.863960E - 04	8.314031E - 01	2.037749E+00	1.012540E - 01
8.314028E - 03	5.940279E+00	8.892729E - 04	8.553271E - 01	1.981203E+00	9.532750E - 02
1.085914E - 02	5.921769E+00	1.494099E - 03	8.795906E - 01	1.925112E+00	8.961141E - 02
1.374360E - 02	5.900676E+00	2.352262E - 03	9.041933E - 01	1.869461E+00	8.412105E - 02
1.696741E - 02	5.876966E+00	3.516675E - 03	9.291354E - 01	1.814233E+00	7.886934E - 02
2.053057E - 02	5.850604E+00	5.040135E - 03	9.544168E - 01	1.759422E+00	7.386631E - 02
2.443307E - 02	5.821579E+00	6.973643E - 03	9.800376E - 01	1.705028E+00	6.911981E - 02
2.867492E - 02	5.789839E+00	9.364780E - 03	1.005998E+00	1.651060E+00	6.463552E - 02
3.325612E - 02	5.755377E+00	1.225613E - 02	1.032297E+00	1.597531E+00	6.041659E - 02
3.817667E - 02	5.718182E+00	1.568390E - 02	1.058935E+00	1.544476E+00	5.646473E - 02
4.343657E - 02	5.678249E+00	1.967660E - 02	1.085914E+00	1.491898E+00	5.277939E - 02
4.903582E - 02	5.635587E+00	2.425397E - 02	1.113232E+00	1.439850E+00	4.935887E - 02
5.497441E - 02	5.590213E+00	2.942621E - 02	1.140888E+00	1.388395E+00	4.620004E - 02
6.125235E - 02	5.542153E+00	3.519320E - 02	1.168884E+00	1.337572E+00	4.329830E - 02
6.786960E - 02	5.491452E+00	4.154436E - 02	1.197220E+00	1.287446E+00	4.064809E - 02
7.482624E - 02	5.438161E+00	4.845838E - 02	1.225895E+00	1.238084E+00	3.824295E - 02
8.212227E - 02	5.382346E+00	5.590355E - 02	1.254910E+00	1.189559E+00	3.607568E - 02
8.975756E - 02	5.324087E+00	6.383777E - 02	1.284263E+00	1.141958E+00	3.413850E - 02
9.773225E - 02	5.263474E+00	7.220995E - 02	1.313956E+00	1.095338E+00	3.242298E - 02
1.060463E - 01	5.200611E+00	8.096004E - 02	1.343988E+00	1.049794E+00	3.092057E - 02
1.146997E - 01	5.135612E+00	9.002066E - 02	1.374360E+00	1.005386E+00	2.962232E - 02
1.236924E - 01	5.068605E+00	9.931791E - 02	1.405071E+00	9.621792E - 01	2.851917E - 02
1.330245E - 01	4.999721E+00	1.087729E - 01	1.436121E+00	9.202393E - 01	2.760208E - 02
1.426959E - 01	4.929100E+00	1.183031E - 01	1.467511E+00	8.795832E - 01	2.686191E - 02
1.527067E - 01	4.856902E+00	1.278236E - 01	1.499240E+00	8.402638E - 01	2.628982E - 02
1.630568E - 01	4.783274E+00	1.372489E - 01	1.531308E+00	8.022841E - 01	2.587697E - 02
1.737463E - 01	4.708379E+00	1.464939E - 01	1.563716E+00	7.656429E - 01	2.561483E - 02
1.847751E - 01	4.632382E+00	1.554756E - 01	1.596463E+00	7.303228E - 01	2.549513E - 02
1.961433E - 01	4.555443E+00	1.641145E - 01	1.629550E+00	6.962931E - 01	2.550988E - 02
2.078508E - 01	4.477732E+00	1.723348E - 01	1.662975E+00	6.635270E - 01	2.565143E - 02
2.198976E - 01	4.399405E+00	1.800669E - 01	1.696740E+00	6.319615E - 01	2.591247E - 02
2.322838E - 01	4.320625E+00	1.872473E - 01	1.730845E+00	6.015308E - 01	2.628611E - 02
2.450094E - 01	4.241552E+00	1.938194E - 01	1.765289E+00	5.721946E - 01	2.676575E - 02
2.580743E - 01	4.162326E+00	1.997346E - 01	1.800072E+00	5.438942E - 01	2.734520E - 02
2.714785E - 01	4.083101E+00	2.049522E - 01	1.835195E+00	5.165442E - 01	2.801871E - 02
2.852222E - 01	4.004002E+00	2.094399E - 01	1.870657E+00	4.900954E - 01	2.878086E - 02
2.993051E - 01	3.925161E+00	2.131739E - 01	1.906458E+00	4.645079E - 01	2.962655E - 02
3.137274E - 01	3.846696E+00	2.161388E - 01	1.942598E+00	4.397179E - 01	3.055114E - 02
3.284891E - 01	3.768714E+00	2.183276E - 01	1.979078E+00	4.156826E - 01	3.155030E - 02
3.435900E - 01	3.691316E+00	2.197410E - 01	2.015898E+00	3.923626E - 01	3.262008E - 02
3.590304E - 01	3.614587E+00	2.203876E - 01	2.053057E+00	3.697281E - 01	3.375677E - 02
3.748101E - 01	3.538604E+00	2.202829E - 01	2.090554E+00	3.477643E - 01	3.495707E - 02
3.909291E - 01	3.463434E+00	2.194488E - 01	2.128391E+00	3.264395E - 01	3.621794E - 02
4.073875E - 01	3.389132E+00	2.179135E - 01	2.166568E+00	3.057423E - 01	3.753662E - 02
4.241852E - 01	3.315742E+00	2.157101E - 01	2.205084E+00	2.856640E - 01	3.891067E - 02
4.413223E - 01	3.243299E+00	2.128764E - 01	2.243939E+00	2.662033E - 01	4.033790E - 02
4.587988E - 01	3.171827E+00	2.094541E - 01	2.283134E+00	2.473618E - 01	4.181628E - 02
4.766145E - 01	3.101343E+00	2.054880E - 01	2.322668E+00	2.291577E - 01	4.334406E - 02
4.947697E - 01	3.031854E+00	2.010252E - 01	2.362542E+00	2.115797E - 01	4.491976E - 02
5.132642E - 01	2.963356E+00	1.961151E - 01	2.402755E+00	1.946663E - 01	4.654196E - 02
5.320980E - 01	2.895843E+00	1.908075E - 01	2.443307E+00	1.784246E - 01	4.820959E - 02
5.512711E - 01	2.829297E+00	1.851534E - 01	2.484199E+00	1.628752E - 01	4.992162E - 02
5.707837E - 01	2.763697E+00	1.792032E - 01	2.525429E+00	1.480551E - 01	5.167713E - 02
5.906355E - 01	2.699018E+00	1.730073E - 01	2.566999E+00	1.339683E - 01	5.347552E - 02
6.108268E - 01	2.635225E+00	1.666147E - 01	2.608909E+00	1.206684E - 01	5.531607E - 02
6.313573E - 01	2.572288E+00	1.600731E - 01	2.651157E+00	1.081736E - 01	5.719841E - 02
6.522272E - 01	2.510165E+00	1.534282E - 01	2.693746E+00	9.651840E - 02	5.912204E - 02
6.734365E - 01	2.448818E+00	1.467239E - 01	2.736673E+00	8.574009E - 02	6.108673E - 02
6.949852E - 01	2.388207E+00	1.400014E - 01	2.779941E+00	7.587230E - 02	6.309205E - 02

Z=3

Moruzzi
Janak
Williams

Beryllium

Atomic Number	4
Lattice	fcc
Lattice constant a_o	5.96 a.u.
Pressure at a_o	-22.5 Kbar
Total energy (solid)	-29.212 Ry
Total energy (atom)	-28.909 Ry
Zero point energy	0.011 Ry
Cohesive energy	0.292 Ry
Bulk modulus	1.34 Mbar
Fermi energy	0.845 Ry
Interstitial electrons (per atom)	0.516
χ/χ_o	1.06
$N(E_f)$	0.054 states/eV-atom
ρ_o	34.13 electrons/$Bohr^3$
$(d\rho(r,E)/dE)_{E=E_f,r=0}$	0.2560 electrons/$Bohr^3$-Ry

Symmetry point	Band energies in Rydbergs (relative to muffin tin zero)	
Γ	-0.026	
X	0.648	
L	0.444	0.856
W	0.913	0.913
K	0.781	0.863

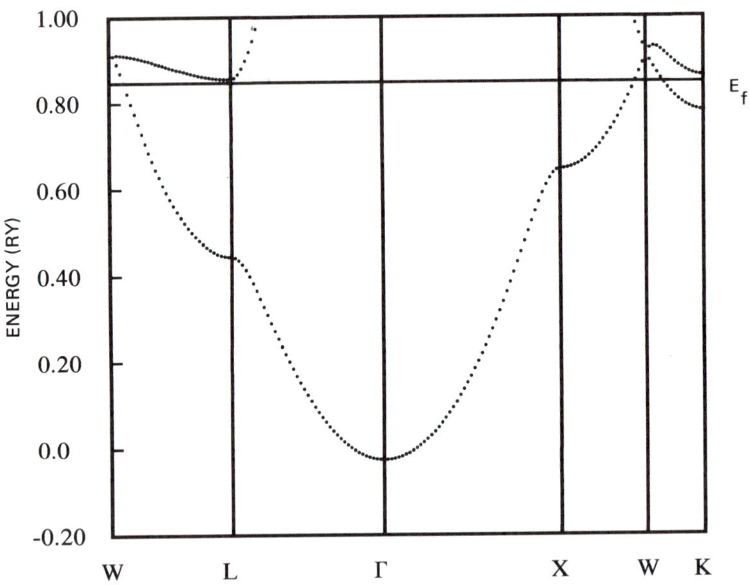

Z=4

Moruzzi
Janak
Williams

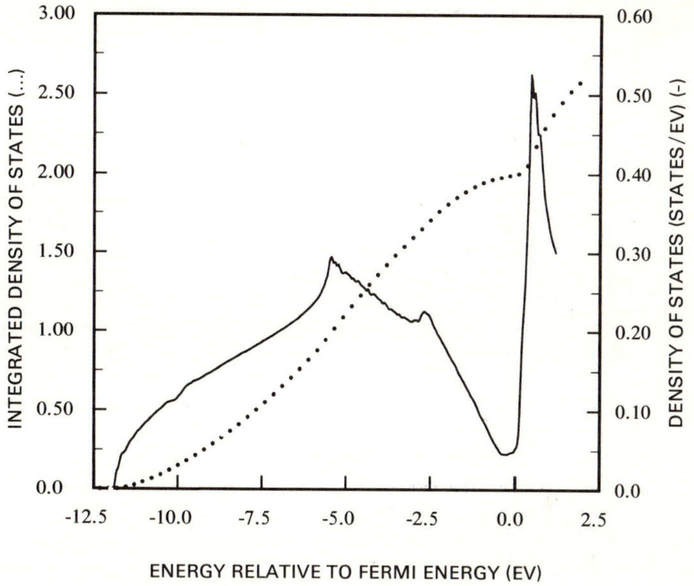

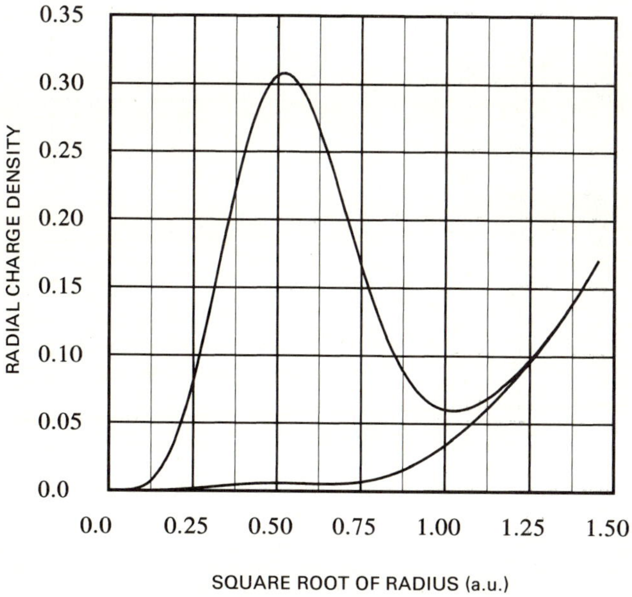

Moruzzi
Janak
Williams

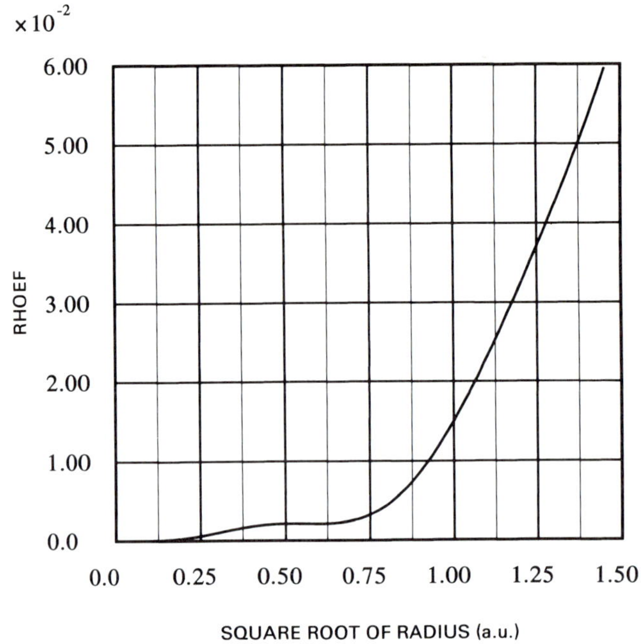

RHOEF=$r^2 \partial \rho(E,r)/\partial E$ (at $E=E_f$)

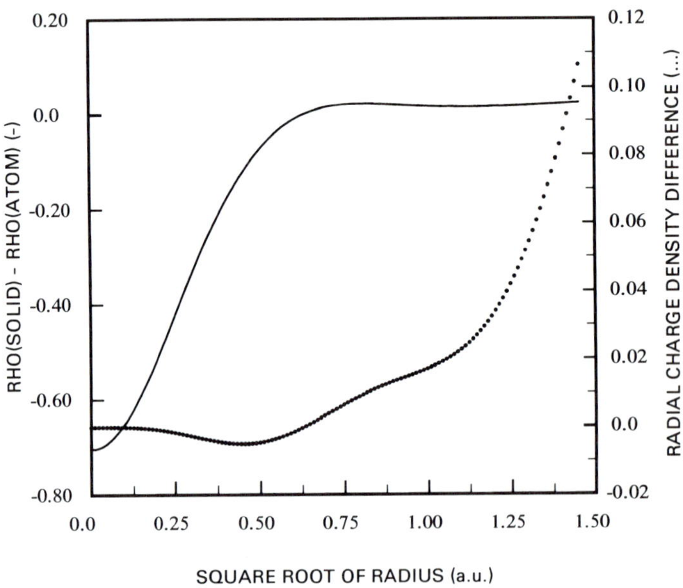

Beryllium

r	2z(r)	$r^2\rho$	r	2z(r)	$r^2\rho$
1.286119E - 04	7.998549E+00	5.640339E - 07	5.433853E - 01	3.499702E+00	1.768097E - 01
5.144477E - 04	7.994195E+00	8.996732E - 06	5.602335E - 01	3.418974E+00	1.677625E - 01
1.157507E - 03	7.986926E+00	4.531226E - 05	5.773389E - 01	3.338718E+00	1.589378E - 01
2.057791E - 03	7.976744E+00	1.421817E - 04	5.947015E - 01	3.258924E+00	1.503767E - 01
3.215298E - 03	7.963602E+00	3.439249E - 04	6.123213E - 01	3.179582E+00	1.421157E - 01
4.630029E - 03	7.947482E+00	7.051406E - 04	6.301984E - 01	3.100695E+00	1.341862E - 01
6.301984E - 03	7.928357E+00	1.289017E - 03	6.483327E - 01	3.022271E+00	1.266150E - 01
8.231163E - 03	7.906190E+00	2.165362E - 03	6.667242E - 01	2.944325E+00	1.194243E - 01
1.041757E - 02	7.880946E+00	3.408435E - 03	6.853729E - 01	2.866884E+00	1.126319E - 01
1.286119E - 02	7.852599E+00	5.094588E - 03	7.042789E - 01	2.789980E+00	1.062513E - 01
1.556204E - 02	7.821108E+00	7.299893E - 03	7.234420E - 01	2.713653E+00	1.002924E - 01
1.852012E - 02	7.786448E+00	1.009765E - 02	7.428625E - 01	2.637952E+00	9.476125E - 02
2.173541E - 02	7.748595E+00	1.355604E - 02	7.625400E - 01	2.562938E+00	8.966064E - 02
2.520794E - 02	7.707536E+00	1.773588E - 02	7.824749E - 01	2.488670E+00	8.499032E - 02
2.893768E - 02	7.663258E+00	2.268846E - 02	8.026670E - 01	2.415220E+00	8.074725E - 02
3.292465E - 02	7.615783E+00	2.845381E - 02	8.231163E - 01	2.342660E+00	7.692611E - 02
3.716885E - 02	7.565085E+00	3.505901E - 02	8.438228E - 01	2.271061E+00	7.351929E - 02
4.167026E - 02	7.511208E+00	4.251714E - 02	8.647866E - 01	2.200497E+00	7.051742E - 02
4.642890E - 02	7.454185E+00	5.082615E - 02	8.860075E - 01	2.131039E+00	6.790966E - 02
5.144477E - 02	7.394064E+00	5.996868E - 02	9.074857E - 01	2.062745E+00	6.568372E - 02
5.671786E - 02	7.330911E+00	6.991154E - 02	9.292211E - 01	1.995675E+00	6.382638E - 02
6.224817E - 02	7.264800E+00	8.060652E - 02	9.512138E - 01	1.929868E+00	6.232372E - 02
6.803566E - 02	7.195824E+00	9.199041E - 02	9.734636E - 01	1.865357E+00	6.116094E - 02
7.408047E - 02	7.124082E+00	1.039866E - 01	9.959707E - 01	1.802159E+00	6.032313E - 02
8.038241E - 02	7.049695E+00	1.165055E - 01	1.018735E+00	1.740282E+00	5.979505E - 02
8.694166E - 02	6.972787E+00	1.294473E - 01	1.041756E+00	1.679726E+00	5.956147E - 02
9.375805E - 02	6.893500E+00	1.427022E - 01	1.065035E+00	1.620438E+00	5.960726E - 02
1.008317E - 01	6.811983E+00	1.561542E - 01	1.088571E+00	1.562419E+00	5.991755E - 02
1.081626E - 01	6.728390E+00	1.696812E - 01	1.112364E+00	1.505614E+00	6.047761E - 02
1.157507E - 01	6.642890E+00	1.831589E - 01	1.136415E+00	1.449977E+00	6.127346E - 02
1.235960E - 01	6.555652E+00	1.964616E - 01	1.160722E+00	1.395478E+00	6.229134E - 02
1.316986E - 01	6.466848E+00	2.094651E - 01	1.185287E+00	1.342033E+00	6.351823E - 02
1.400583E - 01	6.376657E+00	2.220485E - 01	1.210109E+00	1.289643E+00	6.494147E - 02
1.486754E - 01	6.285254E+00	2.340962E - 01	1.235188E+00	1.238214E+00	6.654948E - 02
1.575496E - 01	6.192818E+00	2.454989E - 01	1.260525E+00	1.187736E+00	6.833106E - 02
1.666811E - 01	6.099523E+00	2.561566E - 01	1.286119E+00	1.138157E+00	7.027572E - 02
1.760697E - 01	6.005535E+00	2.659785E - 01	1.311970E+00	1.089445E+00	7.237399E - 02
1.857156E - 01	5.911019E+00	2.748850E - 01	1.338078E+00	1.041597E+00	7.461667E - 02
1.956187E - 01	5.816136E+00	2.828079E - 01	1.364444E+00	9.945615E - 01	7.699585E - 02
2.057791E - 01	5.721035E+00	2.896919E - 01	1.391067E+00	9.483577E - 01	7.950389E - 02
2.161966E - 01	5.625852E+00	2.954940E - 01	1.417946E+00	9.029886E - 01	8.213377E - 02
2.268714E - 01	5.530721E+00	3.001843E - 01	1.445083E+00	8.584399E - 01	8.487988E - 02
2.378034E - 01	5.435764E+00	3.037455E - 01	1.472477E+00	8.147190E - 01	8.773685E - 02
2.489927E - 01	5.341093E+00	3.061730E - 01	1.500129E+00	7.718462E - 01	9.069979E - 02
2.604391E - 01	5.246797E+00	3.074741E - 01	1.528038E+00	7.298440E - 01	9.376484E - 02
2.721428E - 01	5.152969E+00	3.076674E - 01	1.556204E+00	6.887529E - 01	9.692848E - 02
2.841037E - 01	5.059686E+00	3.067824E - 01	1.584627E+00	6.485956E - 01	1.001882E - 01
2.963219E - 01	4.967008E+00	3.048581E - 01	1.613308E+00	6.094051E - 01	1.035418E - 01
3.087972E - 01	4.874986E+00	3.019423E - 01	1.642245E+00	5.712411E - 01	1.069875E - 01
3.215298E - 01	4.783662E+00	2.980904E - 01	1.671440E+00	5.341339E - 01	1.105245E - 01
3.345196E - 01	4.693069E+00	2.933645E - 01	1.700892E+00	4.981369E - 01	1.141524E - 01
3.477666E - 01	4.603218E+00	2.878321E - 01	1.730601E+00	4.633175E - 01	1.178710E - 01
3.612708E - 01	4.514130E+00	2.815652E - 01	1.760568E+00	4.297192E - 01	1.216808E - 01
3.750324E - 01	4.425801E+00	2.746388E - 01	1.790792E+00	3.973993E - 01	1.255831E - 01
3.890510E - 01	4.338232E+00	2.671304E - 01	1.821273E+00	3.664224E - 01	1.295788E - 01
4.033270E - 01	4.251408E+00	2.591186E - 01	1.852012E+00	3.368585E - 01	1.336700E - 01
4.178601E - 01	4.165315E+00	2.506822E - 01	1.883007E+00	3.087895E - 01	1.378586E - 01
4.326505E - 01	4.079926E+00	2.418991E - 01	1.914259E+00	2.822881E - 01	1.421470E - 01
4.476981E - 01	3.995223E+00	2.328464E - 01	1.945769E+00	2.573932E - 01	1.465385E - 01
4.630029E - 01	3.911172E+00	2.235982E - 01	1.977536E+00	2.342174E - 01	1.510356E - 01
4.785649E - 01	3.827746E+00	2.142265E - 01	2.009561E+00	2.128242E - 01	1.556420E - 01
4.943842E - 01	3.744914E+00	2.047994E - 01	2.041842E+00	1.932907E - 01	1.603613E - 01
5.104607E - 01	3.662649E+00	1.953815E - 01	2.074381E+00	1.757177E - 01	1.651970E - 01
5.267944E - 01	3.580918E+00	1.860330E - 01	2.107178E+00	1.601566E - 01	1.701534E - 01

Z=4

Sodium

Atomic Number	11
Lattice	bcc
Lattice constant a_o	7.70 a.u.
Pressure at a_o	-0.11 Kbar
Total energy (solid)	-322.985 Ry
Total energy (atom)	-322.902 Ry
Zero point energy	0.001 Ry
Cohesive energy	0.082 Ry
Bulk modulus	0.09 Mbar
Fermi energy	0.248 Ry
Interstitial electrons (per atom)	0.354
χ/χ_o	1.71
$N(E_f)$	0.45 states/eV-atom
ρ_o	8.269×10^2 electrons/Bohr^3
$(d\rho(r,E)/dE)_{E=E_f,r=0}$	3.6277 electrons/Bohr^3-Ry

Symmetry point	Band energies in Rydbergs (relative to muffin tin zero)	
Γ	-0.007	
H		
N	0.306	0.344
P	0.445	

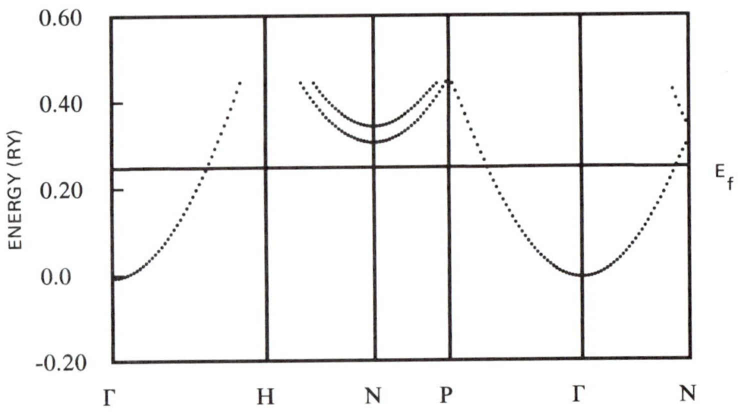

Z=11

Moruzzi
Janak
Williams

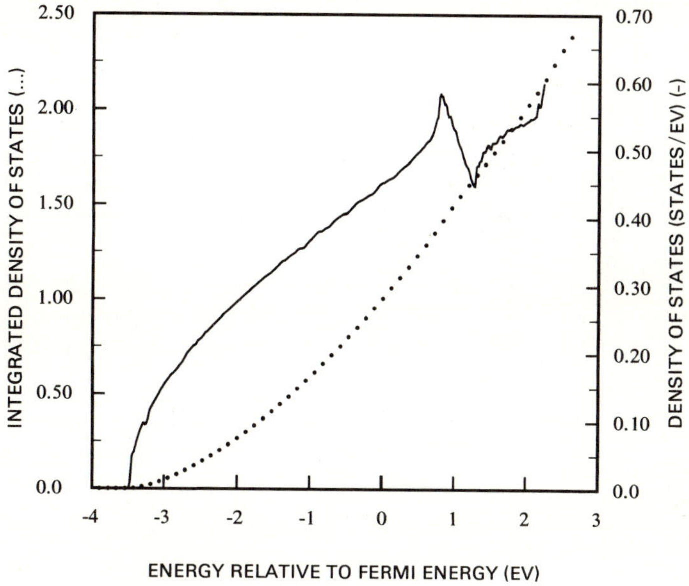

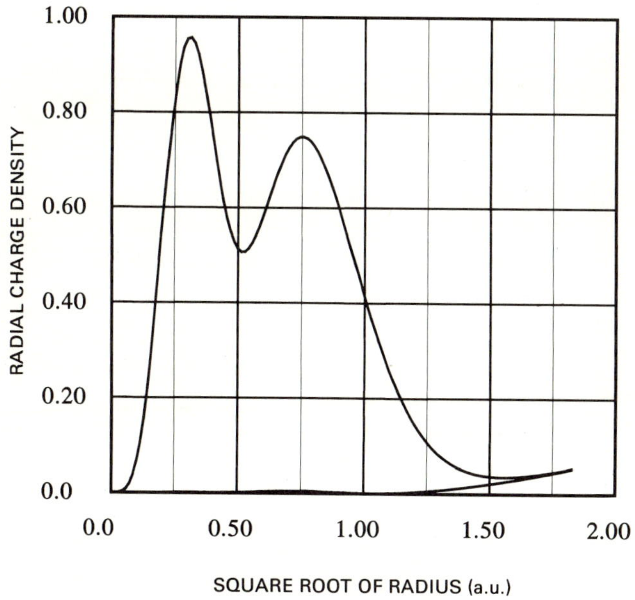

Moruzzi
Janak
Williams

Z=11

Sodium

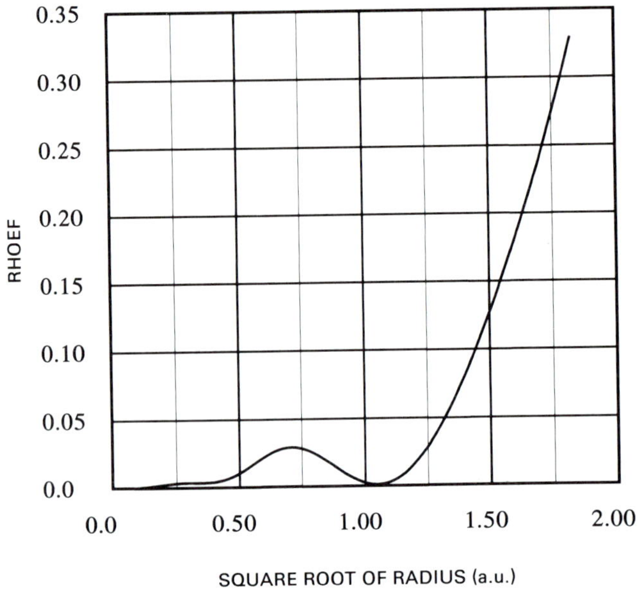

$RHOEF = r^2 \partial\rho(E,r)/\partial E$ (at $E = E_f$)

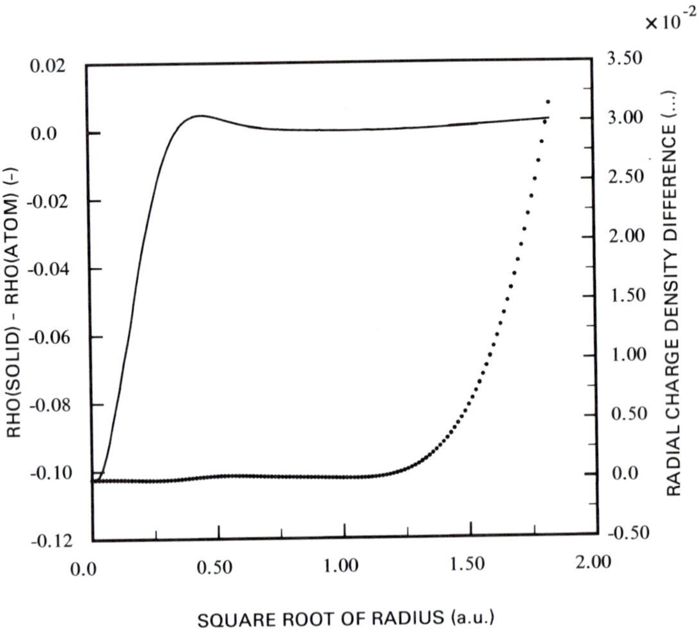

Moruzzi
Janak
Williams

Sodium

r	2z(r)	$r^2\rho$	r	2z(r)	$r^2\rho$
2.035032E - 04	2.198930E+01	3.409119E - 05	8.598012E - 01	4.224504E+00	5.463668E - 01
8.140130E - 04	2.195721E+01	5.381822E - 04	8.864601E - 01	4.060454E+00	5.203066E - 01
1.831529E - 03	2.190352E+01	2.664249E - 03	9.135260E - 01	3.903059E+00	4.940735E - 01
3.256052E - 03	2.182793E+01	8.160654E - 03	9.409989E - 01	3.752038E+00	4.678739E - 01
5.087581E - 03	2.173018E+01	1.913722E - 02	9.688789E - 01	3.607094E+00	4.418938E - 01
7.326115E - 03	2.161000E+01	3.777811E - 02	9.971659E - 01	3.467928E+00	4.162980E - 01
9.971656E - 03	2.146718E+01	6.603765E - 02	1.025859E+00	3.334309E+00	3.912305E - 01
1.302421E - 02	2.130174E+01	1.053566E - 01	1.054960E+00	3.205771E+00	3.668131E - 01
1.648376E - 02	2.111391E+01	1.564304E - 01	1.084468E+00	3.082172E+00	3.431515E - 01
2.035032E - 02	2.090417E+01	2.190595E - 01	1.114384E+00	2.963082E+00	3.203297E - 01
2.462389E - 02	2.067334E+01	2.920940E - 01	1.144705E+00	2.848429E+00	2.984186E - 01
2.930446E - 02	2.042249E+01	3.734788E - 01	1.175434E+00	2.737734E+00	2.774687E - 01
3.439204E - 02	2.015298E+01	4.603912E - 01	1.206571E+00	2.630812E+00	2.575193E - 01
3.988663E - 02	1.986641E+01	5.494545E - 01	1.238113E+00	2.527488E+00	2.385962E - 01
4.578822E - 02	1.956451E+01	6.369991E - 01	1.270063E+00	2.427467E+00	2.207124E - 01
5.209683E - 02	1.924916E+01	7.193487E - 01	1.302421E+00	2.330538E+00	2.038711E - 01
5.881243E - 02	1.892221E+01	7.930942E - 01	1.335184E+00	2.236569E+00	1.880670E - 01
6.593502E - 02	1.858553E+01	8.553345E - 01	1.368356E+00	2.145251E+00	1.732856E - 01
7.346463E - 02	1.824084E+01	9.038666E - 01	1.401934E+00	2.056559E+00	1.595073E - 01
8.140129E - 02	1.788966E+01	9.373055E - 01	1.435919E+00	1.970283E+00	1.467059E - 01
8.974493E - 02	1.753339E+01	9.551353E - 01	1.470310E+00	1.886368E+00	1.348511E - 01
9.849554E - 02	1.717313E+01	9.576925E - 01	1.505110E+00	1.804572E+00	1.239080E - 01
1.076532E - 01	1.680986E+01	9.460863E - 01	1.540316E+00	1.724944E+00	1.138400E - 01
1.172178E - 01	1.644427E+01	9.220730E - 01	1.575929E+00	1.647347E+00	1.046079E - 01
1.271895E - 01	1.607698E+01	8.878983E - 01	1.611949E+00	1.571731E+00	9.617090E - 02
1.375682E - 01	1.570842E+01	8.461227E - 01	1.648376E+00	1.498123E+00	8.848768E - 02
1.483538E - 01	1.533898E+01	7.994466E - 01	1.685210E+00	1.426374E+00	8.151603E - 02
1.595465E - 01	1.496906E+01	7.505449E - 01	1.722451E+00	1.356618E+00	7.521468E - 02
1.711462E - 01	1.459909E+01	7.019265E - 01	1.760099E+00	1.288787E+00	6.954217E - 02
1.831529E - 01	1.422959E+01	6.558183E - 01	1.798154E+00	1.222973E+00	6.445819E - 02
1.955666E - 01	1.386116E+01	6.140817E - 01	1.836617E+00	1.159070E+00	5.992322E - 02
2.083873E - 01	1.349451E+01	5.781651E - 01	1.875485E+00	1.097246E+00	5.589950E - 02
2.216150E - 01	1.313038E+01	5.490825E - 01	1.914762E+00	1.037473E+00	5.235041E - 02
2.352497E - 01	1.276950E+01	5.274203E - 01	1.954445E+00	9.797889E - 01	4.924105E - 02
2.492914E - 01	1.241246E+01	5.133663E - 01	1.994534E+00	9.242954E - 01	4.653831E - 02
2.637402E - 01	1.205973E+01	5.067546E - 01	2.035032E+00	8.708859E - 01	4.421066E - 02
2.785959E - 01	1.171156E+01	5.071203E - 01	2.075936E+00	8.197086E - 01	4.222859E - 02
2.938586E - 01	1.136803E+01	5.137613E - 01	2.117248E+00	7.707086E - 01	4.056432E - 02
3.095284E - 01	1.102905E+01	5.258011E - 01	2.158965E+00	7.239392E - 01	3.919188E - 02
3.256052E - 01	1.069451E+01	5.422480E - 01	2.201091E+00	6.792405E - 01	3.808703E - 02
3.420889E - 01	1.036425E+01	5.620514E - 01	2.243623E+00	6.367105E - 01	3.722739E - 02
3.589797E - 01	1.003820E+01	5.841497E - 01	2.286562E+00	5.962257E - 01	3.659218E - 02
3.762774E - 01	9.716376E+00	6.075119E - 01	2.329908E+00	5.577205E - 01	3.616224E - 02
3.939822E - 01	9.398859E+00	6.311699E - 01	2.373661E+00	5.211869E - 01	3.592001E - 02
4.120940E - 01	9.085843E+00	6.542446E - 01	2.417822E+00	4.864051E - 01	3.584938E - 02
4.306128E - 01	8.777625E+00	6.759620E - 01	2.462389E+00	4.534211E - 01	3.593561E - 02
4.495386E - 01	8.474519E+00	6.956646E - 01	2.507363E+00	4.220788E - 01	3.616536E - 02
4.688714E - 01	8.176918E+00	7.128156E - 01	2.552744E+00	3.923564E - 01	3.652642E - 02
4.886112E - 01	7.885193E+00	7.269986E - 01	2.598533E+00	3.640355E - 01	3.700780E - 02
5.087581E - 01	7.599769E+00	7.379143E - 01	2.644728E+00	3.371704E - 01	3.759959E - 02
5.293119E - 01	7.320986E+00	7.453715E - 01	2.691330E+00	3.116190E - 01	3.829284E - 02
5.502727E - 01	7.049209E+00	7.492791E - 01	2.738339E+00	2.873185E - 01	3.907961E - 02
5.716406E - 01	6.784739E+00	7.496339E - 01	2.785755E+00	2.642790E - 01	3.995270E - 02
5.934154E - 01	6.527838E+00	7.465093E - 01	2.833579E+00	2.423152E - 01	4.090584E - 02
6.155972E - 01	6.278727E+00	7.400432E - 01	2.881809E+00	2.215242E - 01	4.193334E - 02
6.381861E - 01	6.037548E+00	7.304261E - 01	2.930446E+00	2.018750E - 01	4.303031E - 02
6.611820E - 01	5.804407E+00	7.178892E - 01	2.979490E+00	1.832099E - 01	4.419243E - 02
6.845849E - 01	5.579357E+00	7.026950E - 01	3.028942E+00	1.655775E - 01	4.541589E - 02
7.083948E - 01	5.362386E+00	6.851261E - 01	3.078800E+00	1.490391E - 01	4.669741E - 02
7.326116E - 01	5.153453E+00	6.654778E - 01	3.129066E+00	1.335256E - 01	4.803418E - 02
7.572355E - 01	4.952448E+00	6.440496E - 01	3.179738E+00	1.190389E - 01	4.942378E - 02
7.822664E - 01	4.759239E+00	6.211397E - 01	3.230817E+00	1.056577E - 01	5.086418E - 02
8.077043E - 01	4.573657E+00	5.970389E - 01	3.282303E+00	9.332871E - 02	5.235365E - 02
8.335493E - 01	4.395489E+00	5.720265E - 01	3.334197E+00	8.200163E - 02	5.389084E - 02

Z=11

Magnesium

Atomic Number	12
Lattice	fcc
Lattice constant a_o	8.40 a.u.
Pressure at a_o	0.84 Kbar
Total energy (solid)	-398.401 Ry
Total energy (atom)	-398.274 Ry
Zero point energy	0.003 Ry
Cohesive energy	0.124 Ry
Bulk modulus	0.41 Mbar
Fermi energy	0.435 Ry
Interstitial electrons (per atom)	0.536
χ/χ_o	1.47
$N(E_f)$	0.45 states/eV-atom
ρ_o	1.084×10^3 electrons/Bohr^3
$(d\rho(r,E)/dE)_{E=E_f,r=0}$	4.2354 electrons/Bohr^3-Ry

Symmetry point	Band energies in Rydbergs (relative to muffin tin zero)		
Γ	-0.087		
X	0.425	0.484	
L	0.303	0.343	
W	0.550	0.550	0.567
K	0.487	0.496	0.583

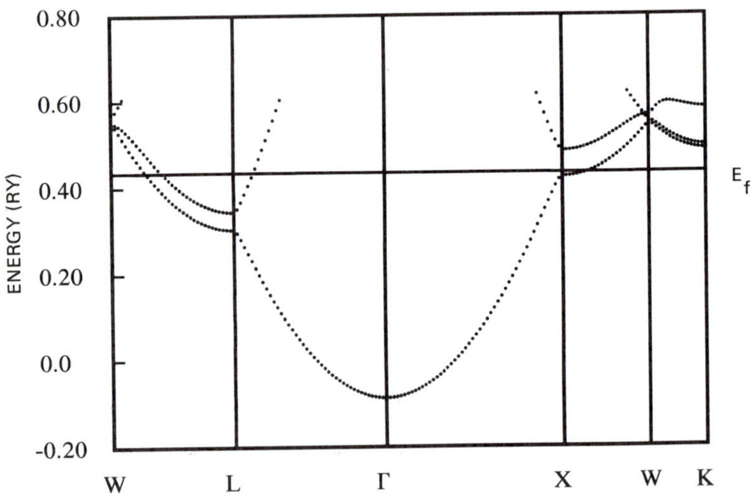

Z=12

Moruzzi
Janak
Williams

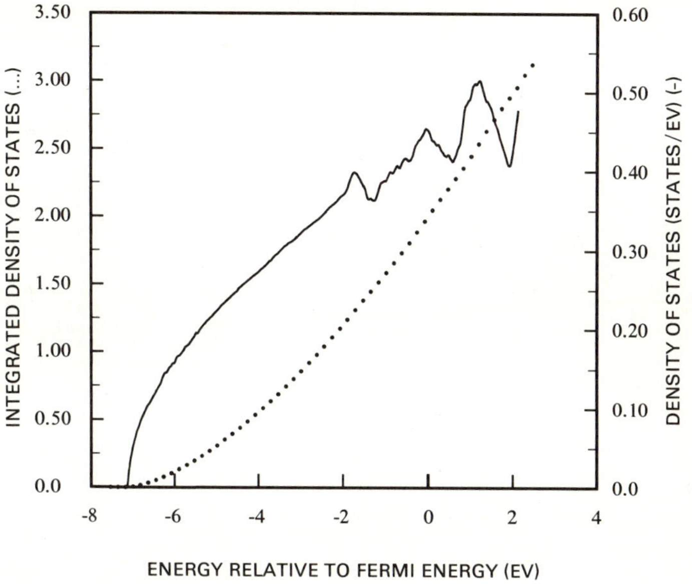

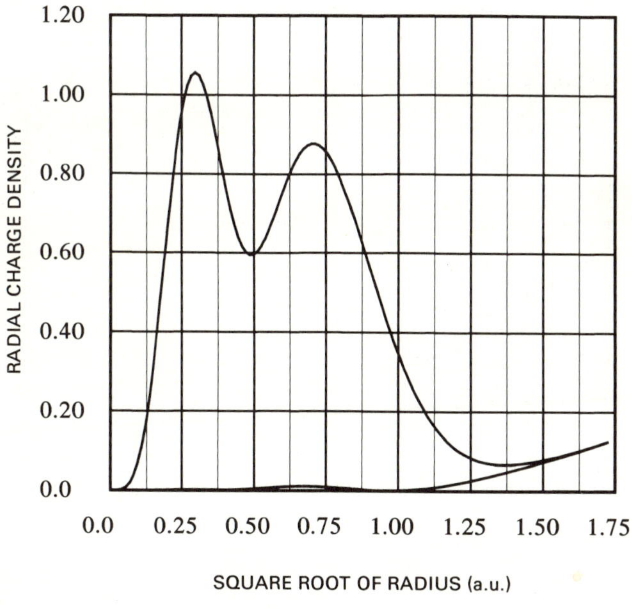

Moruzzi
Janak
Williams

Z=12

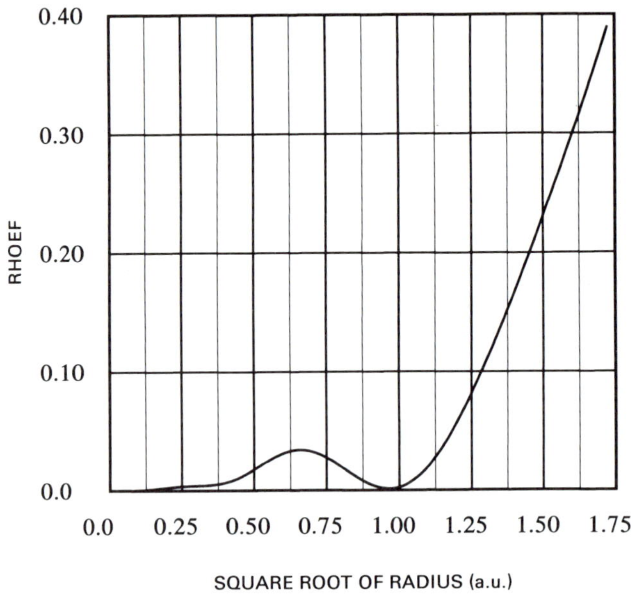

RHOEF$=r^2\partial\rho(E,r)/\partial E$ (at $E=E_f$)

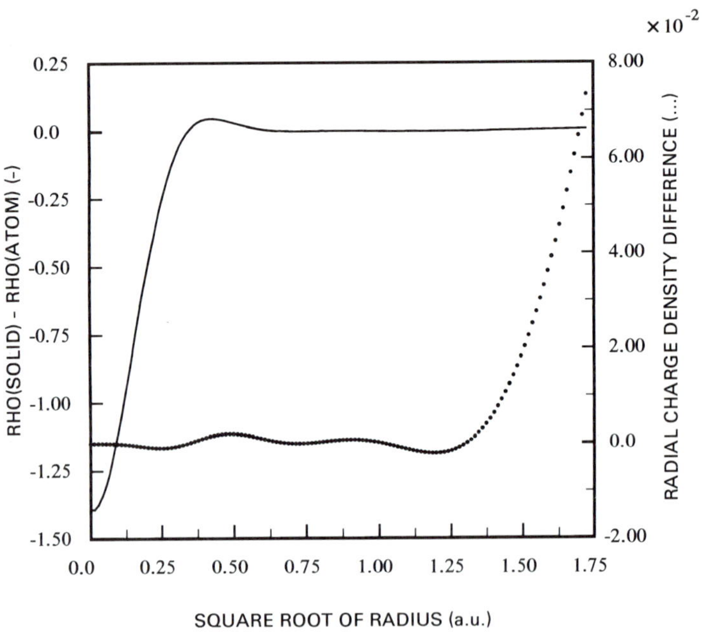

Moruzzi
Janak
Williams

Z=12

Magnesium

r	2z(r)	$r^2\rho$	r	2z(r)	$r^2\rho$
1.812651E - 04	2.398912E+01	3.546906E - 05	7.658451E - 01	5.312310E+00	6.254115E - 01
7.250605E - 04	2.395650E+01	5.601472E - 04	7.895908E - 01	5.129762E+00	5.935821E - 01
1.631386E - 03	2.390190E+01	2.774749E - 03	8.136991E - 01	4.953691E+00	5.616222E - 01
2.900242E - 03	2.382510E+01	8.506622E - 03	8.381699E - 01	4.783772E+00	5.297959E - 01
4.531626E - 03	2.372580E+01	1.997119E - 02	8.630032E - 01	4.619691E+00	4.983405E - 01
6.525543E - 03	2.360374E+01	3.947922E - 02	8.881990E - 01	4.461125E+00	4.674666E - 01
8.881990E - 03	2.345874E+01	6.912398E - 02	9.137574E - 01	4.307754E+00	4.373575E - 01
1.160097E - 02	2.329079E+01	1.104884E - 01	9.396783E - 01	4.159271E+00	4.081705E - 01
1.468247E - 02	2.310008E+01	1.643996E - 01	9.659618E - 01	4.015368E+00	3.800372E - 01
1.812651E - 02	2.288707E+01	2.307647E - 01	9.926078E - 01	3.875756E+00	3.530651E - 01
2.193308E - 02	2.265250E+01	3.085043E - 01	1.019616E+00	3.740181E+00	3.273394E - 01
2.610217E - 02	2.239738E+01	3.955824E - 01	1.046987E+00	3.608407E+00	3.029241E - 01
3.063380E - 02	2.212300E+01	4.891350E - 01	1.074720E+00	3.480097E+00	2.798623E - 01
3.552796E - 02	2.183084E+01	5.856825E - 01	1.102817E+00	3.355119E+00	2.581820E - 01
4.078465E - 02	2.152258E+01	6.813883E - 01	1.131275E+00	3.233282E+00	2.378943E - 01
4.640387E - 02	2.120004E+01	7.723491E - 01	1.160096E+00	3.114422E+00	2.189965E - 01
5.238562E - 02	2.086499E+01	8.548884E - 01	1.189280E+00	2.998399E+00	2.014734E - 01
5.872989E - 02	2.051929E+01	9.258093E - 01	1.218826E+00	2.885020E+00	1.852993E - 01
6.543666E - 02	2.016458E+01	9.826048E - 01	1.248734E+00	2.774354E+00	1.704407E - 01
7.250601E - 02	1.980247E+01	1.023604E+00	1.279006E+00	2.666181E+00	1.568550E - 01
7.993788E - 02	1.943434E+01	1.048040E+00	1.309640E+00	2.560539E+00	1.444952E - 01
8.773232E - 02	1.906134E+01	1.056051E+00	1.340636E+00	2.457414E+00	1.333087E - 01
9.588921E - 02	1.868449E+01	1.048614E+00	1.371995E+00	2.356805E+00	1.232399E - 01
1.044087E - 01	1.830458E+01	1.027428E+00	1.403717E+00	2.258662E+00	1.142302E - 01
1.132907E - 01	1.792224E+01	9.947555E - 01	1.435801E+00	2.163169E+00	1.062205E - 01
1.225352E - 01	1.753802E+01	9.532486E - 01	1.468247E+00	2.070218E+00	9.915024E - 02
1.321422E - 01	1.715236E+01	9.057616E - 01	1.501056E+00	1.980013E+00	9.295946E - 02
1.421118E - 01	1.676570E+01	8.551755E - 01	1.534227E+00	1.892445E+00	8.758855E - 02
1.524439E - 01	1.637852E+01	8.042364E - 01	1.567761E+00	1.807632E+00	8.297956E - 02
1.631386E - 01	1.599139E+01	7.554287E - 01	1.601658E+00	1.725605E+00	7.907605E - 02
1.741958E - 01	1.560491E+01	7.108740E - 01	1.635918E+00	1.646296E+00	7.582366E - 02
1.856155E - 01	1.521980E+01	6.722643E - 01	1.670539E+00	1.569853E+00	7.317066E - 02
1.973977E - 01	1.483681E+01	6.408311E - 01	1.705523E+00	1.496163E+00	7.106751E - 02
2.095425E - 01	1.445666E+01	6.173413E - 01	1.740870E+00	1.425104E+00	6.946748E - 02
2.220497E - 01	1.408002E+01	6.021178E - 01	1.776579E+00	1.356675E+00	6.832683E - 02
2.349195E - 01	1.370735E+01	5.950825E - 01	1.812651E+00	1.290781E+00	6.760448E - 02
2.481519E - 01	1.333897E+01	5.958105E - 01	1.849085E+00	1.227321E+00	6.726223E - 02
2.617468E - 01	1.297503E+01	6.035931E - 01	1.885881E+00	1.166176E+00	6.726480E - 02
2.757042E - 01	1.261552E+01	6.175053E - 01	1.923041E+00	1.107151E+00	6.757969E - 02
2.900242E - 01	1.226036E+01	6.364723E - 01	1.960563E+00	1.050279E+00	6.817698E - 02
3.047066E - 01	1.190949E+01	6.593316E - 01	1.998447E+00	9.952900E - 01	6.902945E - 02
3.197516E - 01	1.156286E+01	6.848893E - 01	2.036695E+00	9.421511E - 01	7.011235E - 02
3.351592E - 01	1.122052E+01	7.119695E - 01	2.075304E+00	8.907644E - 01	7.140344E - 02
3.509293E - 01	1.088259E+01	7.394542E - 01	2.114276E+00	8.410438E - 01	7.288253E - 02
3.670619E - 01	1.054930E+01	7.663143E - 01	2.153610E+00	7.929147E - 01	7.453191E - 02
3.835570E - 01	1.022093E+01	7.916338E - 01	2.193308E+00	7.462479E - 01	7.633567E - 02
4.004146E - 01	9.897845E+00	8.146249E - 01	2.233367E+00	7.011433E - 01	7.827991E - 02
4.176348E - 01	9.580415E+00	8.346351E - 01	2.273789E+00	6.574190E - 01	8.035243E - 02
4.352175E - 01	9.269045E+00	8.511513E - 01	2.314574E+00	6.151354E - 01	8.254272E - 02
4.531628E - 01	8.964133E+00	8.637958E - 01	2.355721E+00	5.743587E - 01	8.484191E - 02
4.714705E - 01	8.666077E+00	8.723201E - 01	2.397231E+00	5.348672E - 01	8.724272E - 02
4.901409E - 01	8.375196E+00	8.765952E - 01	2.439103E+00	4.969059E - 01	8.973891E - 02
5.091737E - 01	8.091816E+00	8.765998E - 01	2.481338E+00	4.604210E - 01	9.232557E - 02
5.285690E - 01	7.816184E+00	8.724074E - 01	2.523935E+00	4.253720E - 01	9.499913E - 02
5.483270E - 01	7.548485E+00	8.641718E - 01	2.566895E+00	3.919487E - 01	9.775668E - 02
5.684474E - 01	7.288874E+00	8.521144E - 01	2.610217E+00	3.600544E - 01	1.005968E - 01
5.889303E - 01	7.037427E+00	8.365096E - 01	2.653902E+00	3.298168E - 01	1.035184E - 01
6.097758E - 01	6.794169E+00	8.176728E - 01	2.697949E+00	3.012990E - 01	1.065216E - 01
6.309838E - 01	6.559078E+00	7.959481E - 01	2.742359E+00	2.745637E - 01	1.096070E - 01
6.525544E - 01	6.332065E+00	7.716979E - 01	2.787132E+00	2.496080E - 01	1.127759E - 01
6.744875E - 01	6.113002E+00	7.452930E - 01	2.832267E+00	2.266564E - 01	1.160301E - 01
6.967831E - 01	5.901721E+00	7.171044E - 01	2.877765E+00	2.056410E - 01	1.193721E - 01
7.194412E - 01	5.698006E+00	6.874963E - 01	2.923625E+00	1.867136E - 01	1.228044E - 01
7.424619E - 01	5.501624E+00	6.568207E - 01	2.969848E+00	1.699645E - 01	1.263303E - 01

Z=12

Aluminum

Atomic Number	13
Lattice	fcc
Lattice constant a_o	7.60 a.u.
Pressure at a_o	-3.92 Kbar
Total energy (solid)	-482.922 Ry
Total energy (atom)	-482.637 Ry
Zero point energy	0.003 Ry
Cohesive energy	0.282 Ry
Bulk modulus	0.80 Mbar
Fermi energy	0.617 Ry
Interstitial electrons (per atom)	0.716
χ/χ_o	1.34
$N(E_f)$	0.41 states/eV-atom
ρ_o	1.391×10^3 electrons/Bohr^3
$(d\rho(r,E)/dE)_{E=E_f,r=0}$	6.1485 electrons/Bohr^3-Ry

Symmetry point	Band energies in Rydbergs (relative to muffin tin zero)			
Γ	-0.206			
X	0.409	0.486		
L	0.273	0.298		
W	0.561	0.561	0.611	0.699
K	0.485	0.513	0.583	

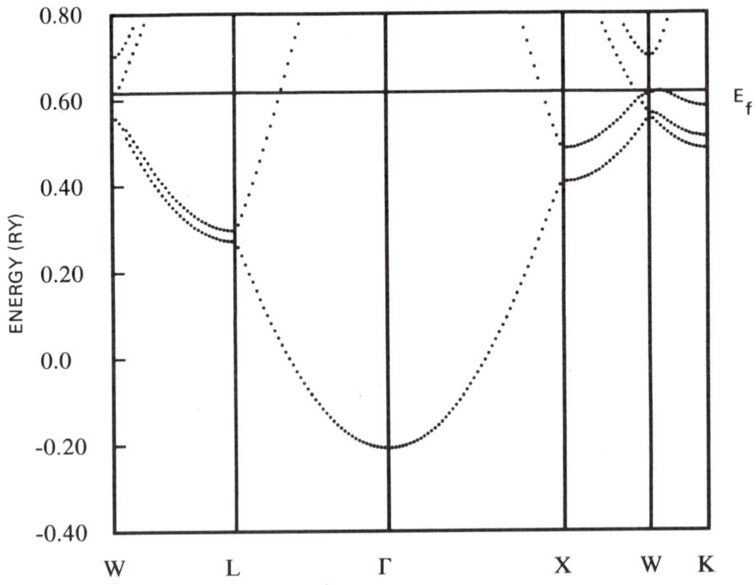

Z=13

Moruzzi
Janak
Williams

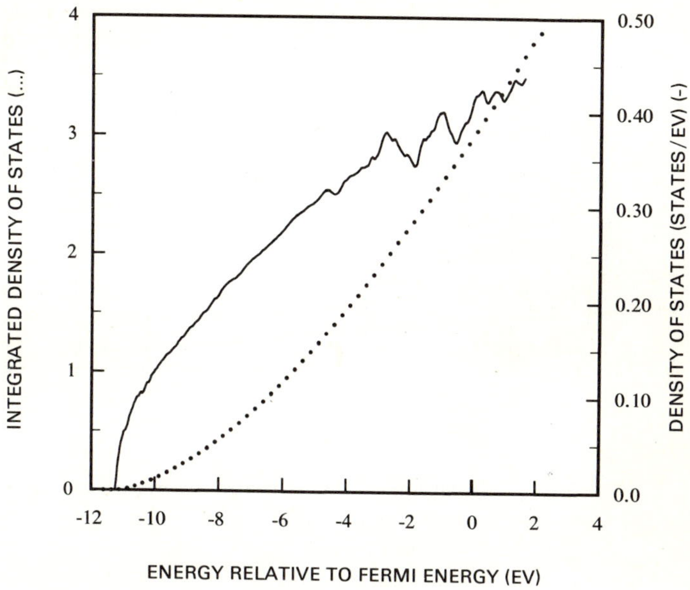

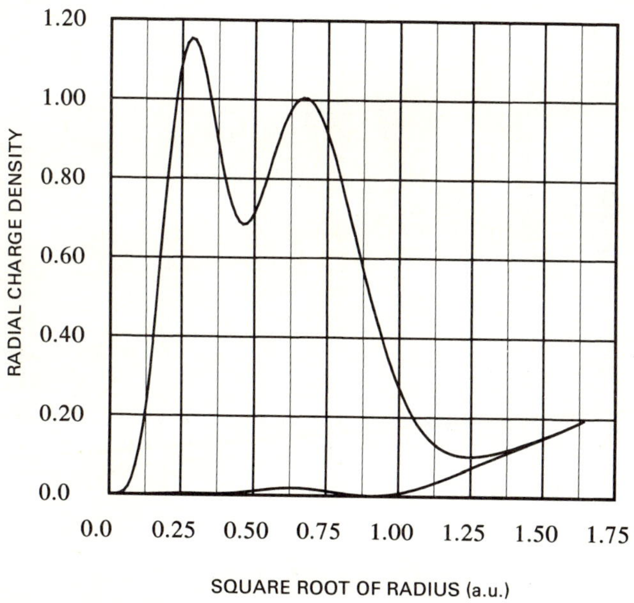

Moruzzi
Janak
Williams

Z=13

Aluminum

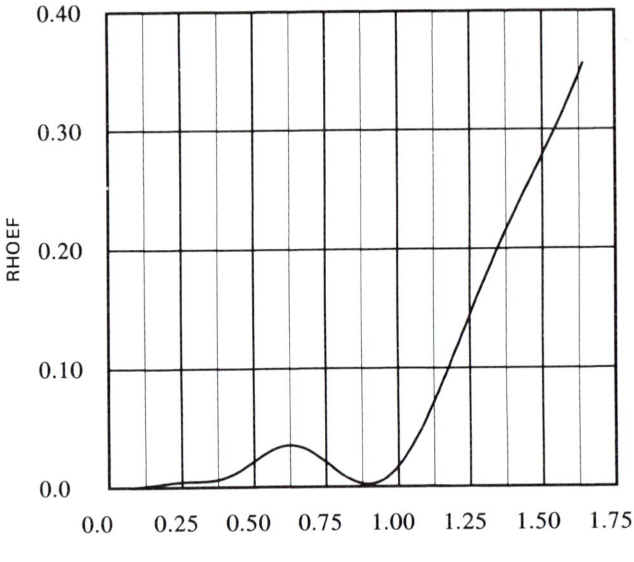

$RHOEF=r^2\partial\rho(E,r)/\partial E$ (at $E=E_f$)

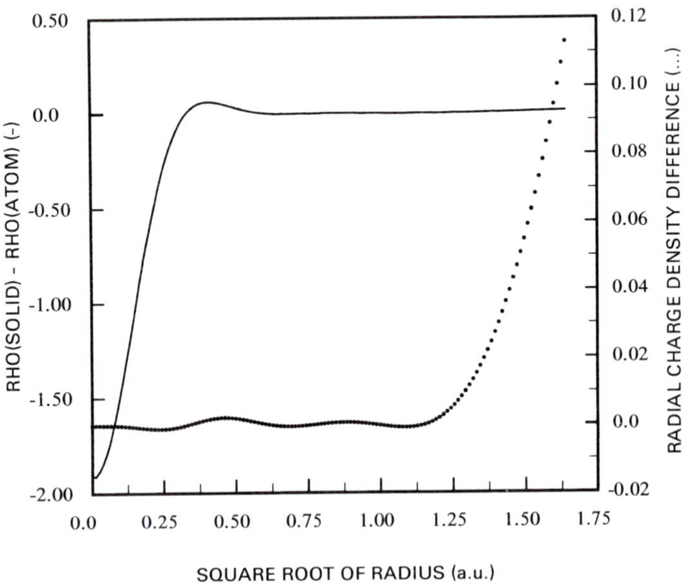

Moruzzi
Janak
Williams

Z=13

Aluminum

r	2z(r)	$r^2\rho$	r	2z(r)	$r^2\rho$
1.640018E - 04	2.598891E+01	3.725776E - 05	6.929075E - 01	6.438246E+00	6.970308E - 01
6.560071E - 04	2.595563E+01	5.885486E - 04	7.143917E - 01	6.237567E+00	6.592512E - 01
1.476016E - 03	2.589995E+01	2.916685E - 03	7.362039E - 01	6.043124E+00	6.214886E - 01
2.624028E - 03	2.582162E+01	8.947179E - 03	7.583442E - 01	5.854572E+00	5.840651E - 01
4.100043E - 03	2.572040E+01	2.102176E - 02	7.808124E - 01	5.671590E+00	5.472701E - 01
5.904064E - 03	2.559599E+01	4.159517E - 02	8.036087E - 01	5.493836E+00	5.113577E - 01
8.036084E - 03	2.544832E+01	7.291001E - 02	8.267329E - 01	5.321001E+00	4.765481E - 01
1.049611E - 02	2.527724E+01	1.166894E - 01	8.501852E - 01	5.152780E+00	4.430276E - 01
1.328414E - 02	2.508296E+01	1.738775E - 01	8.739654E - 01	4.988878E+00	4.109498E - 01
1.640018E - 02	2.486594E+01	2.444622E - 01	8.980737E - 01	4.829037E+00	3.804374E - 01
1.984421E - 02	2.462685E+01	3.273953E - 01	9.225100E - 01	4.673008E+00	3.515840E - 01
2.361625E - 02	2.436667E+01	4.206151E - 01	9.472742E - 01	4.520586E+00	3.244558E - 01
2.771630E - 02	2.408664E+01	5.211736E - 01	9.723665E - 01	4.371578E+00	2.990945E - 01
3.214435E - 02	2.378821E+01	6.254411E - 01	9.977868E - 01	4.225825E+00	2.755191E - 01
3.690040E - 02	2.347299E+01	7.293806E - 01	1.023535E+00	4.083234E+00	2.537290E - 01
4.198445E - 02	2.314273E+01	8.288431E - 01	1.049611E+00	3.943657E+00	2.337055E - 01
4.739651E - 02	2.279922E+01	9.198774E - 01	1.076015E+00	3.807048E+00	2.154146E - 01
5.313657E - 02	2.244424E+01	9.989999E - 01	1.102748E+00	3.673381E+00	1.988101E - 01
5.920464E - 02	2.207948E+01	1.063427E+00	1.129807E+00	3.542727E+00	1.838347E - 01
6.560069E - 02	2.170650E+01	1.111239E+00	1.157196E+00	3.414934E+00	1.704214E - 01
7.232475E - 02	2.132674E+01	1.141465E+00	1.184913E+00	3.290113E+00	1.584977E - 01
7.937682E - 02	2.094136E+01	1.154095E+00	1.212956E+00	3.168402E+00	1.479858E - 01
8.675689E - 02	2.055144E+01	1.150022E+00	1.241329E+00	3.049674E+00	1.388035E - 01
9.446502E - 02	2.015778E+01	1.130938E+00	1.270029E+00	2.934149E+00	1.308676E - 01
1.025011E - 01	1.976108E+01	1.099164E+00	1.299058E+00	2.821693E+00	1.240933E - 01
1.108652E - 01	1.936194E+01	1.057476E+00	1.328414E+00	2.712502E+00	1.183968E - 01
1.195573E - 01	1.896086E+01	1.008903E+00	1.358098E+00	2.606485E+00	1.136950E - 01
1.285774E - 01	1.855835E+01	9.565384E - 01	1.388110E+00	2.503618E+00	1.099067E - 01
1.379254E - 01	1.815489E+01	9.033670E - 01	1.418451E+00	2.403757E+00	1.069537E - 01
1.476015E - 01	1.775110E+01	8.521194E - 01	1.449120E+00	2.306996E+00	1.047610E - 01
1.576056E - 01	1.734761E+01	8.051577E - 01	1.480116E+00	2.213144E+00	1.032570E - 01
1.679378E - 01	1.694514E+01	7.643932E - 01	1.511440E+00	2.122069E+00	1.023740E - 01
1.785979E - 01	1.654451E+01	7.312474E - 01	1.543092E+00	2.033721E+00	1.020485E - 01
1.895860E - 01	1.614641E+01	7.066349E - 01	1.575072E+00	1.947781E+00	1.022212E - 01
2.009022E - 01	1.575153E+01	6.909846E - 01	1.607381E+00	1.864187E+00	1.028371E - 01
2.125463E - 01	1.536037E+01	6.842752E - 01	1.640018E+00	1.782787E+00	1.038454E - 01
2.245184E - 01	1.497327E+01	6.860934E - 01	1.672981E+00	1.703539E+00	1.051998E - 01
2.368186E - 01	1.459042E+01	6.956989E - 01	1.706274E+00	1.626148E+00	1.068580E - 01
2.494467E - 01	1.421187E+01	7.120970E - 01	1.739894E+00	1.550673E+00	1.087818E - 01
2.624028E - 01	1.383759E+01	7.341112E - 01	1.773843E+00	1.476858E+00	1.109372E - 01
2.756870E - 01	1.346755E+01	7.604529E - 01	1.808119E+00	1.404799E+00	1.132938E - 01
2.892991E - 01	1.310177E+01	7.897859E - 01	1.842724E+00	1.334270E+00	1.158249E - 01
3.032392E - 01	1.274031E+01	8.207821E - 01	1.877656E+00	1.265405E+00	1.185071E - 01
3.175074E - 01	1.238334E+01	8.521693E - 01	1.912916E+00	1.198102E+00	1.213205E - 01
3.321036E - 01	1.203111E+01	8.827692E - 01	1.948504E+00	1.132355E+00	1.242483E - 01
3.470277E - 01	1.168392E+01	9.115254E - 01	1.984421E+00	1.068172E+00	1.272764E - 01
3.622799E - 01	1.134215E+01	9.375244E - 01	2.020665E+00	1.005575E+00	1.303937E - 01
3.778601E - 01	1.100619E+01	9.600056E - 01	2.057238E+00	9.446034E - 01	1.335914E - 01
3.937683E - 01	1.067646E+01	9.783675E - 01	2.094138E+00	8.853025E - 01	1.368634E - 01
4.100044E - 01	1.035335E+01	9.921651E - 01	2.131367E+00	8.277281E - 01	1.402057E - 01
4.265686E - 01	1.003724E+01	1.001104E+00	2.168923E+00	7.719449E - 01	1.436162E - 01
4.434608E - 01	9.728474E+00	1.005033E+00	2.206807E+00	7.181091E - 01	1.470945E - 01
4.606810E - 01	9.427362E+00	1.003924E+00	2.245020E+00	6.661355E - 01	1.506430E - 01
4.782292E - 01	9.134114E+00	9.978669E - 01	2.283560E+00	6.162819E - 01	1.542642E - 01
4.961053E - 01	8.848942E+00	9.870458E - 01	2.322429E+00	5.684661E - 01	1.579636E - 01
5.143095E - 01	8.571945E+00	9.717278E - 01	2.361625E+00	5.229619E - 01	1.617469E - 01
5.328417E - 01	8.303185E+00	9.522457E - 01	2.401150E+00	4.796989E - 01	1.656221E - 01
5.517020E - 01	8.042658E+00	9.289824E - 01	2.441002E+00	4.389558E - 01	1.695968E - 01
5.708901E - 01	7.790302E+00	9.023572E - 01	2.481182E+00	4.007636E - 01	1.736814E - 01
5.904064E - 01	7.545998E+00	8.728122E - 01	2.521690E+00	3.652309E - 01	1.778860E - 01
6.102506E - 01	7.309586E+00	8.408011E - 01	2.562528E+00	3.323990E - 01	1.822222E - 01
6.304228E - 01	7.080868E+00	8.067786E - 01	2.603692E+00	3.025584E - 01	1.867020E - 01
6.509230E - 01	6.859587E+00	7.711919E - 01	2.645185E+00	2.757536E - 01	1.913378E - 01
6.717513E - 01	6.645477E+00	7.344723E - 01	2.687005E+00	2.521074E - 01	1.961430E - 01

Z=13

Moruzzi
Janak
Williams

Potassium

Potassium

Atomic Number	19
Lattice	bcc
Lattice constant a_o	9.45 a.u.
Pressure at a_o	1.73 Kbar
Total energy (solid)	-1196.448 Ry
Total energy (atom)	-1196.382 Ry
Zero point energy	0.001 Ry
Cohesive energy	0.066 Ry
Bulk modulus	0.04 Mbar
Fermi energy	0.192 Ry
Interstitial electrons (per atom)	0.386
χ/χ_o	1.95
$N(E_f)$	0.73 states/eV-atom
ρ_o	4.515×10^3 electrons/Bohr^3
$(d\rho(r,E)/dE)_{E=E_f,r=0}$	8.6922 electrons/Bohr^3-Ry

Symmetry point	Band energies in Rydbergs (relative to muffin tin zero)		
Γ	0.016		
H	0.302	0.302	
N	0.221	0.251	
P	0.304	0.304	0.304

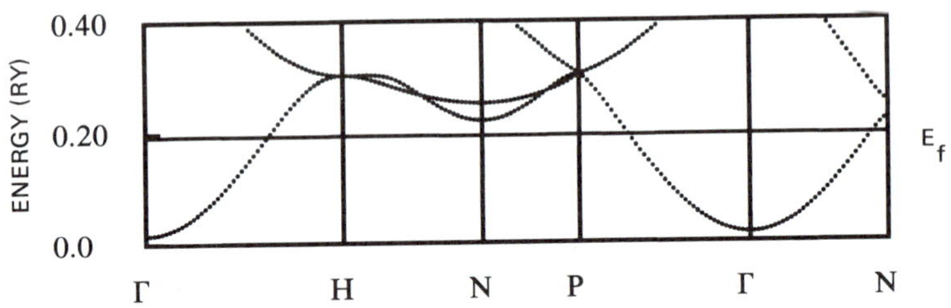

Z=19

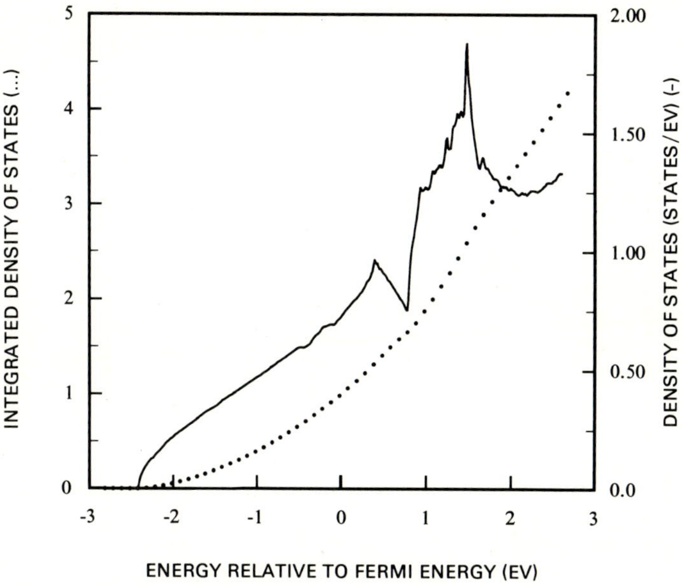

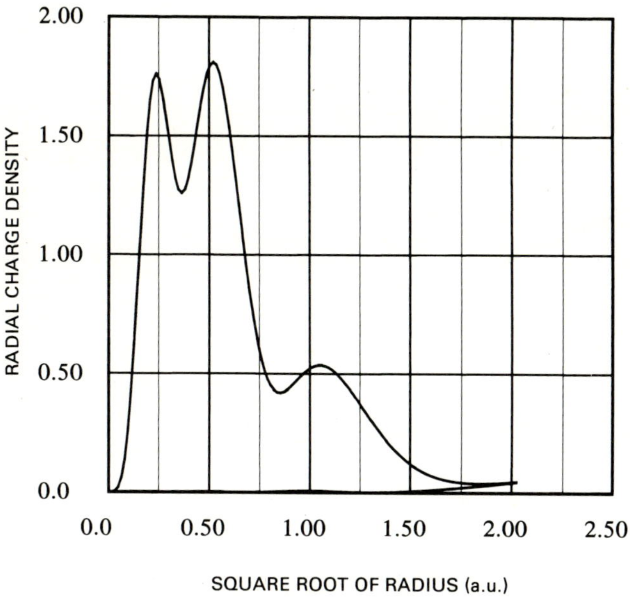

Moruzzi
Janak
Williams

Z=19

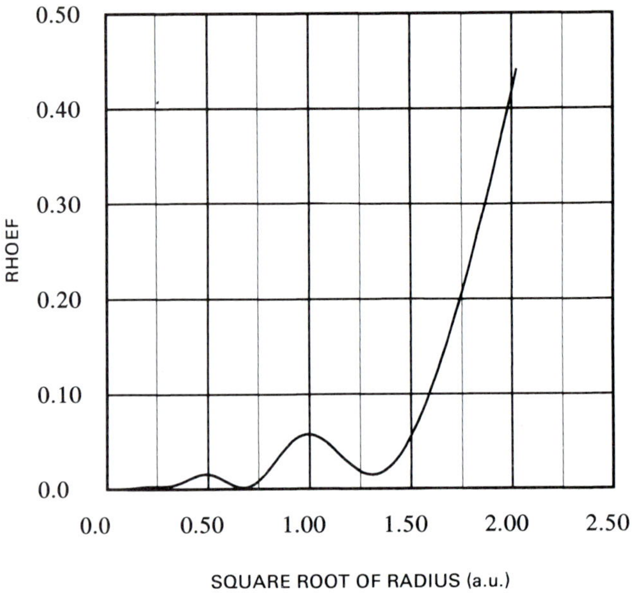

$RHOEF=r^2\partial\rho(E,r)/\partial E$ (at $E=E_f$)

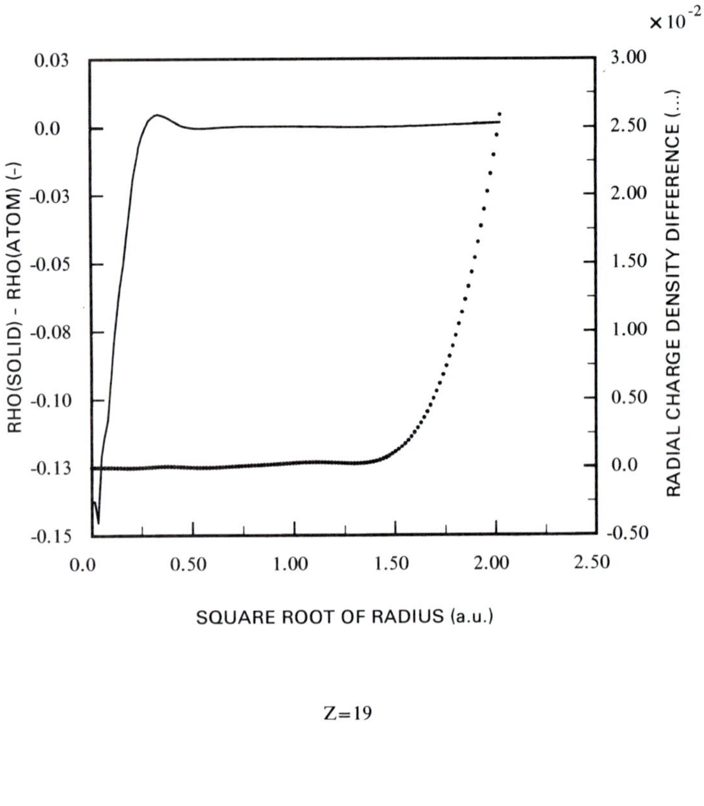

Moruzzi
Janak
Williams

Z=19

r	2z(r)	$r^2\rho$	r	2z(r)	$r^2\rho$
2.497539E - 04	3.797066E+01	2.790000E - 04	1.055210E+00	6.320475E+00	5.331934E - 01
9.990158E - 04	3.788252E+01	4.338700E - 03	1.087928E+00	6.078382E+00	5.358341E - 01
2.247785E - 03	3.773471E+01	2.094710E - 02	1.121145E+00	5.842813E+00	5.357070E - 01
3.996063E - 03	3.752637E+01	6.195050E - 02	1.154861E+00	5.614048E+00	5.328419E - 01
6.243847E - 03	3.725685E+01	1.388770E - 01	1.189078E+00	5.391912E+00	5.273372E - 01
8.991141E - 03	3.692636E+01	2.594801E - 01	1.223794E+00	5.176771E+00	5.193474E - 01
1.223794E - 02	3.653622E+01	4.250816E - 01	1.259009E+00	4.968524E+00	5.090688E - 01
1.598425E - 02	3.608899E+01	6.293631E - 01	1.294724E+00	4.767316E+00	4.967290E - 01
2.023007E - 02	3.558875E+01	8.588611E - 01	1.330938E+00	4.572819E+00	4.825741E - 01
2.497539E - 02	3.504068E+01	1.095003E+00	1.367652E+00	4.385245E+00	4.668629E - 01
3.022023E - 02	3.445074E+01	1.317195E+00	1.404865E+00	4.204339E+00	4.498557E - 01
3.596457E - 02	3.382516E+01	1.506212E+00	1.442578E+00	4.029951E+00	4.318095E - 01
4.220841E - 02	3.316998E+01	1.647238E+00	1.480791E+00	3.861907E+00	4.129729E - 01
4.895177E - 02	3.249069E+01	1.731940E+00	1.519503E+00	3.700156E+00	3.935815E - 01
5.619463E - 02	3.179196E+01	1.759344E+00	1.558714E+00	3.544335E+00	3.738543E - 01
6.393701E - 02	3.107764E+01	1.735408E+00	1.598425E+00	3.394214E+00	3.539926E - 01
7.217884E - 02	3.035080E+01	1.671613E+00	1.638635E+00	3.249692E+00	3.341786E - 01
8.092022E - 02	2.961418E+01	1.582869E+00	1.679345E+00	3.110232E+00	3.145730E - 01
9.016114E - 02	2.887036E+01	1.485202E+00	1.720554E+00	2.975872E+00	2.953186E - 01
9.990156E - 02	2.812213E+01	1.393629E+00	1.762263E+00	2.846212E+00	2.765366E - 01
1.101415E - 01	2.737254E+01	1.320432E+00	1.804472E+00	2.721011E+00	2.583299E - 01
1.208809E - 01	2.662465E+01	1.274093E+00	1.847179E+00	2.600175E+00	2.407832E - 01
1.321198E - 01	2.588127E+01	1.258857E+00	1.890388E+00	2.483179E+00	2.239636E - 01
1.438583E - 01	2.514447E+01	1.274856E+00	1.934094E+00	2.370100E+00	2.079240E - 01
1.560962E - 01	2.441559E+01	1.318697E+00	1.978300E+00	2.260732E+00	1.927021E - 01
1.688336E - 01	2.369539E+01	1.384329E+00	2.023006E+00	2.154599E+00	1.783221E - 01
1.820706E - 01	2.298445E+01	1.464023E+00	2.068212E+00	2.051819E+00	1.647984E - 01
1.958070E - 01	2.228362E+01	1.549340E+00	2.113916E+00	1.952092E+00	1.521344E - 01
2.100430E - 01	2.159395E+01	1.632011E+00	2.160121E+00	1.855289E+00	1.403246E - 01
2.247785E - 01	2.091689E+01	1.704618E+00	2.206825E+00	1.761291E+00	1.293569E - 01
2.400135E - 01	2.025400E+01	1.761088E+00	2.254029E+00	1.670005E+00	1.192122E - 01
2.557480E - 01	1.960684E+01	1.796984E+00	2.301732E+00	1.581496E+00	1.098669E - 01
2.719820E - 01	1.897684E+01	1.809609E+00	2.349935E+00	1.495567E+00	1.012929E - 01
2.887155E - 01	1.836508E+01	1.797969E+00	2.398637E+00	1.412180E+00	9.345889E - 02
3.059486E - 01	1.777229E+01	1.762603E+00	2.447838E+00	1.331456E+00	8.633137E - 02
3.236811E - 01	1.719876E+01	1.705341E+00	2.497539E+00	1.253251E+00	7.987487E - 02
3.419131E - 01	1.664432E+01	1.629033E+00	2.547739E+00	1.177573E+00	7.405305E - 02
3.606446E - 01	1.610849E+01	1.537229E+00	2.598439E+00	1.104445E+00	6.882930E - 02
3.798757E - 01	1.559046E+01	1.433892E+00	2.649639E+00	1.033894E+00	6.416672E - 02
3.996063E - 01	1.508914E+01	1.323138E+00	2.701338E+00	9.661039E - 01	6.002925E - 02
4.198363E - 01	1.460339E+01	1.208987E+00	2.753536E+00	9.009674E - 01	5.638114E - 02
4.405659E - 01	1.413200E+01	1.095179E+00	2.806234E+00	8.385208E - 01	5.318795E - 02
4.617950E - 01	1.367386E+01	9.850323E - 01	2.859432E+00	7.787938E - 01	5.041646E - 02
4.835236E - 01	1.322804E+01	8.813367E - 01	2.913130E+00	7.218098E - 01	4.803488E - 02
5.057517E - 01	1.279383E+01	7.863069E - 01	2.967326E+00	6.677178E - 01	4.601305E - 02
5.284793E - 01	1.237085E+01	7.015573E - 01	3.022022E+00	6.163662E - 01	4.432229E - 02
5.517064E - 01	1.195902E+01	6.281191E - 01	3.077218E+00	5.677334E - 01	4.293576E - 02
5.754331E - 01	1.155854E+01	5.664802E - 01	3.132913E+00	5.219182E - 01	4.182823E - 02
5.996592E - 01	1.116977E+01	5.166433E - 01	3.189108E+00	4.785774E - 01	4.097623E - 02
6.243848E - 01	1.079314E+01	4.781981E - 01	3.245802E+00	4.379241E - 01	4.035799E - 02
6.496100E - 01	1.042889E+01	4.503992E - 01	3.302996E+00	3.997276E - 01	3.995333E - 02
6.753346E - 01	1.007705E+01	4.322464E - 01	3.360688E+00	3.640351E - 01	3.974365E - 02
7.015588E - 01	9.737244E+00	4.225602E - 01	3.418881E+00	3.304595E - 01	3.971189E - 02
7.282825E - 01	9.408785E+00	4.200537E - 01	3.477573E+00	2.991810E - 01	3.984239E - 02
7.555056E - 01	9.090761E+00	4.233955E - 01	3.536765E+00	2.699527E - 01	4.012093E - 02
7.832283E - 01	8.782147E+00	4.312627E - 01	3.596457E+00	2.426705E - 01	4.053451E - 02
8.114505E - 01	8.481972E+00	4.423856E - 01	3.656647E+00	2.173836E - 01	4.107133E - 02
8.401722E - 01	8.189384E+00	4.555818E - 01	3.717338E+00	1.937143E - 01	4.172067E - 02
8.693934E - 01	7.903728E+00	4.697818E - 01	3.778527E+00	1.718695E - 01	4.247290E - 02
8.991141E - 01	7.624536E+00	4.840457E - 01	3.840216E+00	1.517755E - 01	4.331927E - 02
9.293344E - 01	7.351524E+00	4.975729E - 01	3.902405E+00	1.330866E - 01	4.425195E - 02
9.600541E - 01	7.084588E+00	5.097045E - 01	3.965093E+00	1.160365E - 01	4.526383E - 02
9.912733E - 01	6.823699E+00	5.199217E - 01	4.028281E+00	1.002966E - 01	4.634855E - 02
1.022992E+00	6.568936E+00	5.278389E - 01	4.091969E+00	8.611995E - 02	4.750047E - 02

Z=19

Calcium

Calcium

Atomic Number	20
Lattice	fcc
Lattice constant a_o	10.0 a.u.
Pressure at a_o	0.92 Kbar
Total energy (solid)	-1351.609 Ry
Total energy (atom)	-1351.442 Ry
Zero point energy	0.002 Ry
Cohesive energy	0.165 Ry
Bulk modulus	0.17 Mbar
Fermi energy	0.330 Ry
Interstitial electrons (per atom)	0.599
χ/χ_o	4.49
$N(E_f)$	1.56 states/eV-atom
ρ_o	5.292×10^3 electrons/Bohr^3
$(d\rho(r,E)/dE)_{E=E_f,r=0}$	3.3957 electrons/Bohr^3-Ry

Symmetry point	Band energies in Rydbergs (relative to muffin tin zero)			
Γ	0.028			
X	0.280	0.371	0.439	
L	0.264	0.333	0.498	0.498
W	0.331	0.395	0.395	
K	0.320	0.366	0.415	

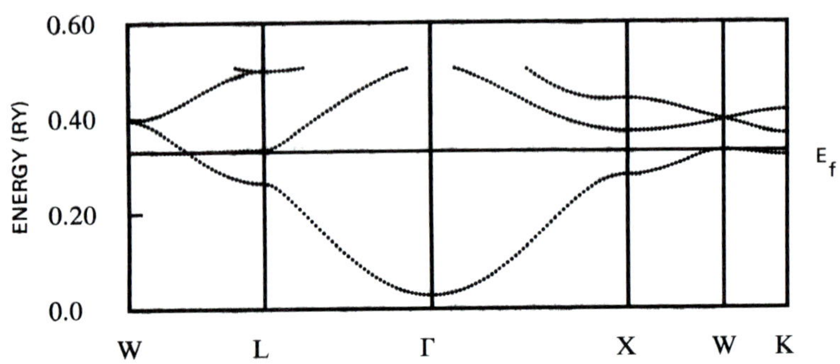

Z=20

Moruzzi
Janak
Williams

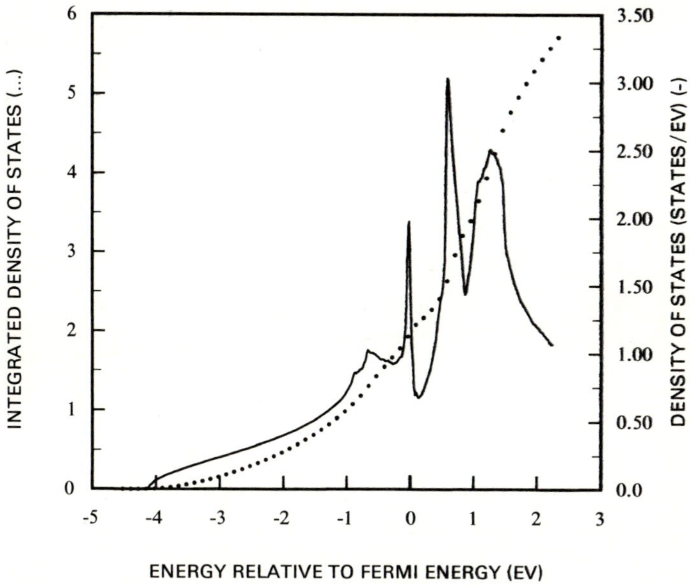

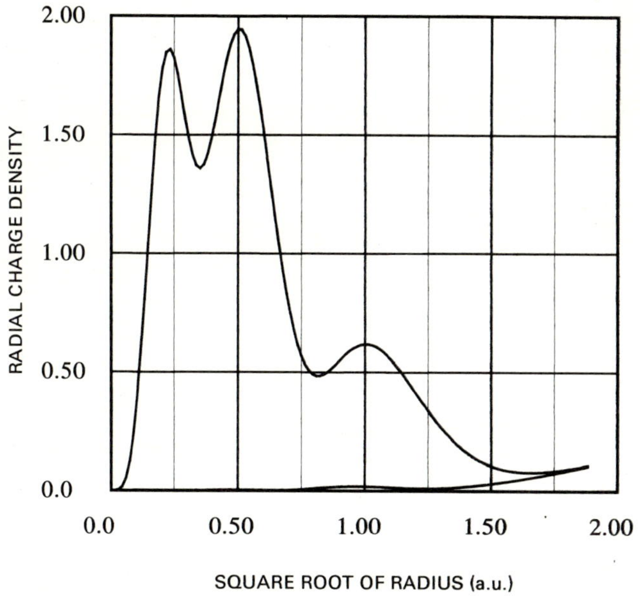

Moruzzi
Janak
Williams

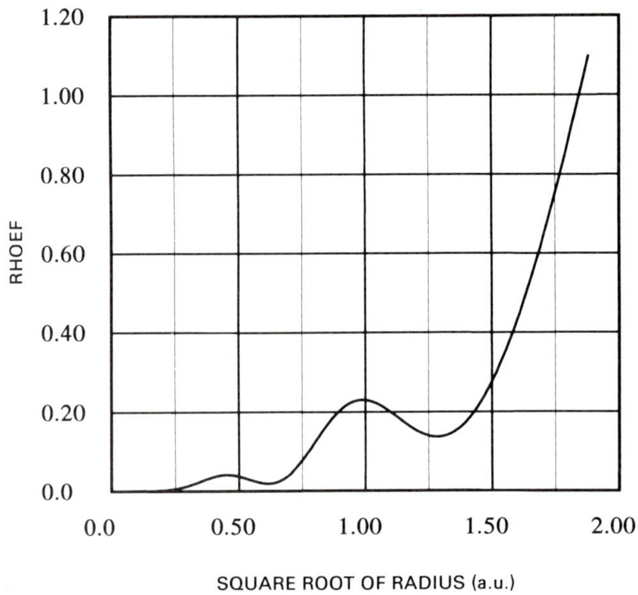

RHOEF=$r^2\partial\rho(E,r)/\partial E$ (at $E=E_f$)

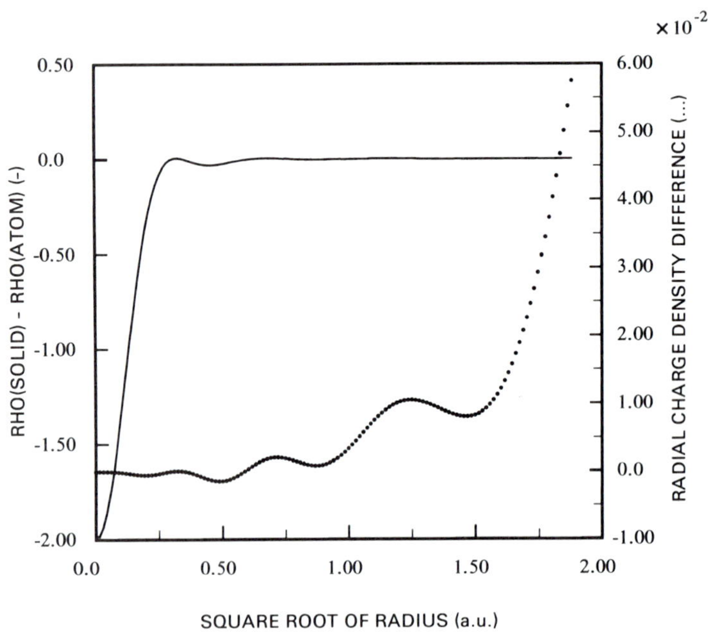

Z=20

Calcium

r	2z(r)	$r^2\rho$	r	2z(r)	$r^2\rho$
2.157918E - 04	3.997270E+01	2.443092E - 04	9.117203E - 01	7.779718E+00	6.014583E - 01
8.631672E - 04	3.989067E+01	3.809035E - 03	9.399891E - 01	7.507431E+00	6.100018E - 01
1.942126E - 03	3.975319E+01	1.846890E - 02	9.686894E - 01	7.241258E+00	6.156318E - 01
3.452669E - 03	3.955951E+01	5.495021E - 02	9.978213E - 01	6.981323E+00	6.181883E - 01
5.394794E - 03	3.930904E+01	1.241376E - 01	1.027385E+00	6.727774E+00	6.176158E - 01
7.768504E - 03	3.900168E+01	2.341321E - 01	1.057380E+00	6.480742E+00	6.139501E - 01
1.057380E - 02	3.863841E+01	3.878304E - 01	1.087806E+00	6.240401E+00	6.073025E - 01
1.381068E - 02	3.822105E+01	5.815695E - 01	1.118664E+00	6.006658E+00	5.978456E - 01
1.747913E - 02	3.775279E+01	8.051093E - 01	1.149954E+00	5.779545E+00	5.858001E - 01
2.157918E - 02	3.723778E+01	1.042931E+00	1.181676E+00	5.559050E+00	5.714220E - 01
2.611081E - 02	3.668097E+01	1.276547E+00	1.213828E+00	5.345435E+00	5.549923E - 01
3.107402E - 02	3.608774E+01	1.487312E+00	1.246413E+00	5.138158E+00	5.368025E - 01
3.646882E - 02	3.546353E+01	1.659184E+00	1.279429E+00	4.937409E+00	5.171520E - 01
4.229519E - 02	3.481345E+01	1.780953E+00	1.312877E+00	4.743048E+00	4.963360E - 01
4.855315E - 02	3.414204E+01	1.847555E+00	1.346756E+00	4.554745E+00	4.746408E - 01
5.524270E - 02	3.345319E+01	1.860387E+00	1.381067E+00	4.372313E+00	4.523389E - 01
6.236383E - 02	3.275002E+01	1.826639E+00	1.415810E+00	4.195702E+00	4.296869E - 01
6.991649E - 02	3.203528E+01	1.757904E+00	1.450984E+00	4.024538E+00	4.069195E - 01
7.790083E - 02	3.131131E+01	1.668329E+00	1.486589E+00	3.858753E+00	3.842508E - 01
8.631670E - 02	3.058049E+01	1.572670E+00	1.522627E+00	3.697968E+00	3.618711E - 01
9.516418E - 02	2.984540E+01	1.484522E+00	1.559095E+00	3.542110E+00	3.399485E - 01
1.044432E - 01	2.910872E+01	1.414945E+00	1.595996E+00	3.390810E+00	3.186266E - 01
1.141539E - 01	2.837326E+01	1.371567E+00	1.633327E+00	3.244012E+00	2.980282E - 01
1.242961E - 01	2.764140E+01	1.358247E+00	1.671091E+00	3.101363E+00	2.782532E - 01
1.348699E - 01	2.691504E+01	1.375171E+00	1.709287E+00	2.962682E+00	2.593822E - 01
1.458752E - 01	2.619543E+01	1.419334E+00	1.747913E+00	2.827958E+00	2.414773E - 01
1.573122E - 01	2.548334E+01	1.485260E+00	1.786971E+00	2.697050E+00	2.245831E - 01
1.691808E - 01	2.477934E+01	1.565842E+00	1.826462E+00	2.569517E+00	2.087277E - 01
1.814809E - 01	2.408417E+01	1.653182E+00	1.866383E+00	2.445718E+00	1.939272E - 01
1.942126E - 01	2.339877E+01	1.739380E+00	1.906736E+00	2.325112E+00	1.801837E - 01
2.073759E - 01	2.272440E+01	1.817161E+00	1.947520E+00	2.208102E+00	1.674904E - 01
2.209708E - 01	2.206238E+01	1.880363E+00	1.988737E+00	2.094182E+00	1.558294E - 01
2.349973E - 01	2.141415E+01	1.924258E+00	2.030385E+00	1.983643E+00	1.451771E - 01
2.494553E - 01	2.078102E+01	1.945690E+00	2.072464E+00	1.876485E+00	1.355029E - 01
2.643449E - 01	2.016402E+01	1.943101E+00	2.114975E+00	1.772564E+00	1.267706E - 01
2.796661E - 01	1.956393E+01	1.916423E+00	2.157918E+00	1.671908E+00	1.189409E - 01
2.954190E - 01	1.898116E+01	1.866904E+00	2.201292E+00	1.574708E+00	1.119717E - 01
3.116034E - 01	1.841573E+01	1.796867E+00	2.245097E+00	1.481002E+00	1.058189E - 01
3.282193E - 01	1.786736E+01	1.709452E+00	2.289335E+00	1.390516E+00	1.004368E - 01
3.452669E - 01	1.733542E+01	1.608335E+00	2.334003E+00	1.303744E+00	9.578037E - 02
3.627460E - 01	1.681912E+01	1.497485E+00	2.379105E+00	1.220246E+00	9.180379E - 02
3.806567E - 01	1.631749E+01	1.380920E+00	2.424636E+00	1.140493E+00	8.846265E - 02
3.989990E - 01	1.582948E+01	1.262503E+00	2.470600E+00	1.064013E+00	8.571333E - 02
4.177729E - 01	1.535412E+01	1.145788E+00	2.516995E+00	9.910849E - 01	8.351386E - 02
4.369784E - 01	1.489048E+01	1.033891E+00	2.563822E+00	9.216651E - 01	8.182400E - 02
4.566154E - 01	1.443791E+01	9.294122E - 01	2.611080E+00	8.555338E - 01	8.060551E - 02
4.766841E - 01	1.399594E+01	8.343803E - 01	2.658771E+00	7.926145E - 01	7.982200E - 02
4.971843E - 01	1.356441E+01	7.502560E - 01	2.706892E+00	7.329642E - 01	7.943958E - 02
5.181161E - 01	1.314338E+01	6.779377E - 01	2.755445E+00	6.763328E - 01	7.942629E - 02
5.394795E - 01	1.273318E+01	6.178048E - 01	2.804430E+00	6.227638E - 01	7.975245E - 02
5.612745E - 01	1.233418E+01	5.697726E - 01	2.853847E+00	5.720049E - 01	8.039069E - 02
5.835010E - 01	1.194672E+01	5.333611E - 01	2.903694E+00	5.241002E - 01	8.131576E - 02
6.061592E - 01	1.157104E+01	5.077680E - 01	2.953974E+00	4.787996E - 01	8.250445E - 02
6.292489E - 01	1.120699E+01	4.919443E - 01	3.004684E+00	4.361693E - 01	8.393532E - 02
6.527702E - 01	1.085417E+01	4.846677E - 01	3.055827E+00	3.959798E - 01	8.558935E - 02
6.767231E - 01	1.051189E+01	4.846103E - 01	3.107402E+00	3.581623E - 01	8.744884E - 02
7.011076E - 01	1.017927E+01	4.903993E - 01	3.159408E+00	3.228171E - 01	8.949786E - 02
7.259236E - 01	9.855392E+00	5.006707E - 01	3.211844E+00	2.899055E - 01	9.172219E - 02
7.511712E - 01	9.539389E+00	5.141122E - 01	3.264713E+00	2.592482E - 01	9.410906E - 02
7.768505E - 01	9.230536E+00	5.294986E - 01	3.318014E+00	2.308277E - 01	9.664661E - 02
8.029613E - 01	8.928272E+00	5.457178E - 01	3.371746E+00	2.047964E - 01	9.932476E - 02
8.295037E - 01	8.632224E+00	5.617892E - 01	3.425910E+00	1.810079E - 01	1.021340E - 01
8.564777E - 01	8.342150E+00	5.768741E - 01	3.480506E+00	1.594852E - 01	1.050661E - 01
8.838832E - 01	8.057979E+00	5.902796E - 01	3.535533E+00	1.404052E - 01	1.081131E - 01

Z=20

Scandium

Atomic Number	21
Lattice	fcc
Lattice constant a_o	8.49 a.u.
Pressure at a_o	-8.3 Kbar
Total energy (solid)	-1517.685 Ry
Total energy (atom)	-1517.322 Ry
Zero point energy	0.003 Ry
Cohesive energy	0.360 Ry
Bulk modulus	0.57 Mbar
Fermi energy	0.488 Ry
Interstitial electrons (per atom)	0.748
χ/χ_o	2.47
$N(E_f)$	1.73 states/eV-atom
ρ_o	6.152×10^3 electrons/Bohr^3
$(d\rho(r,E)/dE)_{E=E_f,r=0}$	3.0394 electrons/Bohr^3-Ry

Symmetry point	Band energies in Rydbergs (relative to muffin tin zero)			
Γ	0.108	0.608	0.608	0.608
X	0.349	0.437		
L	0.362	0.520	0.594	0.594
W	0.417	0.512	0.512	
K	0.406	0.460	0.571	

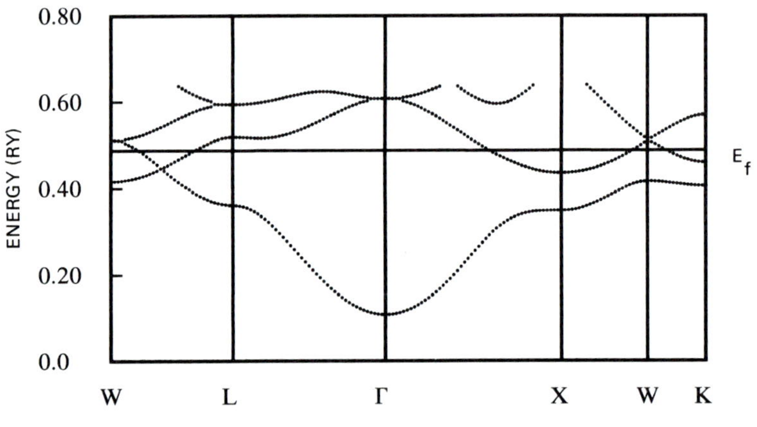

Z=21

Moruzzi
Janak
Williams

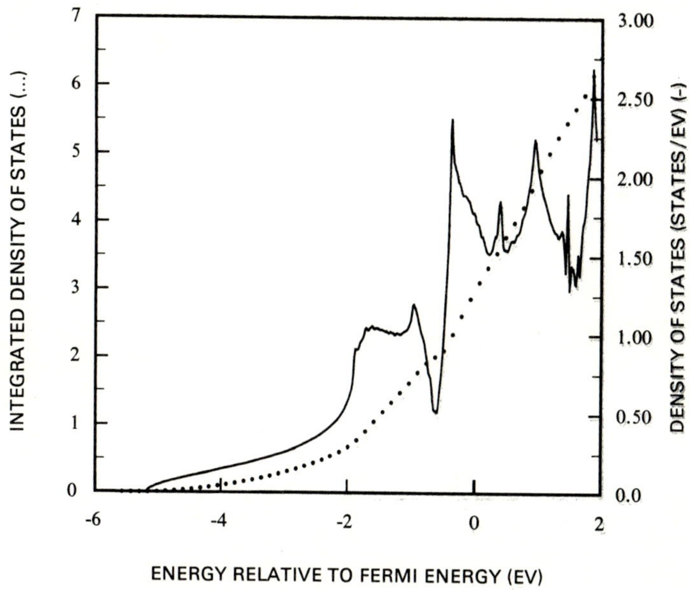

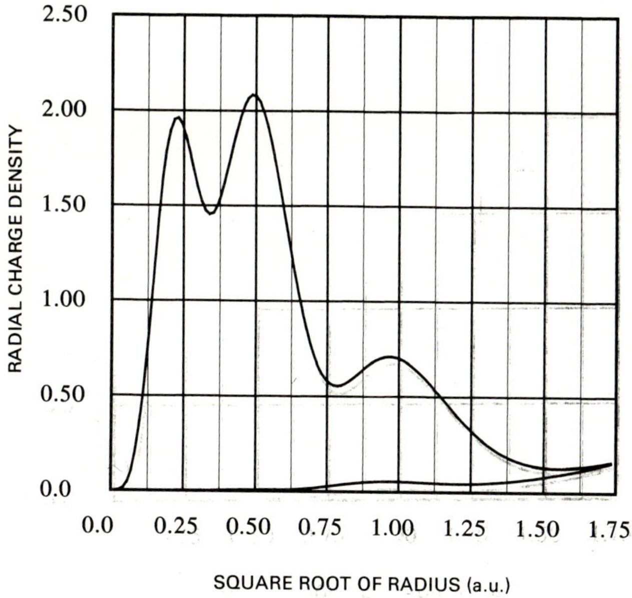

Moruzzi
Janak
Williams

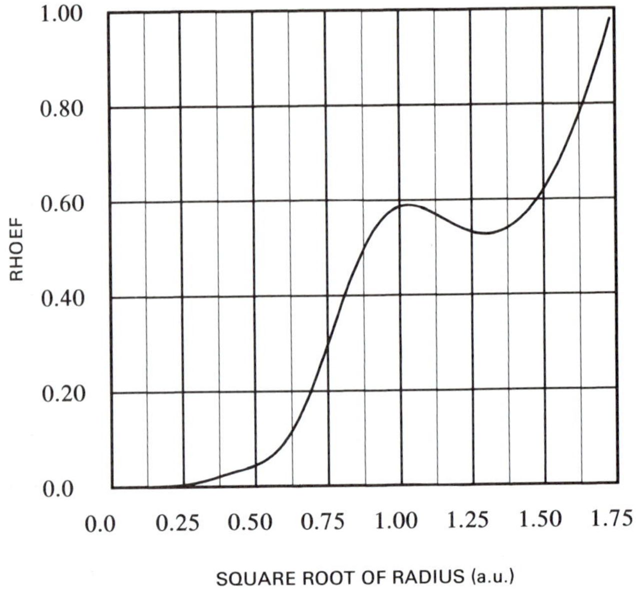

RHOEF=$r^2\partial\rho(E,r)/\partial E$ (at $E=E_f$)

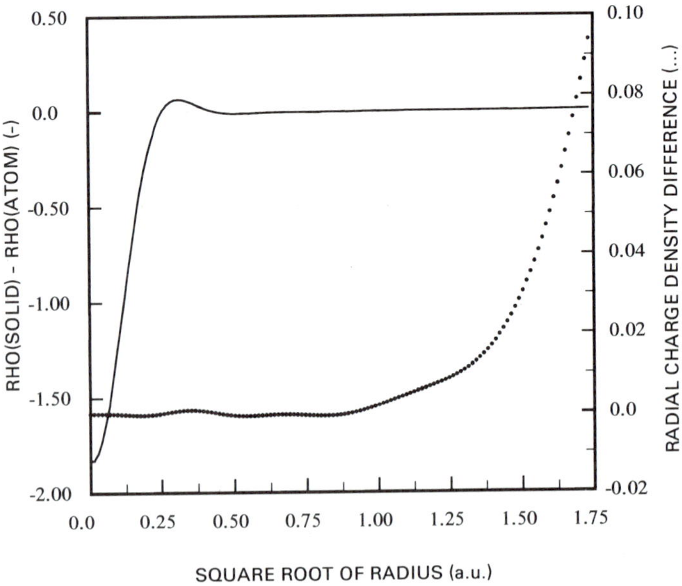

Moruzzi
Janak
Williams

Z=21

r	2z(r)	$r^2\rho$	r	2z(r)	$r^2\rho$
1.832504E - 04	4.197508E+01	2.050084E - 04	7.742329E - 01	9.377714E+00	6.492659E - 01
7.330016E - 04	4.190021E+01	3.205271E - 03	7.982388E - 01	9.071157E+00	6.656396E - 01
1.649254E - 03	4.177481E+01	1.561423E - 02	8.226110E - 01	8.770273E+00	6.801476E - 01
2.932006E - 03	4.159822E+01	4.676148E - 02	8.473498E - 01	8.475098E+00	6.922626E - 01
4.581258E - 03	4.136992E+01	1.065292E - 01	8.724552E - 01	8.185715E+00	7.015890E - 01
6.597012E - 03	4.108969E+01	2.029907E - 01	8.979270E - 01	7.902277E+00	7.078552E - 01
8.979268E - 03	4.075795E+01	3.403331E - 01	9.237652E - 01	7.624926E+00	7.109028E - 01
1.172803E - 02	4.037619E+01	5.174843E - 01	9.499701E - 01	7.353780E+00	7.106748E - 01
1.484328E - 02	3.994647E+01	7.277048E - 01	9.765414E - 01	7.088975E+00	7.072023E - 01
1.832504E - 02	3.947197E+01	9.592044E - 01	1.003479E+00	6.830655E+00	7.005914E - 01
2.217330E - 02	3.895657E+01	1.196637E+00	1.030783E+00	6.578926E+00	6.910100E - 01
2.638806E - 02	3.840468E+01	1.423198E+00	1.058454E+00	6.333645E+00	6.786744E - 01
3.096932E - 02	3.782089E+01	1.622887E+00	1.086492E+00	6.094820E+00	6.638379E - 01
3.591708E - 02	3.720975E+01	1.782593E+00	1.114895E+00	5.862756E+00	6.467797E - 01
4.123134E - 02	3.657552E+01	1.893677E+00	1.143665E+00	5.637069E+00	6.277948E - 01
4.691210E - 02	3.592192E+01	1.952827E+00	1.172802E+00	5.417840E+00	6.071856E - 01
5.295936E - 02	3.525220E+01	1.962125E+00	1.202306E+00	5.204790E+00	5.852539E - 01
5.937313E - 02	3.456902E+01	1.928432E+00	1.232175E+00	4.998116E+00	5.622960E - 01
6.615335E - 02	3.387468E+01	1.862222E+00	1.262411E+00	4.797338E+00	5.385957E - 01
7.330012E - 02	3.317125E+01	1.776097E+00	1.293015E+00	4.602284E+00	5.144209E - 01
8.081341E - 02	3.246085E+01	1.683241E+00	1.323984E+00	4.412948E+00	4.900227E - 01
8.869314E - 02	3.174557E+01	1.595968E+00	1.355320E+00	4.229140E+00	4.656302E - 01
9.693944E - 02	3.102783E+01	1.524578E+00	1.387022E+00	4.050676E+00	4.414508E - 01
1.055522E - 01	3.031001E+01	1.476577E+00	1.419090E+00	3.877367E+00	4.176691E - 01
1.145315E - 01	2.959430E+01	1.456282E+00	1.451526E+00	3.708874E+00	3.944461E - 01
1.238772E - 01	2.888251E+01	1.464828E+00	1.484328E+00	3.545013E+00	3.719209E - 01
1.335895E - 01	2.817581E+01	1.500454E+00	1.517496E+00	3.385958E+00	3.502119E - 01
1.436683E - 01	2.747507E+01	1.559017E+00	1.551031E+00	3.231220E+00	3.294146E - 01
1.541136E - 01	2.678088E+01	1.634659E+00	1.584932E+00	3.080834E+00	3.096063E - 01
1.649253E - 01	2.609384E+01	1.720478E+00	1.619200E+00	2.934674E+00	2.908460E - 01
1.761036E - 01	2.541466E+01	1.809225E+00	1.653834E+00	2.792473E+00	2.731755E - 01
1.876484E - 01	2.474426E+01	1.893881E+00	1.688835E+00	2.654306E+00	2.566223E - 01
1.995597E - 01	2.408380E+01	1.968133E+00	1.724202E+00	2.520105E+00	2.412003E - 01
2.118374E - 01	2.343449E+01	2.026740E+00	1.759936E+00	2.389653E+00	2.269113E - 01
2.244817E - 01	2.279749E+01	2.065756E+00	1.796037E+00	2.263080E+00	2.137471E - 01
2.374925E - 01	2.217386E+01	2.082617E+00	1.832503E+00	2.140362E+00	2.016908E - 01
2.508698E - 01	2.156451E+01	2.076147E+00	1.869337E+00	2.021324E+00	1.907176E - 01
2.646136E - 01	2.097000E+01	2.046468E+00	1.906537E+00	1.906138E+00	1.807972E - 01
2.787238E - 01	2.039067E+01	1.994832E+00	1.944103E+00	1.794806E+00	1.718941E - 01
2.932006E - 01	1.982655E+01	1.923445E+00	1.982036E+00	1.687342E+00	1.639687E - 01
3.080439E - 01	1.927734E+01	1.835219E+00	2.020335E+00	1.583595E+00	1.569794E - 01
3.232537E - 01	1.874258E+01	1.733573E+00	2.059001E+00	1.483744E+00	1.508821E - 01
3.388300E - 01	1.822160E+01	1.622192E+00	2.098033E+00	1.387789E+00	1.456317E - 01
3.547727E - 01	1.771361E+01	1.504836E+00	2.137432E+00	1.295563E+00	1.411827E - 01
3.710821E - 01	1.721773E+01	1.385150E+00	2.177197E+00	1.207222E+00	1.374897E - 01
3.877578E - 01	1.673318E+01	1.266544E+00	2.217329E+00	1.122737E+00	1.345081E - 01
4.048001E - 01	1.625923E+01	1.152040E+00	2.257828E+00	1.041910E+00	1.321941E - 01
4.222089E - 01	1.579528E+01	1.044229E+00	2.298693E+00	9.648596E - 01	1.305054E - 01
4.399842E - 01	1.534098E+01	9.451910E - 01	2.339924E+00	8.915260E - 01	1.294013E - 01
4.581260E - 01	1.489613E+01	8.564900E - 01	2.381522E+00	8.216875E - 01	1.288431E - 01
4.766343E - 01	1.446082E+01	7.791769E - 01	2.423486E+00	7.555994E - 01	1.287938E - 01
4.955091E - 01	1.403527E+01	7.138095E - 01	2.465817E+00	6.926956E - 01	1.292185E - 01
5.147504E - 01	1.361982E+01	6.604992E - 01	2.508514E+00	6.334016E - 01	1.300846E - 01
5.343581E - 01	1.321481E+01	6.189617E - 01	2.551579E+00	5.773186E - 01	1.313613E - 01
5.543324E - 01	1.282046E+01	5.885812E - 01	2.595009E+00	5.245453E - 01	1.330202E - 01
5.746732E - 01	1.243678E+01	5.684742E - 01	2.638805E+00	4.750290E - 01	1.350347E - 01
5.953805E - 01	1.206354E+01	5.575569E - 01	2.682969E+00	4.285535E - 01	1.373805E - 01
6.164544E - 01	1.170023E+01	5.546063E - 01	2.727499E+00	3.852432E - 01	1.400348E - 01
6.378946E - 01	1.134621E+01	5.583199E - 01	2.772395E+00	3.450745E - 01	1.429766E - 01
6.597014E - 01	1.100070E+01	5.673683E - 01	2.817657E+00	3.080259E - 01	1.461872E - 01
6.818748E - 01	1.066293E+01	5.804402E - 01	2.863287E+00	2.739304E - 01	1.496485E - 01
7.044145E - 01	1.033224E+01	5.962794E - 01	2.909283E+00	2.429586E - 01	1.533446E - 01
7.273208E - 01	1.000805E+01	6.137166E - 01	2.955646E+00	2.149591E - 01	1.572605E - 01
7.505937E - 01	9.689959E+00	6.316909E - 01	3.002375E+00	1.901256E - 01	1.613823E - 01

Z=21

Titanium

Atomic Number	22
Lattice	fcc
Lattice constant a_o	7.56 a.u.
Pressure at a_o	3.89 Kbar
Total energy (solid)	-1695.044 Ry
Total energy (atom)	-1694.591 Ry
Zero point energy	0.003 Ry
Cohesive energy	0.450 Ry
Bulk modulus	1.16 Mbar
Fermi energy	0.643 Ry
Interstitial electrons (per atom)	0.843
χ/χ_o	2.17
$N(E_f)$	1.59 states/eV-atom
ρ_o	7.100×10^3 electrons/Bohr^3
$(d\rho(r,E)/dE)_{E=E_f,r=0}$	2.9368 electrons/Bohr^3-Ry

Symmetry point	Band energies in Rydbergs (relative to muffin tin zero)			
Γ	0.200	0.707	0.707	0.707
X	0.415	0.504		
L	0.451	0.692	0.692	0.706
W	0.499	0.619	0.619	
K	0.484	0.549	0.709	

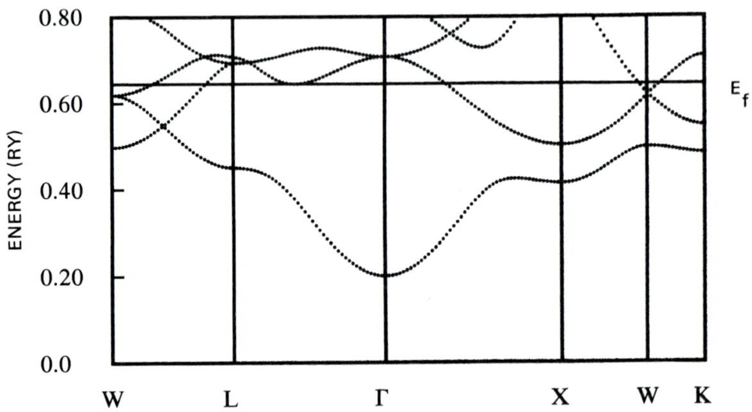

Z=22

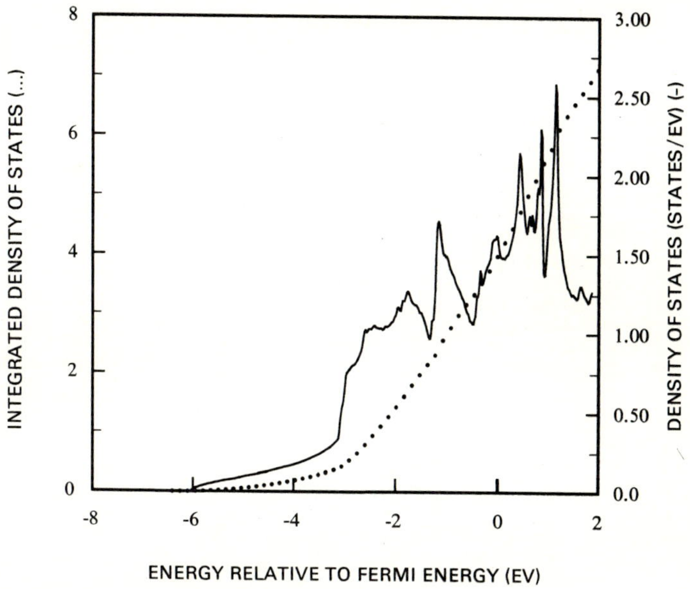

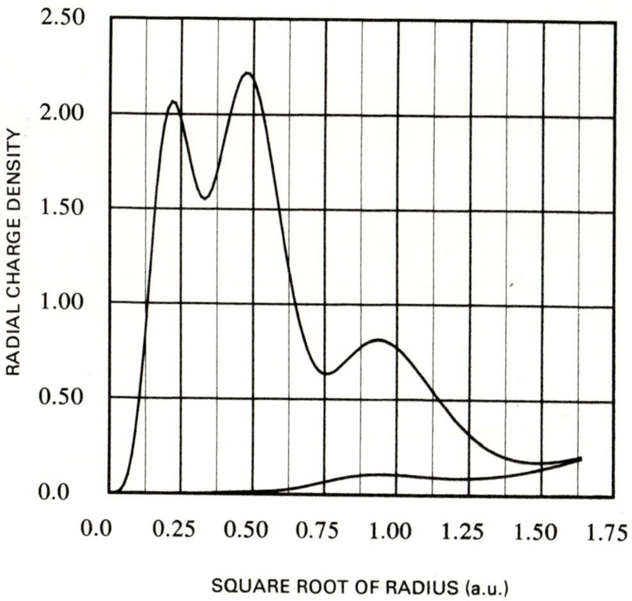

Moruzzi
Janak
Williams

Z=22

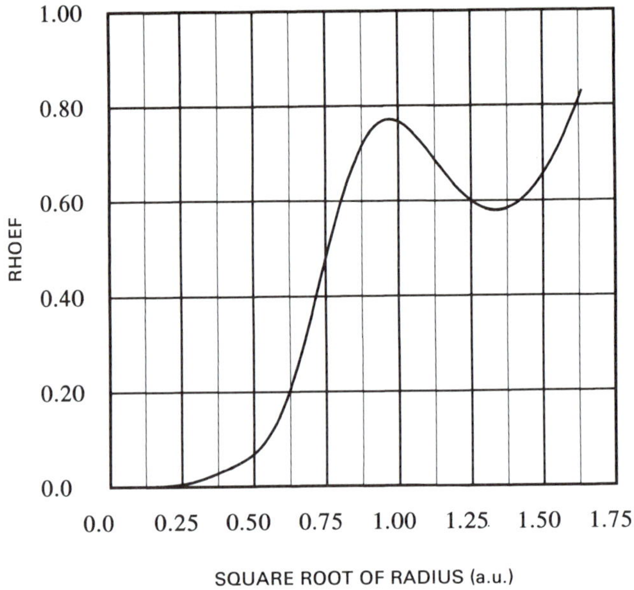

RHOEF=$r^2\partial\rho(E,r)/\partial E$ (at $E=E_f$)

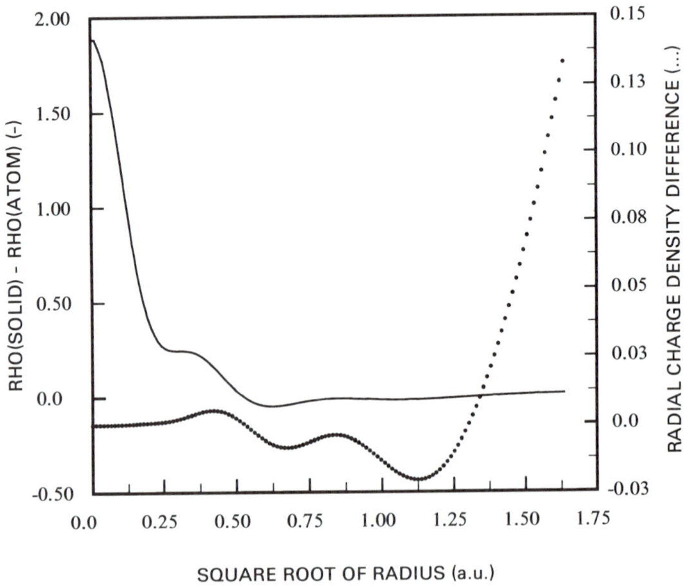

Titanium

r	2z(r)	$r^2\rho$	r	2z(r)	$r^2\rho$
1.631386E - 04	4.397621E+01	1.876169E - 04	6.892606E - 01	1.065113E+01	7.128007E - 01
6.525544E - 04	4.390468E+01	2.937924E - 03	7.106318E - 01	1.031335E+01	7.327533E - 01
1.468248E - 03	4.378514E+01	1.434899E - 02	7.323292E - 01	9.981374E+00	7.518561E - 01
2.610218E - 03	4.361674E+01	4.312842E - 02	7.543529E - 01	9.655141E+00	7.693397E - 01
4.078463E - 03	4.339909E+01	9.871107E - 02	7.767029E - 01	9.334702E+00	7.845734E - 01
5.872987E - 03	4.313188E+01	1.891652E - 01	7.993792E - 01	9.020134E+00	7.970630E - 01
7.993791E - 03	4.281546E+01	3.192861E - 01	8.223817E - 01	8.711576E+00	8.064450E - 01
1.044087E - 02	4.245088E+01	4.892392E - 01	8.457105E - 01	8.409192E+00	8.124781E - 01
1.321423E - 02	4.203986E+01	6.939989E - 01	8.693656E - 01	8.113116E+00	8.150323E - 01
1.631386E - 02	4.158504E+01	9.236658E - 01	8.933470E - 01	7.823505E+00	8.140771E - 01
1.973977E - 02	4.108974E+01	1.164588E+00	9.176546E - 01	7.540499E+00	8.096678E - 01
2.349196E - 02	4.055780E+01	1.401093E+00	9.422886E - 01	7.264182E+00	8.019331E - 01
2.757042E - 02	3.999335E+01	1.617457E+00	9.672488E - 01	6.994638E+00	7.910617E - 01
3.197517E - 02	3.940057E+01	1.799891E+00	9.925353E - 01	6.731915E+00	7.772889E - 01
3.670619E - 02	3.878345E+01	1.938137E+00	1.018147E+00	6.476135E+00	7.608860E - 01
4.176348E - 02	3.814563E+01	2.026538E+00	1.044086E+00	6.227071E+00	7.421489E - 01
4.714706E - 02	3.749028E+01	2.064459E+00	1.070352E+00	5.984792E+00	7.213882E - 01
5.285691E - 02	3.682014E+01	2.056026E+00	1.096944E+00	5.749048E+00	6.989210E - 01
5.889304E - 02	3.613750E+01	2.009342E+00	1.123861E+00	5.520087E+00	6.750639E - 01
6.525540E - 02	3.544435E+01	1.935323E+00	1.151106E+00	5.297435E+00	6.501259E - 01
7.194412E - 02	3.474266E+01	1.846301E+00	1.178676E+00	5.081305E+00	6.244040E - 01
7.895905E - 02	3.403432E+01	1.754686E+00	1.206573E+00	4.871189E+00	5.981792E - 01
8.630031E - 02	3.332141E+01	1.671721E+00	1.234796E+00	4.667103E+00	5.717130E - 01
9.396780E - 02	3.260616E+01	1.606561E+00	1.263345E+00	4.468870E+00	5.452467E - 01
1.019616E - 01	3.189066E+01	1.565642E+00	1.292220E+00	4.276324E+00	5.189980E - 01
1.102816E - 01	3.117691E+01	1.552436E+00	1.321423E+00	4.089116E+00	4.931615E - 01
1.189280E - 01	3.046648E+01	1.567480E+00	1.350950E+00	3.907426E+00	4.679093E - 01
1.279007E - 01	2.976041E+01	1.608687E+00	1.380805E+00	3.730742E+00	4.433888E - 01
1.371995E - 01	2.905945E+01	1.671834E+00	1.410985E+00	3.559261E+00	4.197264E - 01
1.468247E - 01	2.836417E+01	1.751148E+00	1.441492E+00	3.392492E+00	3.970259E - 01
1.567762E - 01	2.767513E+01	1.839937E+00	1.472325E+00	3.230476E+00	3.753718E - 01
1.670539E - 01	2.699306E+01	1.931202E+00	1.503485E+00	3.073102E+00	3.548296E - 01
1.776579E - 01	2.631883E+01	2.018166E+00	1.534970E+00	2.920261E+00	3.354477E - 01
1.885882E - 01	2.565347E+01	2.094716E+00	1.566783E+00	2.771698E+00	3.172585E - 01
1.998448E - 01	2.499811E+01	2.155740E+00	1.598921E+00	2.627689E+00	3.002809E - 01
2.114276E - 01	2.435382E+01	2.197323E+00	1.631386E+00	2.487826E+00	2.845210E - 01
2.233368E - 01	2.372160E+01	2.216870E+00	1.664177E+00	2.352237E+00	2.699743E - 01
2.355721E - 01	2.310228E+01	2.213100E+00	1.697293E+00	2.221066E+00	2.566268E - 01
2.481338E - 01	2.249641E+01	2.185980E+00	1.730737E+00	2.093946E+00	2.444564E - 01
2.610217E - 01	2.190436E+01	2.136586E+00	1.764506E+00	1.971035E+00	2.334346E - 01
2.742360E - 01	2.132617E+01	2.066936E+00	1.798603E+00	1.852145E+00	2.235276E - 01
2.877765E - 01	2.076169E+01	1.979775E+00	1.833025E+00	1.737631E+00	2.146971E - 01
3.016433E - 01	2.021053E+01	1.878378E+00	1.867773E+00	1.627305E+00	2.069020E - 01
3.158363E - 01	1.967213E+01	1.766337E+00	1.902848E+00	1.520990E+00	2.000986E - 01
3.303556E - 01	1.914584E+01	1.647357E+00	1.938250E+00	1.418857E+00	1.942420E - 01
3.452013E - 01	1.863092E+01	1.525097E+00	1.973977E+00	1.320888E+00	1.892862E - 01
3.603731E - 01	1.812666E+01	1.403004E+00	2.010031E+00	1.227063E+00	1.851852E - 01
3.758714E - 01	1.763243E+01	1.284214E+00	2.046411E+00	1.137363E+00	1.818935E - 01
3.916958E - 01	1.714772E+01	1.171444E+00	2.083117E+00	1.051748E+00	1.793664E - 01
4.078465E - 01	1.667216E+01	1.066947E+00	2.120149E+00	9.701850E - 01	1.775603E - 01
4.243235E - 01	1.620561E+01	9.724740E - 01	2.157508E+00	8.924478E - 01	1.764332E - 01
4.411268E - 01	1.574810E+01	8.892639E - 01	2.195192E+00	8.188444E - 01	1.759444E - 01
4.582563E - 01	1.529979E+01	8.180698E - 01	2.233204E+00	7.489707E - 01	1.760558E - 01
4.757122E - 01	1.486100E+01	7.591841E - 01	2.271542E+00	6.829574E - 01	1.767307E - 01
4.934943E - 01	1.443206E+01	7.124876E - 01	2.310205E+00	6.207555E - 01	1.779345E - 01
5.116026E - 01	1.401319E+01	6.775038E - 01	2.349195E+00	5.623230E - 01	1.796347E - 01
5.300373E - 01	1.360445E+01	6.534615E - 01	2.388512E+00	5.076213E - 01	1.818014E - 01
5.487983E - 01	1.320573E+01	6.393573E - 01	2.428155E+00	4.564481E - 01	1.844059E - 01
5.678855E - 01	1.281666E+01	6.340181E - 01	2.468123E+00	4.091253E - 01	1.874218E - 01
5.872990E - 01	1.243667E+01	6.361613E - 01	2.508419E+00	3.652956E - 01	1.908249E - 01
6.070387E - 01	1.206514E+01	6.444473E - 01	2.549041E+00	3.251253E - 01	1.945927E - 01
6.271048E - 01	1.170140E+01	6.575305E - 01	2.589988E+00	2.887976E - 01	1.987032E - 01
6.474971E - 01	1.134482E+01	6.740974E - 01	2.631262E+00	2.559827E - 01	2.031382E - 01
6.682158E - 01	1.099486E+01	6.929040E - 01	2.672863E+00	2.267063E - 01	2.078791E - 01

Z=22

Vanadium

Atomic Number	23
Lattice	bcc
Lattice constant a_o	5.54 a.u.
Pressure at a_o	4.90 Kbar
Total energy (solid)	-1883.982 Ry
Total energy (atom)	-1883.529 Ry
Zero point energy	0.003 Ry
Cohesive energy	0.450 Ry
Bulk modulus	1.64 Mbar
Fermi energy	0.762 Ry
Interstitial electrons (per atom)	1.106
χ/χ_o	2.34
$N(E_f)$	1.64 states/eV-atom
ρ_o	8.141×10^3 electrons/Bohr^3
$(d\rho(r,E)/dE)_{E=E_f,r=0}$	4.1629 electrons/Bohr^3-Ry

Symmetry point	Band energies in Rydbergs (relative to muffin tin zero)			
Γ	0.288	0.793	0.793	0.793
H	0.504	0.504		
N	0.512	0.670	0.911	
P	0.678	0.678	0.678	

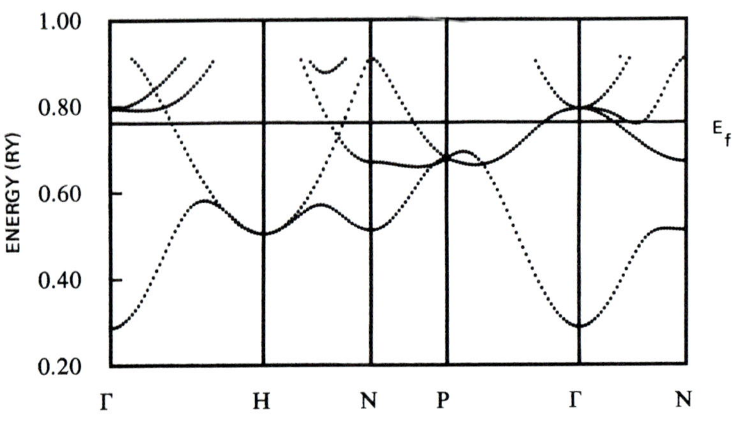

Z=23

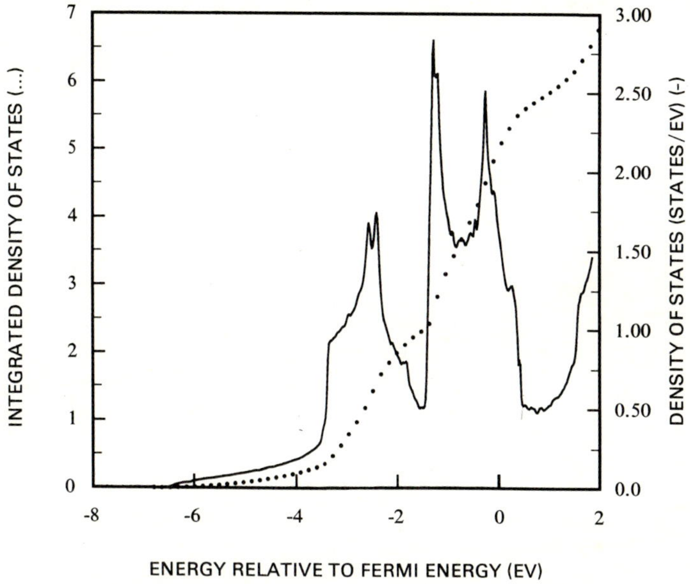

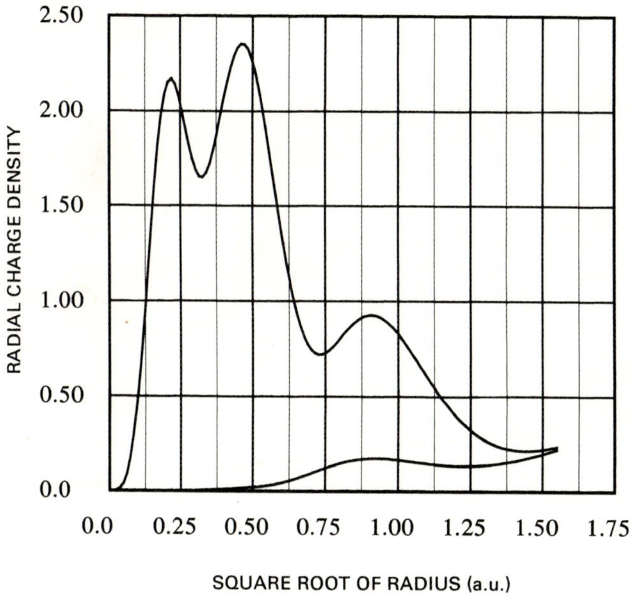

Moruzzi
Janak
Williams

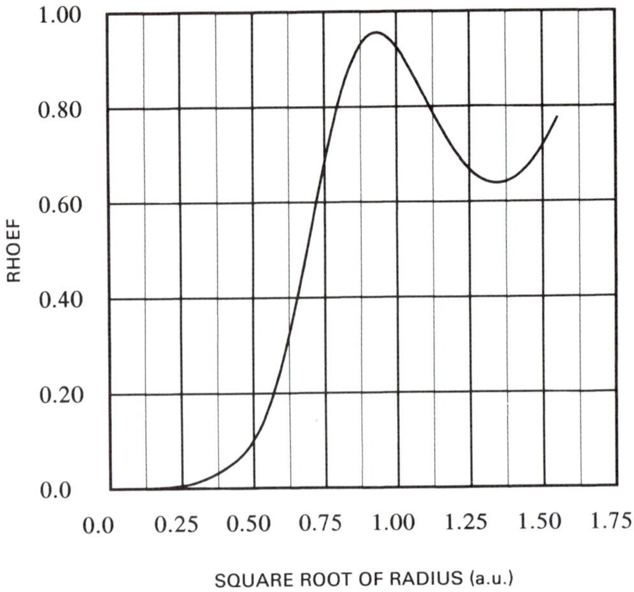

RHOEF=$r^2 \partial\rho(E,r)/\partial E$ (at $E=E_f$)

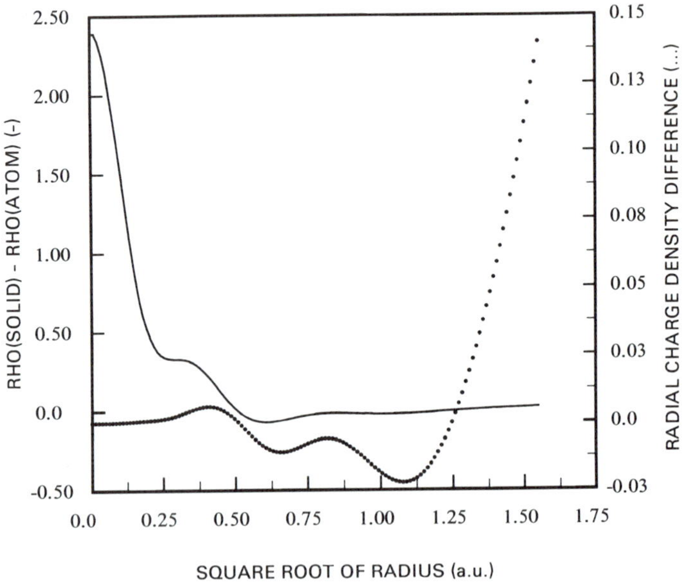

Moruzzi
Janak
Williams

Z=23

Vanadium

r	2z(r)	$r^2\rho$	r	2z(r)	$r^2\rho$
1.464166E - 04	4.597719E+01	1.733550E - 04	6.186101E - 01	1.196737E+01	7.805382E - 01
5.856664E - 04	4.590855E+01	2.718203E - 03	6.377907E - 01	1.159949E+01	8.021646E - 01
1.317749E - 03	4.579390E+01	1.330528E - 02	6.572641E - 01	1.123770E+01	8.242836E - 01
2.342666E - 03	4.563246E+01	4.011542E - 02	6.770303E - 01	1.088186E+01	8.459044E - 01
3.660415E - 03	4.542381E+01	9.218109E - 02	6.970894E - 01	1.053190E+01	8.661680E - 01
5.270995E - 03	4.516768E+01	1.775115E - 01	7.174413E - 01	1.018784E+01	8.843532E - 01
7.174410E - 03	4.486427E+01	3.013374E - 01	7.380860E - 01	9.849802E+00	8.998768E - 01
9.370662E - 03	4.451433E+01	4.647917E - 01	7.590236E - 01	9.517900E+00	9.122902E - 01
1.185974E - 02	4.411943E+01	6.642464E - 01	7.802540E - 01	9.192294E+00	9.212729E - 01
1.464166E - 02	4.368164E+01	8.914180E - 01	8.017772E - 01	8.873153E+00	9.266232E - 01
1.771641E - 02	4.320386E+01	1.134192E+00	8.235933E - 01	8.560637E+00	9.282469E - 01
2.108399E - 02	4.268947E+01	1.378045E+00	8.457022E - 01	8.254896E+00	9.261459E - 01
2.474440E - 02	4.214215E+01	1.607775E+00	8.681040E - 01	7.956060E+00	9.204041E - 01
2.869765E - 02	4.156575E+01	1.809303E+00	8.907986E - 01	7.664230E+00	9.111753E - 01
3.294373E - 02	4.096399E+01	1.971289E+00	9.137859E - 01	7.379492E+00	8.986700E - 01
3.748265E - 02	4.034039E+01	2.086292E+00	9.370662E - 01	7.101864E+00	8.831429E - 01
4.231440E - 02	3.969806E+01	2.151423E+00	9.606393E - 01	6.831365E+00	8.648823E - 01
4.743898E - 02	3.903972E+01	2.168391E+00	9.845052E - 01	6.567990E+00	8.441991E - 01
5.285639E - 02	3.836766E+01	2.143041E+00	1.008663E+00	6.311816E+00	8.214163E - 01
5.856664E - 02	3.768391E+01	2.084446E+00	1.033115E+00	6.062359E+00	7.968637E - 01
6.456971E - 02	3.699028E+01	2.003757E+00	1.057859E+00	5.819986E+00	7.708684E - 01
7.086563E - 02	3.628854E+01	1.912951E+00	1.082896E+00	5.584395E+00	7.437487E - 01
7.745433E - 02	3.558058E+01	1.823623E+00	1.108227E+00	5.355260E+00	7.158111E - 01
8.433592E - 02	3.486829E+01	1.745967E+00	1.133850E+00	5.132617E+00	6.873446E - 01
9.151036E - 02	3.415375E+01	1.688019E+00	1.159765E+00	4.916497E+00	6.586201E - 01
9.897757E - 02	3.343895E+01	1.655173E+00	1.185974E+00	4.706356E+00	6.298860E - 01
1.067377E - 01	3.272554E+01	1.650048E+00	1.212476E+00	4.502213E+00	6.013672E - 01
1.147906E - 01	3.201489E+01	1.672552E+00	1.239269E+00	4.304089E+00	5.732670E - 01
1.231363E - 01	3.130803E+01	1.720223E+00	1.266356E+00	4.111439E+00	5.457628E - 01
1.317749E - 01	3.060548E+01	1.788680E+00	1.293736E+00	3.924290E+00	5.190099E - 01
1.407063E - 01	2.990785E+01	1.872175E+00	1.321409E+00	3.742486E+00	4.931408E - 01
1.499306E - 01	2.921568E+01	1.964168E+00	1.349375E+00	3.565883E+00	4.682665E - 01
1.594477E - 01	2.852968E+01	2.057872E+00	1.377633E+00	3.394346E+00	4.444767E - 01
1.692576E - 01	2.785071E+01	2.146749E+00	1.406184E+00	3.227753E+00	4.218426E - 01
1.793603E - 01	2.717970E+01	2.224905E+00	1.435029E+00	3.065804E+00	4.004186E - 01
1.897559E - 01	2.651773E+01	2.287395E+00	1.464166E+00	2.908775E+00	3.802418E - 01
2.004443E - 01	2.586575E+01	2.330420E+00	1.493595E+00	2.756393E+00	3.613359E - 01
2.114255E - 01	2.522473E+01	2.351424E+00	1.523317E+00	2.608591E+00	3.437110E - 01
2.226996E - 01	2.459540E+01	2.349107E+00	1.553333E+00	2.465117E+00	3.273661E - 01
2.342665E - 01	2.397835E+01	2.323370E+00	1.583641E+00	2.326297E+00	3.122912E - 01
2.461263E - 01	2.337389E+01	2.275188E+00	1.614243E+00	2.191717E+00	2.984667E - 01
2.582788E - 01	2.278214E+01	2.206461E+00	1.645137E+00	2.061534E+00	2.858673E - 01
2.707242E - 01	2.220299E+01	2.119809E+00	1.676323E+00	1.935905E+00	2.744606E - 01
2.834625E - 01	2.163612E+01	2.018399E+00	1.707803E+00	1.814444E+00	2.642106E - 01
2.964936E - 01	2.108110E+01	1.905733E+00	1.739575E+00	1.697312E+00	2.550771E - 01
3.098175E - 01	2.053737E+01	1.785458E+00	1.771641E+00	1.584497E+00	2.470175E - 01
3.234342E - 01	2.000429E+01	1.661214E+00	1.803998E+00	1.476165E+00	2.399871E - 01
3.373438E - 01	1.948128E+01	1.536468E+00	1.836649E+00	1.371929E+00	2.339408E - 01
3.515462E - 01	1.896776E+01	1.414404E+00	1.869593E+00	1.271953E+00	2.288325E - 01
3.660415E - 01	1.846330E+01	1.297830E+00	1.902829E+00	1.176209E+00	2.246161E - 01
3.808295E - 01	1.796759E+01	1.189106E+00	1.936359E+00	1.084487E+00	2.212469E - 01
3.959104E - 01	1.748047E+01	1.090117E+00	1.970181E+00	9.971272E - 01	2.186802E - 01
4.112842E - 01	1.700198E+01	1.002250E+00	2.004296E+00	9.139103E - 01	2.168731E - 01
4.269508E - 01	1.653229E+01	9.264048E - 01	2.038704E+00	8.347972E - 01	2.157845E - 01
4.429102E - 01	1.607164E+01	8.630207E - 01	2.073405E+00	7.595689E - 01	2.153751E - 01
4.591624E - 01	1.562033E+01	8.121158E - 01	2.108398E+00	6.885631E - 01	2.156066E - 01
4.757075E - 01	1.517859E+01	7.733356E - 01	2.143685E+00	6.213745E - 01	2.164441E - 01
4.925454E - 01	1.474651E+01	7.460108E - 01	2.179264E+00	5.583460E - 01	2.178535E - 01
5.096762E - 01	1.432403E+01	7.292176E - 01	2.215137E+00	4.990828E - 01	2.198041E - 01
5.270997E - 01	1.391088E+01	7.218392E - 01	2.251301E+00	4.439371E - 01	2.222661E - 01
5.448161E - 01	1.350662E+01	7.226225E - 01	2.287759E+00	3.925255E - 01	2.252127E - 01
5.628254E - 01	1.311069E+01	7.302359E - 01	2.324510E+00	3.450309E - 01	2.286189E - 01
5.811275E - 01	1.272253E+01	7.433168E - 01	2.361553E+00	3.014617E - 01	2.324612E - 01
5.997224E - 01	1.234157E+01	7.605170E - 01	2.398890E+00	2.618270E - 01	2.367184E - 01

Z=23

Chromium

Atomic Number	24
Lattice	bcc
Lattice constant a_o	5.30 a.u.
Pressure at a_o	0.08 Kbar
Total energy (solid)	-2084.745 Ry
Total energy (atom)	-2084.386 Ry
Zero point energy	0.005 Ry
Cohesive energy	0.354 Ry
Bulk modulus	2.70 Mbar
Fermi energy	0.859 Ry
Interstitial electrons (per atom)	1.134
χ/χ_o	1.36
$N(E_f)$	0.70 states/eV-atom
ρ_o	9.279×10^3 electrons/Bohr^3
$(d\rho(r,E)/dE)_{E=E_f,r=0}$	0.7981 electrons/Bohr^3-Ry

Symmetry point	Band energies in Rydbergs (relative to muffin tin zero)					
Γ	0.291	0.790	0.790	0.790	0.936	0.936
H	0.506	0.506				
N	0.517	0.668	0.943	0.958	0.966	
P	0.683	0.683	0.683	0.971	0.971	

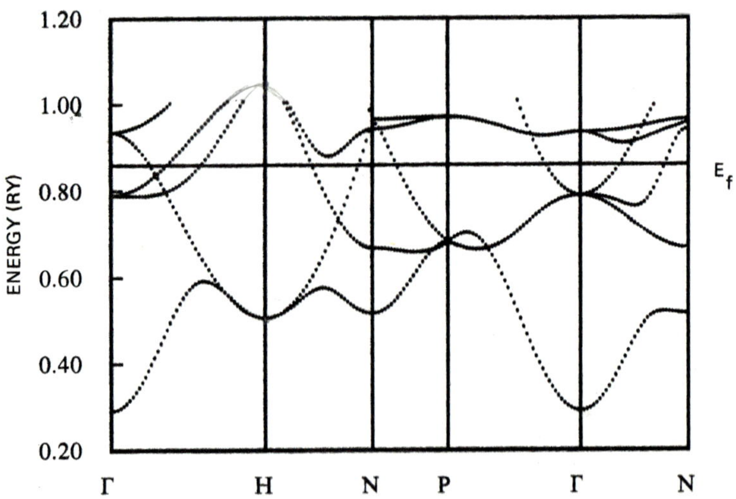

Z=24

Moruzzi
Janak
Williams

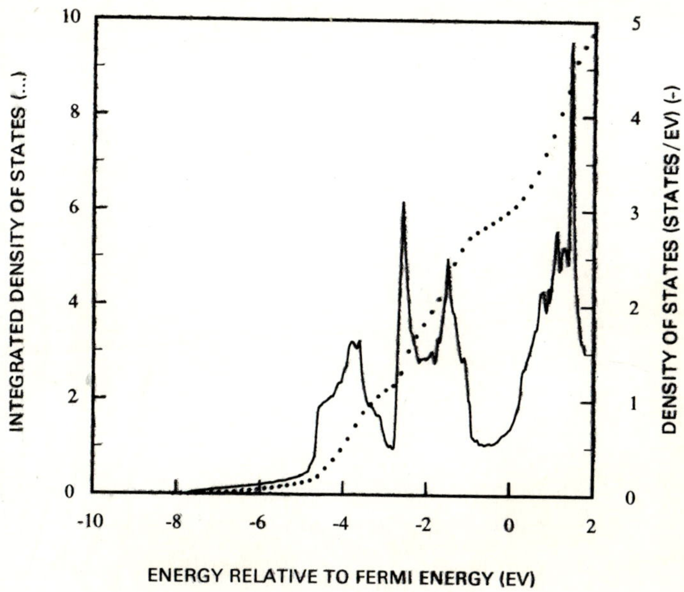

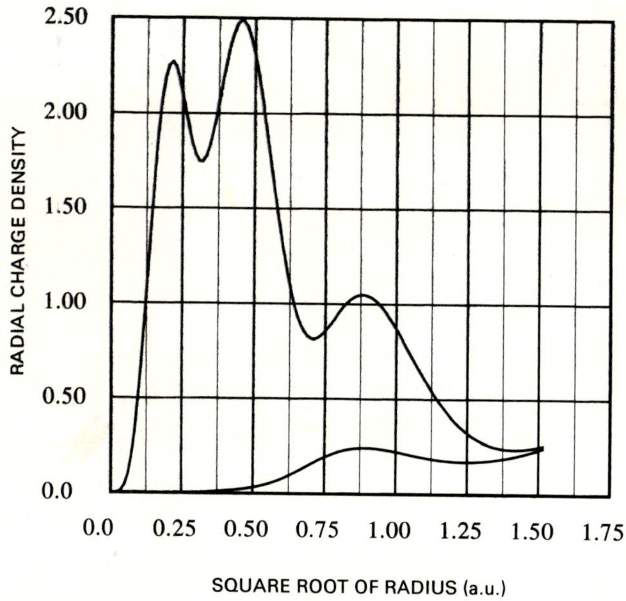

Moruzzi
Janak
Williams

Z=24

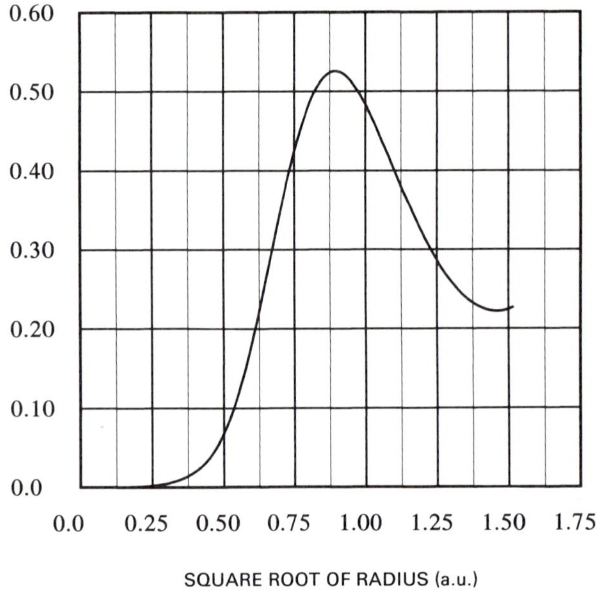

$RHOEF = r^2 \partial\rho(E,r)/\partial E$ (at $E=E_f$)

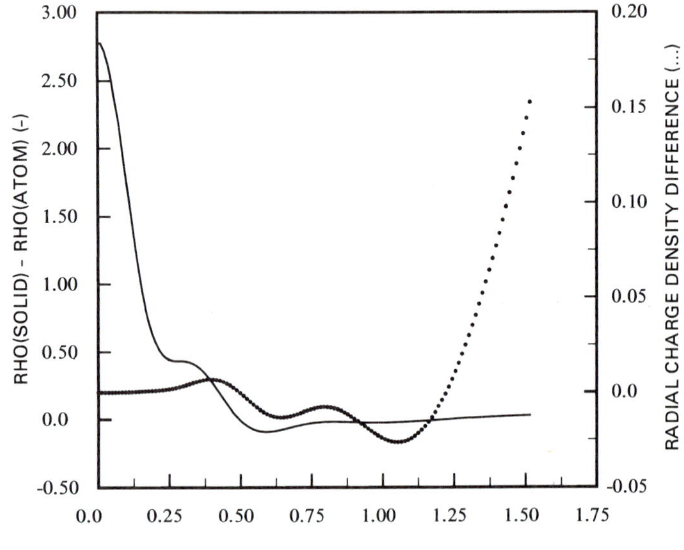

Moruzzi
Janak
Williams

Z=24

Chromium

r	2z(r)	$r^2\rho$	r	2z(r)	$r^2\rho$
1.400737E - 04	4.797673E+01	1.808335E - 04	5.918111E - 01	1.263979E+01	8.972753E - 01
5.602946E - 04	4.790675E+01	2.835565E - 03	6.101608E - 01	1.224525E+01	9.220389E - 01
1.260663E - 03	4.778986E+01	1.388055E - 02	6.287906E - 01	1.185718E+01	9.467141E - 01
2.241178E - 03	4.762527E+01	4.185325E - 02	6.477005E - 01	1.147551E+01	9.702721E - 01
3.501841E - 03	4.741261E+01	9.618449E - 02	6.668906E - 01	1.110031E+01	9.918351E - 01
5.042650E - 03	4.715157E+01	1.852439E - 01	6.863608E - 01	1.073164E+01	1.010678E+00
6.863605E - 03	4.684233E+01	3.145100E - 01	7.061113E - 01	1.036965E+01	1.026233E+00
8.964714E - 03	4.648567E+01	4.851911E - 01	7.261418E - 01	1.001454E+01	1.038073E+00
1.134596E - 02	4.608319E+01	6.935319E - 01	7.464525E - 01	9.666476E+00	1.045915E+00
1.400736E - 02	4.563690E+01	9.309150E - 01	7.670433E - 01	9.325652E+00	1.049601E+00
1.694891E - 02	4.514975E+01	1.184721E+00	7.879143E - 01	8.992249E+00	1.049086E+00
2.017060E - 02	4.462508E+01	1.439804E+00	8.090654E - 01	8.666415E+00	1.044427E+00
2.367245E - 02	4.406660E+01	1.680298E+00	8.304967E - 01	8.348285E+00	1.035763E+00
2.745443E - 02	4.347816E+01	1.891500E+00	8.522081E - 01	8.037946E+00	1.023306E+00
3.151657E - 02	4.286351E+01	2.061553E+00	8.741996E - 01	7.735463E+00	1.007321E+00
3.585885E - 02	4.222614E+01	2.182672E+00	8.964713E - 01	7.440858E+00	9.881200E - 01
4.048128E - 02	4.156920E+01	2.251824E+00	9.190232E - 01	7.154111E+00	9.660406E - 01
4.538386E - 02	4.089543E+01	2.270785E+00	9.418552E - 01	6.875204E+00	9.414421E - 01
5.056659E - 02	4.020715E+01	2.245661E+00	9.649674E - 01	6.604039E+00	9.146932E - 01
5.602946E - 02	3.950645E+01	2.185938E+00	9.883596E - 01	6.340544E+00	8.861623E - 01
6.177248E - 02	3.879514E+01	2.103278E+00	1.012032E+00	6.084678E+00	8.562114E - 01
6.779563E - 02	3.807503E+01	2.010214E+00	1.035984E+00	5.836114E+00	8.251916E - 01
7.409894E - 02	3.734804E+01	1.918873E+00	1.060217E+00	5.594727E+00	7.934350E - 01
8.068240E - 02	3.661613E+01	1.839910E+00	1.084730E+00	5.360341E+00	7.612528E - 01
8.754599E - 02	3.588141E+01	1.781700E+00	1.109523E+00	5.132975E+00	7.289340E - 01
9.468979E - 02	3.514581E+01	1.749858E+00	1.134596E+00	4.912238E+00	6.967415E - 01
1.021137E - 01	3.441110E+01	1.747051E+00	1.159949E+00	4.697941E+00	6.649115E - 01
1.098177E - 01	3.367857E+01	1.773111E+00	1.185583E+00	4.489891E+00	6.336544E - 01
1.178019E - 01	3.294926E+01	1.825367E+00	1.211496E+00	4.288097E+00	6.031529E - 01
1.260663E - 01	3.222383E+01	1.899138E+00	1.237690E+00	4.092182E+00	5.735657E - 01
1.346108E - 01	3.150282E+01	1.988302E+00	1.264164E+00	3.901973E+00	5.450251E - 01
1.434354E - 01	3.078683E+01	2.085925E+00	1.290918E+00	3.717303E+00	5.176402E - 01
1.525401E - 01	3.007661E+01	2.184828E+00	1.317952E+00	3.538216E+00	4.914989E - 01
1.619251E - 01	2.937303E+01	2.278114E+00	1.345267E+00	3.364178E+00	4.666685E - 01
1.715902E - 01	2.867708E+01	2.359591E+00	1.372861E+00	3.195652E+00	4.431977E - 01
1.815354E - 01	2.798982E+01	2.424091E+00	1.400736E+00	3.031927E+00	4.211181E - 01
1.917608E - 01	2.731227E+01	2.467677E+00	1.428891E+00	2.873107E+00	4.004470E - 01
2.022663E - 01	2.664532E+01	2.487748E+00	1.457326E+00	2.719291E+00	3.811883E - 01
2.130520E - 01	2.598981E+01	2.483054E+00	1.486041E+00	2.570210E+00	3.633344E - 01
2.241178E - 01	2.534621E+01	2.453613E+00	1.515037E+00	2.425797E+00	3.468677E - 01
2.354638E - 01	2.471490E+01	2.400591E+00	1.544312E+00	2.286190E+00	3.317624E - 01
2.470899E - 01	2.409598E+01	2.326125E+00	1.573867E+00	2.151149E+00	3.179860E - 01
2.589961E - 01	2.348933E+01	2.233123E+00	1.603703E+00	2.020630E+00	3.055000E - 01
2.711825E - 01	2.289464E+01	2.125040E+00	1.633819E+00	1.894601E+00	2.942624E - 01
2.836491E - 01	2.231149E+01	2.005673E+00	1.664214E+00	1.773227E+00	2.842267E - 01
2.963958E - 01	2.173935E+01	1.878956E+00	1.694891E+00	1.656082E+00	2.753453E - 01
3.094227E - 01	2.117764E+01	1.748782E+00	1.725847E+00	1.543535E+00	2.675682E - 01
3.227296E - 01	2.062582E+01	1.618837E+00	1.757083E+00	1.435560E+00	2.608455E - 01
3.363168E - 01	2.008339E+01	1.492479E+00	1.788600E+00	1.331730E+00	2.551269E - 01
3.501841E - 01	1.955002E+01	1.372644E+00	1.820396E+00	1.232411E+00	2.503628E - 01
3.643315E - 01	1.902544E+01	1.261769E+00	1.852473E+00	1.137378E+00	2.465041E - 01
3.787591E - 01	1.850961E+01	1.161761E+00	1.884830E+00	1.046591E+00	2.435039E - 01
3.934668E - 01	1.800269E+01	1.073993E+00	1.917468E+00	9.600245E - 01	2.413167E - 01
4.084547E - 01	1.750482E+01	9.992988E - 01	1.950385E+00	8.778360E - 01	2.398987E - 01
4.237227E - 01	1.701630E+01	9.380245E - 01	1.983582E+00	7.997991E - 01	2.392086E - 01
4.392709E - 01	1.653740E+01	8.900638E - 01	2.017060E+00	7.258826E - 01	2.392076E - 01
4.550993E - 01	1.606828E+01	8.549184E - 01	2.050817E+00	6.562603E - 01	2.398583E - 01
4.712077E - 01	1.560901E+01	8.317614E - 01	2.084856E+00	5.905107E - 01	2.411276E - 01
4.875963E - 01	1.515938E+01	8.195057E - 01	2.119174E+00	5.290148E - 01	2.429829E - 01
5.042651E - 01	1.471906E+01	8.168700E - 01	2.153772E+00	4.715642E - 01	2.453951E - 01
5.212140E - 01	1.428760E+01	8.224437E - 01	2.188650E+00	4.183550E - 01	2.483371E - 01
5.384431E - 01	1.386443E+01	8.347470E - 01	2.223809E+00	3.689976E - 01	2.517853E - 01
5.559523E - 01	1.344906E+01	8.522852E - 01	2.259248E+00	3.238983E - 01	2.557165E - 01
5.737417E - 01	1.304097E+01	8.735929E - 01	2.294967E+00	2.828924E - 01	2.601116E - 01

Z=24

Manganese

Atomic Number	25
Lattice	fcc
Lattice constant a_o	6.543 a.u.
Pressure at a_o	-14.66 Kbar
Total energy (solid)	-2297.579 Ry
Total energy (atom)	-2297.211 Ry
Zero point energy	0.004 Ry
Cohesive energy	0.364 Ry
Bulk modulus	2.91 Mbar
Fermi energy	0.821 Ry
Interstitial electrons (per atom)	0.877
χ/χ_o	2.74
$N(E_f)$	1.56 states/eV-atom
ρ_o	1.052×10^4 electrons/Bohr^3
$(d\rho(r,E)/dE)_{E=E_f,r=0}$	2.2478 electrons/Bohr^3-Ry

Symmetry point	Band energies in Rydbergs (relative to muffin tin zero)					
Γ	0.222	0.734	0.734	0.734	0.860	0.860
X	0.444	0.524				
L	0.487	0.721	0.721	0.863		
W	0.537	0.660	0.660	0.844		
K	0.520	0.582	0.770	0.874		

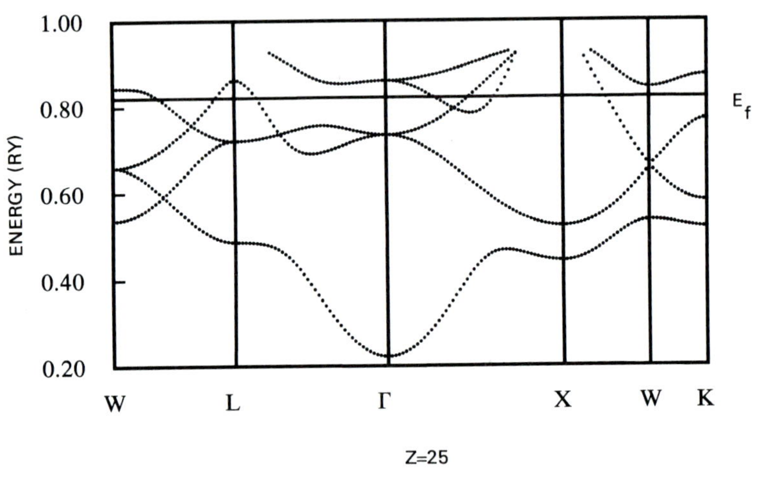

Z=25

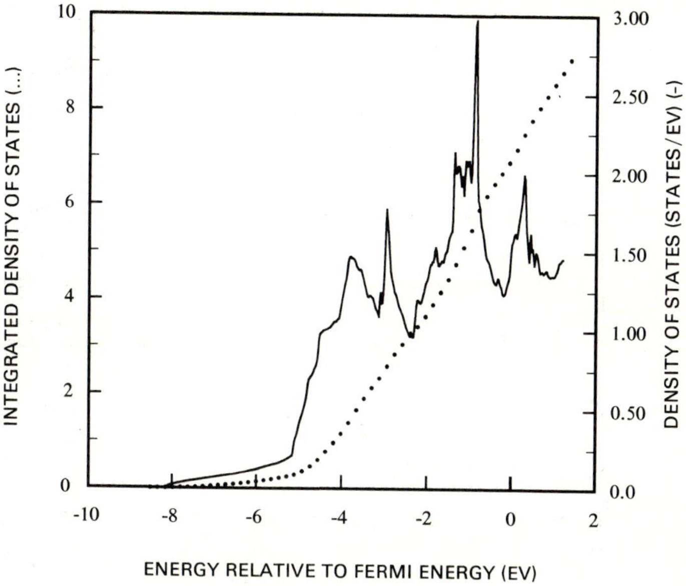

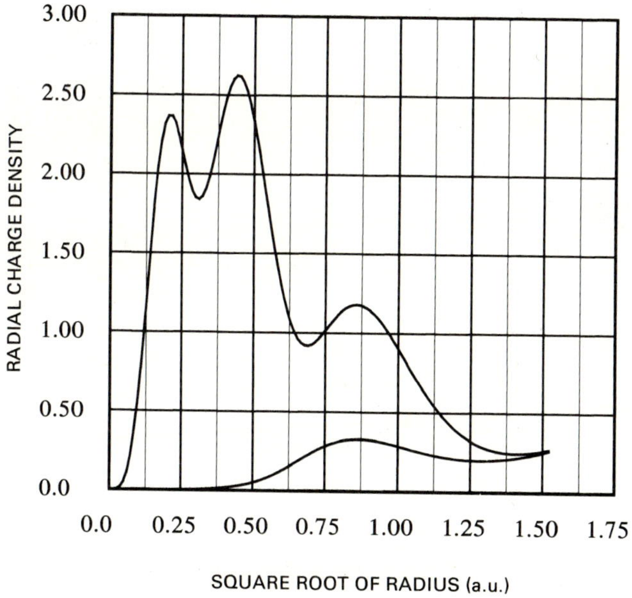

Moruzzi
Janak
Williams

Manganese

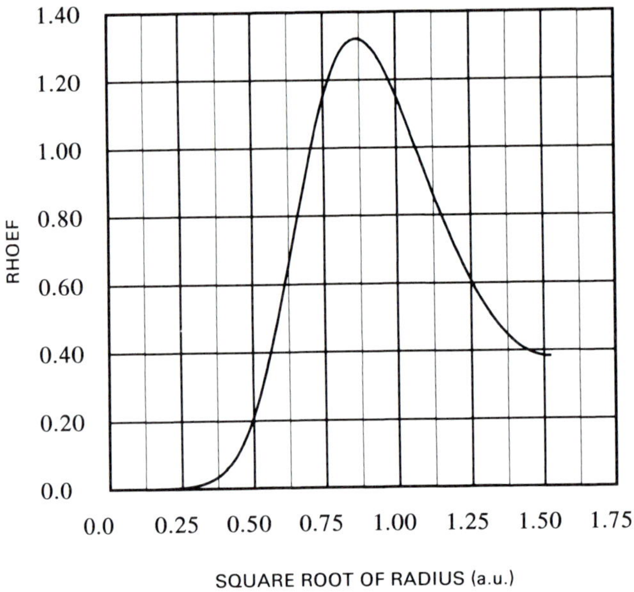

RHOEF=$r^2\partial\rho(E,r)/\partial E$ (at $E=E_f$)

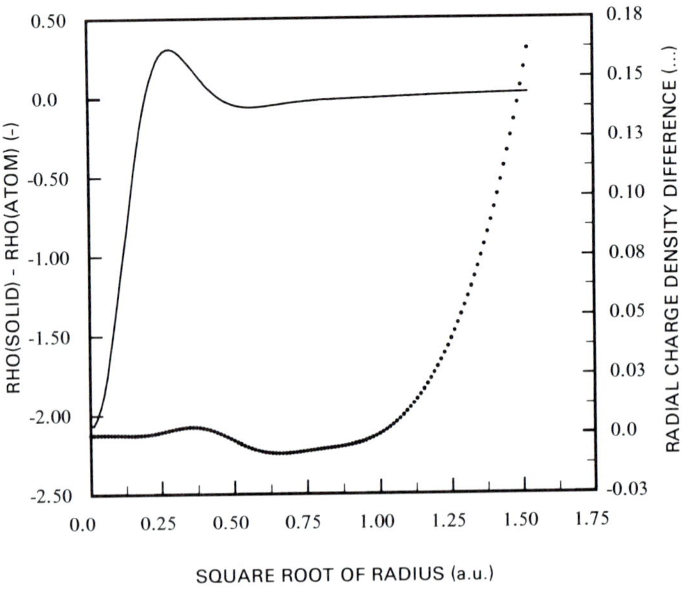

Moruzzi
Janak
Williams

Z=25

Manganese

r	2z(r)	$r^2\rho$	r	2z(r)	$r^2\rho$
1.411926E - 04	4.997505E+01	2.081931E - 04	5.965386E - 01	1.262725E+01	1.069750E+00
5.647703E - 04	4.990005E+01	3.261288E - 03	6.150348E - 01	1.220833E+01	1.096277E+00
1.270733E - 03	4.977480E+01	1.593773E - 02	6.338134E - 01	1.179685E+01	1.120325E+00
2.259081E - 03	4.959839E+01	4.794329E - 02	6.528745E - 01	1.139298E+01	1.141060E+00
3.529814E - 03	4.937044E+01	1.098478E - 01	6.722178E - 01	1.099690E+01	1.157837E+00
5.082931E - 03	4.909070E+01	2.107795E - 01	6.918436E - 01	1.060886E+01	1.170186E+00
6.918434E - 03	4.875945E+01	3.563097E - 01	7.117518E - 01	1.022906E+01	1.177809E+00
9.036325E - 03	4.837766E+01	5.469263E - 01	7.319423E - 01	9.857740E+00	1.180552E+00
1.143660E - 02	4.794704E+01	7.773609E - 01	7.524152E - 01	9.495101E+00	1.178401E+00
1.411926E - 02	4.747023E+01	1.036883E+00	7.731705E - 01	9.141325E+00	1.171456E+00
1.708430E - 02	4.695032E+01	1.310491E+00	7.942082E - 01	8.796534E+00	1.159916E+00
2.033173E - 02	4.639110E+01	1.580761E+00	8.155283E - 01	8.460834E+00	1.144058E+00
2.386154E - 02	4.579663E+01	1.830045E+00	8.371308E - 01	8.134254E+00	1.124224E+00
2.767374E - 02	4.517108E+01	2.042651E+00	8.590156E - 01	7.816846E+00	1.100798E+00
3.176833E - 02	4.451840E+01	2.206699E+00	8.811828E - 01	7.508536E+00	1.074194E+00
3.614530E - 02	4.384227E+01	2.315336E+00	9.036325E - 01	7.209281E+00	1.044848E+00
4.080465E - 02	4.314594E+01	2.367311E+00	9.263645E - 01	6.918967E+00	1.013198E+00
4.574639E - 02	4.243214E+01	2.366751E+00	9.493788E - 01	6.637466E+00	9.796799E - 01
5.097052E - 02	4.170334E+01	2.322362E+00	9.726756E - 01	6.364607E+00	9.447144E - 01
5.647703E - 02	4.096158E+01	2.246157E+00	9.962548E - 01	6.100196E+00	9.087060E - 01
6.226593E - 02	4.020894E+01	2.151926E+00	1.020116E+00	5.844150E+00	8.720340E - 01
6.833720E - 02	3.944737E+01	2.053702E+00	1.044260E+00	5.595975E+00	8.350506E - 01
7.469082E - 02	3.867909E+01	1.964349E+00	1.068686E+00	5.355620E+00	7.980763E - 01
8.132690E - 02	3.790623E+01	1.894460E+00	1.093395E+00	5.122855E+00	7.614014E - 01
8.824533E - 02	3.713109E+01	1.851611E+00	1.118386E+00	4.897440E+00	7.252833E - 01
9.544617E - 02	3.635564E+01	1.839997E+00	1.143660E+00	4.679140E+00	6.899455E - 01
1.029294E - 01	3.558144E+01	1.860453E+00	1.169215E+00	4.467719E+00	6.555826E - 01
1.106949E - 01	3.480974E+01	1.910745E+00	1.195053E+00	4.262958E+00	6.223566E - 01
1.187429E - 01	3.404135E+01	1.986119E+00	1.221174E+00	4.064424E+00	5.904019E - 01
1.270733E - 01	3.327696E+01	2.079964E+00	1.247577E+00	3.872334E+00	5.598256E - 01
1.356860E - 01	3.251721E+01	2.184548E+00	1.274262E+00	3.686077E+00	5.307100E - 01
1.445812E - 01	3.176283E+01	2.291724E+00	1.301230E+00	3.505680E+00	5.031154E - 01
1.537587E - 01	3.101483E+01	2.393565E+00	1.328481E+00	3.330979E+00	4.770812E - 01
1.632186E - 01	3.027428E+01	2.482915E+00	1.356013E+00	3.161822E+00	4.526287E - 01
1.729609E - 01	2.954233E+01	2.553779E+00	1.383828E+00	2.998074E+00	4.297634E - 01
1.829855E - 01	2.882008E+01	2.601593E+00	1.411925E+00	2.839610E+00	4.084771E - 01
1.932926E - 01	2.810860E+01	2.623363E+00	1.440305E+00	2.686320E+00	3.887497E - 01
2.038820E - 01	2.740871E+01	2.617683E+00	1.468967E+00	2.537900E+00	3.705515E - 01
2.147539E - 01	2.672099E+01	2.584644E+00	1.497911E+00	2.394681E+00	3.538438E - 01
2.259081E - 01	2.604585E+01	2.525685E+00	1.527139E+00	2.256169E+00	3.385822E - 01
2.373447E - 01	2.538339E+01	2.443375E+00	1.556647E+00	2.122716E+00	3.247167E - 01
2.490637E - 01	2.473351E+01	2.341156E+00	1.586439E+00	1.993843E+00	3.121940E - 01
2.610651E - 01	2.409586E+01	2.223092E+00	1.616513E+00	1.869705E+00	3.009581E - 01
2.733488E - 01	2.347003E+01	2.093606E+00	1.646870E+00	1.750253E+00	2.909508E - 01
2.859150E - 01	2.285547E+01	1.957229E+00	1.677508E+00	1.635440E+00	2.821143E - 01
2.987635E - 01	2.225163E+01	1.818390E+00	1.708429E+00	1.525221E+00	2.743903E - 01
3.118944E - 01	2.165800E+01	1.681231E+00	1.739634E+00	1.419344E+00	2.677217E - 01
3.253077E - 01	2.107416E+01	1.549453E+00	1.771119E+00	1.318185E+00	2.620518E - 01
3.390033E - 01	2.049982E+01	1.426221E+00	1.802888E+00	1.221285E+00	2.573271E - 01
3.529814E - 01	1.993491E+01	1.314084E+00	1.834938E+00	1.129023E+00	2.534950E - 01
3.672419E - 01	1.937946E+01	1.214952E+00	1.867271E+00	1.040941E+00	2.505070E - 01
3.817847E - 01	1.883369E+01	1.130085E+00	1.899887E+00	9.572135E - 01	2.483154E - 01
3.966099E - 01	1.829788E+01	1.060132E+00	1.932785E+00	8.778031E - 01	2.468765E - 01
4.117175E - 01	1.777229E+01	1.005174E+00	1.965965E+00	8.026827E - 01	2.461498E - 01
4.271075E - 01	1.725719E+01	9.647886E - 01	1.999428E+00	7.318242E - 01	2.460970E - 01
4.427799E - 01	1.675258E+01	9.381319E - 01	2.033173E+00	6.652096E - 01	2.466834E - 01
4.587346E - 01	1.625835E+01	9.240168E - 01	2.067200E+00	6.028198E - 01	2.478775E - 01
4.749718E - 01	1.577419E+01	9.209984E - 01	2.101510E+00	5.444432E - 01	2.496507E - 01
4.914913E - 01	1.529962E+01	9.274558E - 01	2.136102E+00	4.904979E - 01	2.519768E - 01
5.082933E - 01	1.483411E+01	9.416682E - 01	2.170977E+00	4.405740E - 01	2.548339E - 01
5.253776E - 01	1.437717E+01	9.618832E - 01	2.206134E+00	3.949022E - 01	2.582018E - 01
5.427442E - 01	1.392832E+01	9.863793E - 01	2.241573E+00	3.535104E - 01	2.620633E - 01
5.603933E - 01	1.348717E+01	1.013513E+00	2.277294E+00	3.166460E - 01	2.664039E - 01
5.783248E - 01	1.305353E+01	1.041760E+00	2.313299E+00	2.837312E - 01	2.712124E - 01

Z=25

Iron

Atomic Number	26
Lattice	bcc
Lattice constant a_o	5.15 a.u.
Pressure at a_o	-8.28 Kbar
Total energy (solid)	-2522.812 Ry
Total energy (atom)	-2522.369 Ry
Zero point energy	0.004 Ry
Cohesive energy	0.439 Ry
Bulk modulus	3.06 Mbar
Fermi energy	0.833 Ry
Interstitial electrons (per atom)	1.044
χ/χ_o	-2.34
$N(E_f)$	3.06 states/eV-atom
ρ_o	1.186×10^4 electrons/Bohr^3
$(d\rho(r,E)/dE)_{E=E_f,r=0}$	1.9133 electrons/Bohr^3-Ry

Symmetry point	Band energies in Rydbergs (relative to muffin tin zero)					
Γ	0.186	0.699	0.699	0.699	0.826	0.826
H	0.463	0.463	0.916	0.916	0.916	
N	0.460	0.598	0.828	0.848	0.878	
P	0.605	0.605	0.605	0.851	0.851	

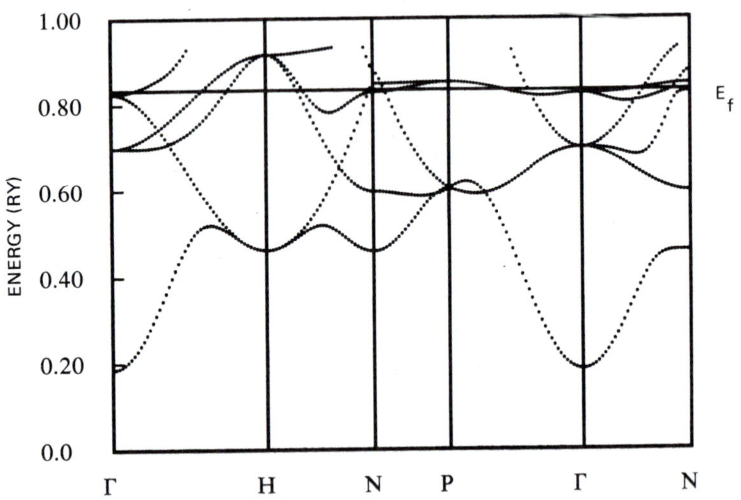

Z=26

Moruzzi
Janak
Williams

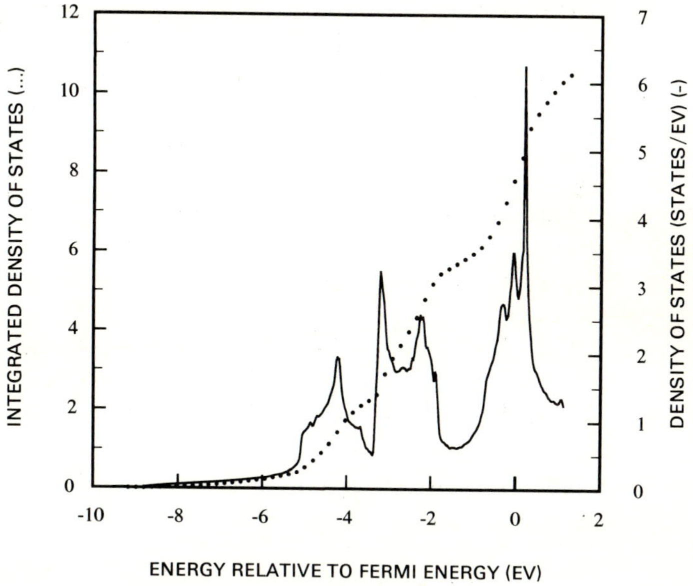

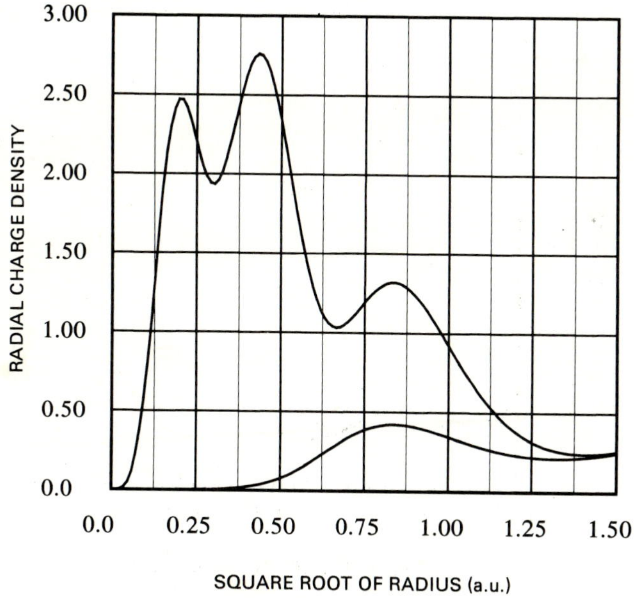

Moruzzi
Janak
Williams

Z=26

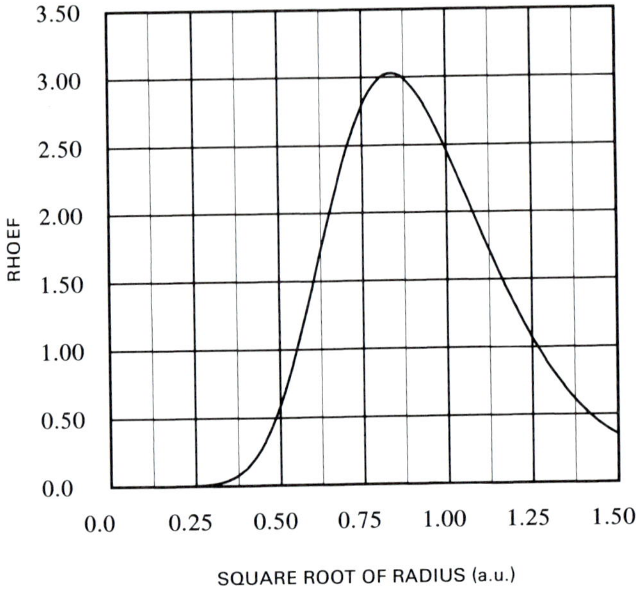

RHOEF=$r^2\partial\rho(E,r)/\partial E$ (at $E=E_f$)

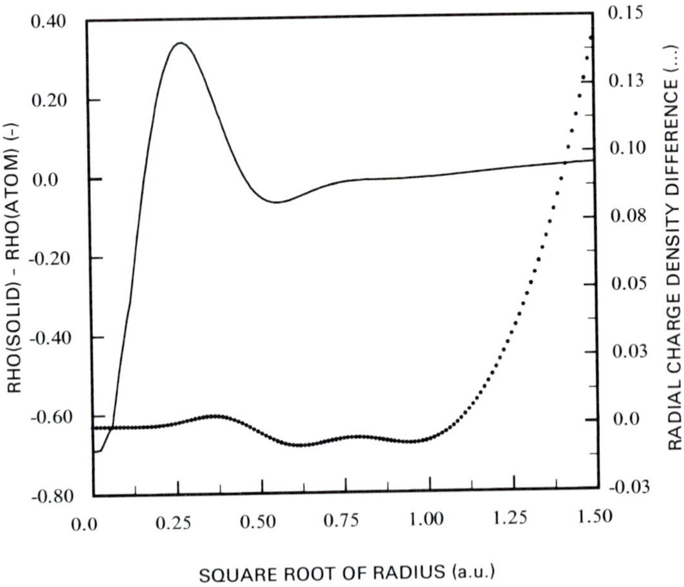

Iron

r	2z(r)	$r^2\rho$	r	2z(r)	$r^2\rho$
1.361093E - 04	5.197447E+01	2.181906E - 04	5.750618E - 01	1.317289E+01	1.214734E+00
5.444372E - 04	5.189778E+01	3.417710E - 03	5.928921E - 01	1.272852E+01	1.242575E+00
1.224984E - 03	5.176965E+01	1.670066E - 02	6.109946E - 01	1.229237E+01	1.267129E+00
2.177749E - 03	5.158926E+01	5.023189E - 02	6.293694E - 01	1.186469E+01	1.287592E+00
3.402733E - 03	5.135616E+01	1.150727E - 01	6.480163E - 01	1.144574E+01	1.303360E+00
4.899934E - 03	5.107014E+01	2.207608E - 01	6.669356E - 01	1.103577E+01	1.314021E+00
6.669354E - 03	5.073148E+01	3.730943E - 01	6.861269E - 01	1.063505E+01	1.319335E+00
8.710995E - 03	5.034111E+01	5.725338E - 01	7.055906E - 01	1.024378E+01	1.319222E+00
1.102485E - 02	4.990088E+01	8.135056E - 01	7.253264E - 01	9.862201E+00	1.313738E+00
1.361093E - 02	4.941339E+01	1.084720E+00	7.453345E - 01	9.490438E+00	1.303058E+00
1.646923E - 02	4.888173E+01	1.370435E+00	7.656148E - 01	9.128634E+00	1.287450E+00
1.959974E - 02	4.830977E+01	1.652397E+00	7.861673E - 01	8.776867E+00	1.267262E+00
2.300247E - 02	4.770160E+01	1.912157E+00	8.069920E - 01	8.435174E+00	1.242900E+00
2.667742E - 02	4.706139E+01	2.133368E+00	8.280889E - 01	8.103513E+00	1.214808E+00
3.062459E - 02	4.639314E+01	2.303693E+00	8.494581E - 01	7.781835E+00	1.183455E+00
3.484398E - 02	4.570055E+01	2.416120E+00	8.710995E - 01	7.470019E+00	1.149325E+00
3.933559E - 02	4.498688E+01	2.469497E+00	8.930131E - 01	7.167960E+00	1.112899E+00
4.409941E - 02	4.425494E+01	2.468319E+00	9.151989E - 01	6.875468E+00	1.074646E+00
4.913546E - 02	4.350714E+01	2.421854E+00	9.376569E - 01	6.592342E+00	1.035021E+00
5.444372E - 02	4.274564E+01	2.342782E+00	9.603872E - 01	6.318355E+00	9.944513E - 01
6.002420E - 02	4.197249E+01	2.245605E+00	9.833897E - 01	6.053279E+00	9.533358E - 01
6.587690E - 02	4.118977E+01	2.145009E+00	1.006664E+00	5.796864E+00	9.120406E - 01
7.200181E - 02	4.039964E+01	2.054392E+00	1.030210E+00	5.548967E+00	8.708989E - 01
7.839894E - 02	3.960442E+01	1.984720E+00	1.054030E+00	5.308837E+00	8.302078E - 01
8.506829E - 02	3.880629E+01	1.943734E+00	1.078121E+00	5.076861E+00	7.902262E - 01
9.200984E - 02	3.800729E+01	1.935612E+00	1.102485E+00	4.852324E+00	7.511817E - 01
9.922367E - 02	3.720898E+01	1.960981E+00	1.127121E+00	4.634962E+00	7.132654E - 01
1.067097E - 01	3.641266E+01	2.017265E+00	1.152029E+00	4.424731E+00	6.766374E - 01
1.144679E - 01	3.561902E+01	2.099266E+00	1.177209E+00	4.221382E+00	6.414270E - 01
1.224983E - 01	3.482884E+01	2.199870E+00	1.202662E+00	4.024454E+00	6.077364E - 01
1.308010E - 01	3.404285E+01	2.310843E+00	1.228386E+00	3.833939E+00	5.756406E - 01
1.393759E - 01	3.326184E+01	2.423569E+00	1.254383E+00	3.649401E+00	5.451943E - 01
1.482230E - 01	3.248679E+01	2.529735E+00	1.280652E+00	3.470860E+00	5.164275E - 01
1.573423E - 01	3.171881E+01	2.621885E+00	1.307193E+00	3.298134E+00	4.893554E - 01
1.667339E - 01	3.095912E+01	2.693853E+00	1.334007E+00	3.130609E+00	4.639767E - 01
1.763976E - 01	3.020886E+01	2.741012E+00	1.361093E+00	2.968786E+00	4.402736E - 01
1.863336E - 01	2.946906E+01	2.760425E+00	1.388451E+00	2.812087E+00	4.182206E - 01
1.965418E - 01	2.874055E+01	2.750847E+00	1.416080E+00	2.660600E+00	3.977804E - 01
2.070222E - 01	2.802396E+01	2.712619E+00	1.443983E+00	2.513990E+00	3.789087E - 01
2.177749E - 01	2.731961E+01	2.647491E+00	1.472157E+00	2.372373E+00	3.615535E - 01
2.287997E - 01	2.662766E+01	2.558386E+00	1.500605E+00	2.235435E+00	3.456610E - 01
2.400968E - 01	2.594795E+01	2.449122E+00	1.529324E+00	2.103308E+00	3.311715E - 01
2.516661E - 01	2.528024E+01	2.324134E+00	1.558315E+00	1.975698E+00	3.180241E - 01
2.635076E - 01	2.462407E+01	2.188182E+00	1.587579E+00	1.852758E+00	3.061571E - 01
2.756213E - 01	2.397897E+01	2.046101E+00	1.617114E+00	1.734418E+00	2.955078E - 01
2.880073E - 01	2.334444E+01	1.902569E+00	1.646922E+00	1.620405E+00	2.860150E - 01
3.006654E - 01	2.272006E+01	1.761910E+00	1.677002E+00	1.510878E+00	2.776173E - 01
3.135958E - 01	2.210550E+01	1.627950E+00	1.707355E+00	1.405571E+00	2.702568E - 01
3.267984E - 01	2.150061E+01	1.503902E+00	1.737979E+00	1.304652E+00	2.638758E - 01
3.402732E - 01	2.090535E+01	1.392306E+00	1.768876E+00	1.207858E+00	2.584211E - 01
3.540202E - 01	2.031987E+01	1.295001E+00	1.800045E+00	1.115366E+00	2.538404E - 01
3.680395E - 01	1.974442E+01	1.213133E+00	1.831487E+00	1.026915E+00	2.500862E - 01
3.823310E - 01	1.917929E+01	1.147193E+00	1.863200E+00	9.426896E - 01	2.471122E - 01
3.968947E - 01	1.862474E+01	1.097073E+00	1.895185E+00	8.626562E - 01	2.448766E - 01
4.117306E - 01	1.808090E+01	1.062152E+00	1.927444E+00	7.865660E - 01	2.433414E - 01
4.268388E - 01	1.754774E+01	1.041371E+00	1.959973E+00	7.146141E - 01	2.424700E - 01
4.422191E - 01	1.702507E+01	1.033331E+00	1.992776E+00	6.465609E - 01	2.422311E - 01
4.578717E - 01	1.651253E+01	1.036385E+00	2.025850E+00	5.826201E - 01	2.425954E - 01
4.737964E - 01	1.600963E+01	1.048724E+00	2.059197E+00	5.225603E - 01	2.435378E - 01
4.899935E - 01	1.551592E+01	1.068460E+00	2.092816E+00	4.666053E - 01	2.450348E - 01
5.064627E - 01	1.503094E+01	1.093699E+00	2.126707E+00	4.147660E - 01	2.470674E - 01
5.232041E - 01	1.455434E+01	1.122602E+00	2.160871E+00	3.668419E - 01	2.496194E - 01
5.402178E - 01	1.408586E+01	1.153436E+00	2.195307E+00	3.228537E - 01	2.526767E - 01
5.575037E - 01	1.362538E+01	1.184616E+00	2.230015E+00	2.830612E - 01	2.562286E - 01

Z=26

Cobalt

Atomic Number	27
Lattice	fcc
Lattice constant a_o	6.448 a.u.
Pressure at a_o	-0.7 Kbar
Total energy (solid)	-2760.760 Ry
Total energy (atom)	-2760.288 Ry
Zero point energy	0.004 Ry
Cohesive energy	0.468 Ry
Bulk modulus	2.84 Mbar
Fermi energy	0.772 Ry
Interstitial electrons (per atom)	0.787
χ/χ_o	38.2
$N(E_f)$	2.01 states/eV-atom
ρ_o	1.332×10^4 electrons/Bohr^3
$(d\rho(r,E)/dE)_{E=E_f,r=0}$	1.7662 electrons/Bohr^3-Ry

Symmetry point	Band energies in Rydbergs (relative to muffin tin zero)					
Γ	0.103	0.624	0.624	0.624	0.725	0.725
X	0.385	0.453	0.794	0.818	0.818	
L	0.408	0.615	0.615	0.758	0.798	0.798
W	0.465	0.562	0.562	0.715	0.818	
K	0.451	0.497	0.653	0.738	0.792	

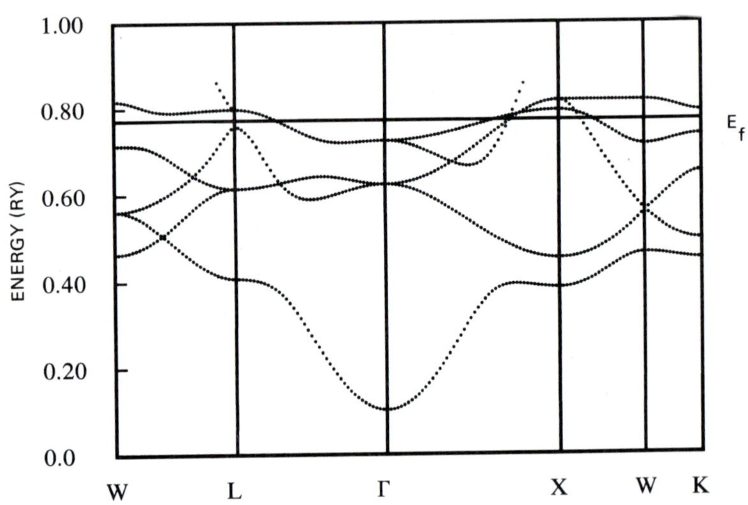

Z=27

Moruzzi
Janak
Williams

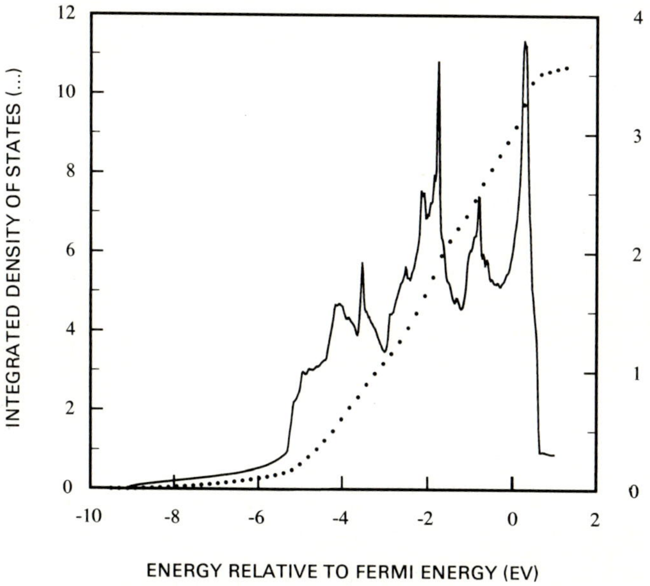
12
10
8
6
4
2
0
4
3
2
1
0
-10
-8
-6
-4
-2
0
2
INTEGRATED DENSITY OF STATES (...)
DENSITY OF STATES (STATES/EV) (-)
ENERGY RELATIVE TO FERMI ENERGY (EV)

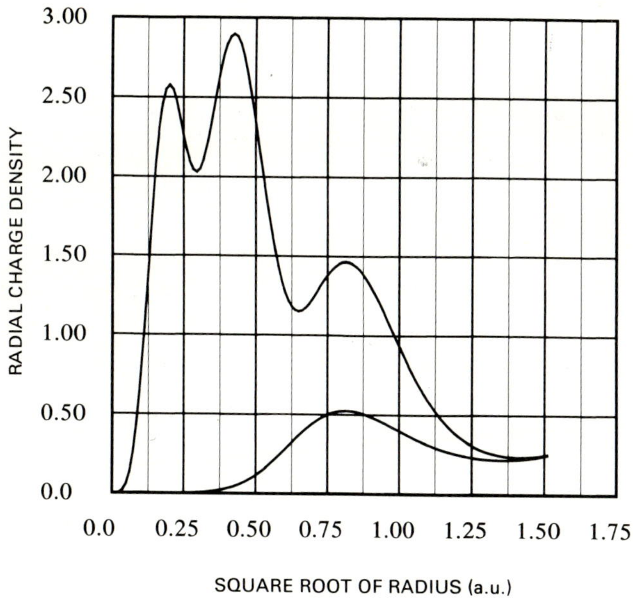
3.00
2.50
2.00
1.50
1.00
0.50
0.0
0.0
0.25
0.50
0.75
1.00
1.25
1.50
1.75
RADIAL CHARGE DENSITY
SQUARE ROOT OF RADIUS (a.u.)

Z=27

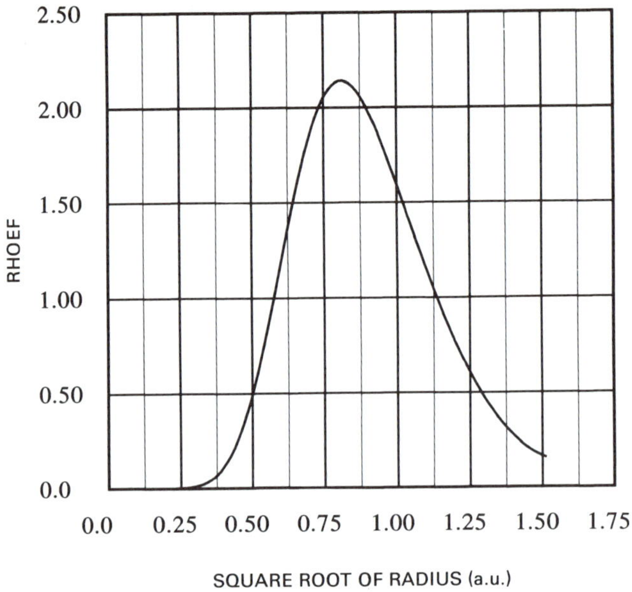

RHOEF$=r^2\partial\rho(E,r)/\partial E$ (at $E=E_f$)

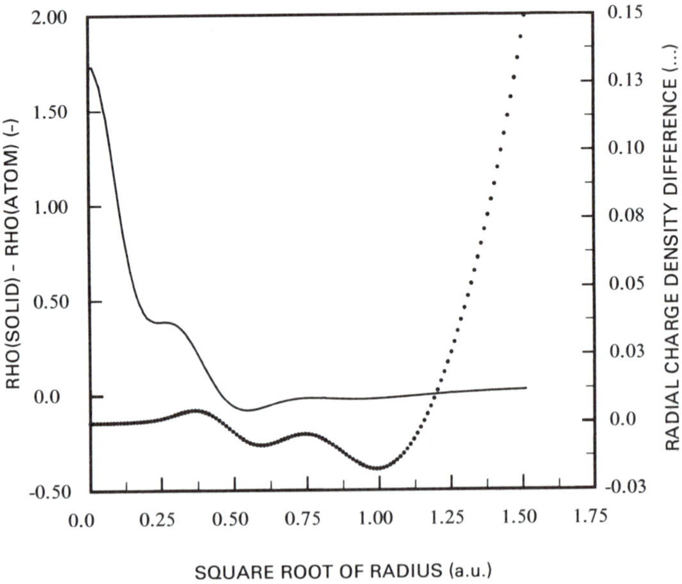

Moruzzi
Janak
Williams

Z=27

Cobalt

r	2z(r)	$r^2\rho$	r	2z(r)	$r^2\rho$
1.391426E - 04	5.397238E+01	2.558676E - 04	5.878773E - 01	1.285420E+01	1.418939E+00
5.565702E - 04	5.388939E+01	4.002638E - 03	6.061050E - 01	1.238994E+01	1.439755E+00
1.252283E - 03	5.375066E+01	1.951636E - 02	6.246109E - 01	1.193595E+01	1.454862E+00
2.226281E - 03	5.355534E+01	5.852217E - 02	6.433952E - 01	1.149248E+01	1.463860E+00
3.478564E - 03	5.330298E+01	1.335399E - 01	6.624577E - 01	1.105986E+01	1.466557E+00
5.009130E - 03	5.299338E+01	2.549660E - 01	6.817985E - 01	1.063834E+01	1.462945E+00
6.817985E - 03	5.262698E+01	4.284767E - 01	7.014176E - 01	1.022814E+01	1.453169E+00
8.905124E - 03	5.220496E+01	6.532613E - 01	7.213150E - 01	9.829413E+00	1.437507E+00
1.127055E - 02	5.172955E+01	9.214242E - 01	7.414907E - 01	9.442279E+00	1.416340E+00
1.391425E - 02	5.120390E+01	1.218637E+00	7.619447E - 01	9.066764E+00	1.390130E+00
1.683625E - 02	5.063147E+01	1.525922E+00	7.826769E - 01	8.702890E+00	1.359395E+00
2.003653E - 02	5.001669E+01	1.822152E+00	8.036874E - 01	8.350581E+00	1.324690E+00
2.351509E - 02	4.936409E+01	2.086910E+00	8.249762E - 01	8.009727E+00	1.286588E+00
2.727194E - 02	4.867824E+01	2.303129E+00	8.465433E - 01	7.680167E+00	1.245665E+00
3.130708E - 02	4.796338E+01	2.459159E+00	8.683887E - 01	7.361703E+00	1.202487E+00
3.562050E - 02	4.722337E+01	2.549976E+00	8.905123E - 01	7.054111E+00	1.157598E+00
4.021220E - 02	4.646156E+01	2.577427E+00	9.129143E - 01	6.757100E+00	1.111512E+00
4.508219E - 02	4.568085E+01	2.549570E+00	9.355946E - 01	6.470408E+00	1.064713E+00
5.023046E - 02	4.488373E+01	2.479282E+00	9.585531E - 01	6.193713E+00	1.017638E+00
5.565702E - 02	4.407249E+01	2.382433E+00	9.817899E - 01	5.926695E+00	9.706871E - 01
6.136187E - 02	4.324944E+01	2.275886E+00	1.005304E+00	5.669181E+00	9.242111E - 01
6.734496E - 02	4.241693E+01	2.175615E+00	1.029098E+00	5.420559E+00	8.785235E - 01
7.360637E - 02	4.157750E+01	2.095143E+00	1.053169E+00	5.180571E+00	8.338910E - 01
8.014607E - 02	4.073358E+01	2.044446E+00	1.077519E+00	4.948886E+00	7.905384E - 01
8.696407E - 02	3.988753E+01	2.029353E+00	1.102148E+00	4.725182E+00	7.486538E - 01
9.406036E - 02	3.904121E+01	2.051461E+00	1.127054E+00	4.509375E+00	7.083851E - 01
1.014349E - 01	3.819608E+01	2.108471E+00	1.152239E+00	4.300935E+00	6.698508E - 01
1.090878E - 01	3.735301E+01	2.194845E+00	1.177702E+00	4.099571E+00	6.331354E - 01
1.170189E - 01	3.651291E+01	2.302647E+00	1.203444E+00	3.905009E+00	5.982969E - 01
1.252283E - 01	3.567657E+01	2.422511E+00	1.229464E+00	3.716994E+00	5.653695E - 01
1.337160E - 01	3.484486E+01	2.544548E+00	1.255761E+00	3.535520E+00	5.343629E - 01
1.424820E - 01	3.401898E+01	2.659207E+00	1.282337E+00	3.360126E+00	5.052720E - 01
1.515262E - 01	3.320012E+01	2.757954E+00	1.309192E+00	3.190607E+00	4.780736E - 01
1.608488E - 01	3.238960E+01	2.833789E+00	1.336325E+00	3.026771E+00	4.527317E - 01
1.704496E - 01	3.158875E+01	2.881576E+00	1.363736E+00	2.868443E+00	4.292009E - 01
1.803287E - 01	3.079865E+01	2.898199E+00	1.391425E+00	2.715698E+00	4.074249E - 01
1.904861E - 01	3.002023E+01	2.882550E+00	1.419393E+00	2.568156E+00	3.873410E - 01
2.009218E - 01	2.925414E+01	2.835385E+00	1.447639E+00	2.425683E+00	3.688829E - 01
2.116358E - 01	2.850073E+01	2.759091E+00	1.476163E+00	2.288160E+00	3.519802E - 01
2.226281E - 01	2.776013E+01	2.657381E+00	1.504966E+00	2.155478E+00	3.365605E - 01
2.338986E - 01	2.703218E+01	2.534935E+00	1.534046E+00	2.027771E+00	3.225501E - 01
2.454475E - 01	2.631657E+01	2.397061E+00	1.563405E+00	1.904709E+00	3.098757E - 01
2.572746E - 01	2.561287E+01	2.249347E+00	1.593042E+00	1.786217E+00	2.984656E - 01
2.693800E - 01	2.492059E+01	2.097340E+00	1.622958E+00	1.672211E+00	2.882493E - 01
2.817637E - 01	2.423930E+01	1.946292E+00	1.653152E+00	1.562627E+00	2.791584E - 01
2.944256E - 01	2.356866E+01	1.800930E+00	1.683624E+00	1.457623E+00	2.711272E - 01
3.073659E - 01	2.290848E+01	1.665309E+00	1.714375E+00	1.356681E+00	2.640945E - 01
3.205844E - 01	2.225871E+01	1.542683E+00	1.745404E+00	1.260206E+00	2.580003E - 01
3.340812E - 01	2.161951E+01	1.435481E+00	1.776711E+00	1.168148E+00	2.527894E - 01
3.478564E - 01	2.099110E+01	1.345272E+00	1.808296E+00	1.079994E+00	2.484114E - 01
3.619097E - 01	2.037387E+01	1.272831E+00	1.840160E+00	9.959343E - 01	2.448181E - 01
3.762414E - 01	1.976802E+01	1.218170E+00	1.872302E+00	9.161667E - 01	2.419650E - 01
3.908514E - 01	1.917377E+01	1.180665E+00	1.904722E+00	8.404218E - 01	2.398129E - 01
4.057397E - 01	1.859106E+01	1.159139E+00	1.937421E+00	7.686791E - 01	2.383254E - 01
4.209062E - 01	1.801971E+01	1.151987E+00	1.970397E+00	7.011465E - 01	2.374696E - 01
4.363511E - 01	1.745938E+01	1.157290E+00	2.003653E+00	6.373495E - 01	2.372179E - 01
4.520742E - 01	1.690959E+01	1.172923E+00	2.037186E+00	5.777466E - 01	2.375437E - 01
4.680756E - 01	1.636989E+01	1.196669E+00	2.070997E+00	5.221074E - 01	2.384263E - 01
4.843552E - 01	1.583986E+01	1.226300E+00	2.105087E+00	4.704425E - 01	2.398468E - 01
5.009132E - 01	1.531917E+01	1.259662E+00	2.139456E+00	4.227747E - 01	2.417906E - 01
5.177494E - 01	1.480768E+01	1.294736E+00	2.174102E+00	3.793622E - 01	2.442449E - 01
5.348639E - 01	1.430531E+01	1.329691E+00	2.209027E+00	3.397748E - 01	2.472020E - 01
5.522568E - 01	1.381213E+01	1.362914E+00	2.244230E+00	3.045288E - 01	2.506549E - 01
5.699279E - 01	1.332835E+01	1.393036E+00	2.279712E+00	2.734467E - 01	2.546006E - 01

Z=27

Nickel

Atomic Number	28
Lattice	fcc
Lattice constant a_o	6.55 a.u.
Pressure at a_o	3.01 Kbar
Total energy (solid)	-3011.649 Ry
Total energy (atom)	-3011.233 Ry
Zero point energy	0.003 Ry
Cohesive energy	0.413 Ry
Bulk modulus	2.20 Mbar
Fermi energy	0.677 Ry
Interstitial electrons (per atom)	0.715
χ/χ_o	-0.98
$N(E_f)$	4.06 states/eV-atom
ρ_o	1.488×10^4 electrons/Bohr^3
$(d\rho(r,E)/dE)_{E=E_f,r=0}$	4.4742 electrons/Bohr^3-Ry

Symmetry point	Band energies in Rydbergs (relative to muffin tin zero)					
Γ	0.023	0.532	0.532	0.532	0.615	0.615
X	0.331	0.390	0.671	0.691	0.691	
L	0.342	0.525	0.525	0.661	0.675	0.675
W	0.401	0.480	0.480	0.608	0.691	
K	0.389	0.425	0.556	0.626	0.670	

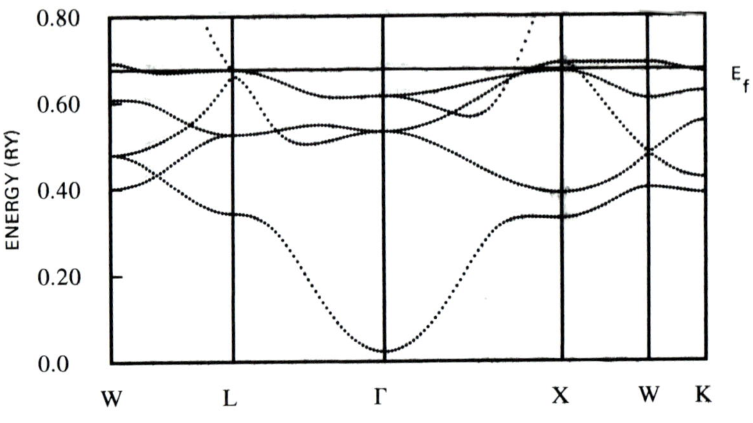

Z=28

Moruzzi
Janak
Williams

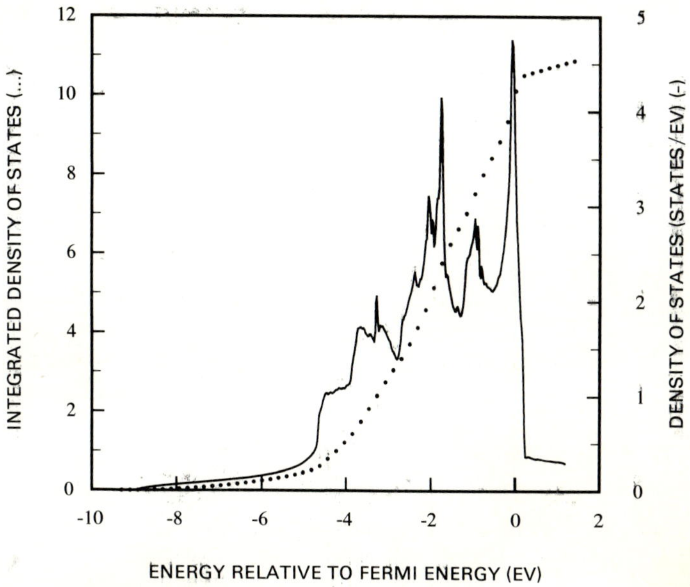

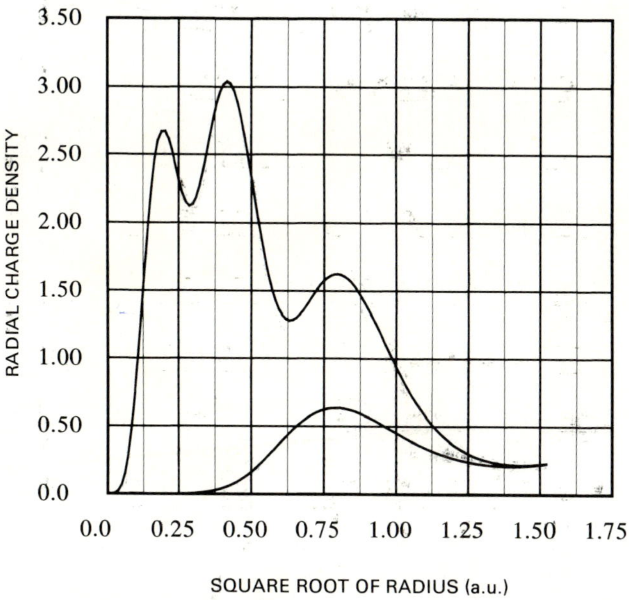

Moruzzi
Janak
Williams

Z=28

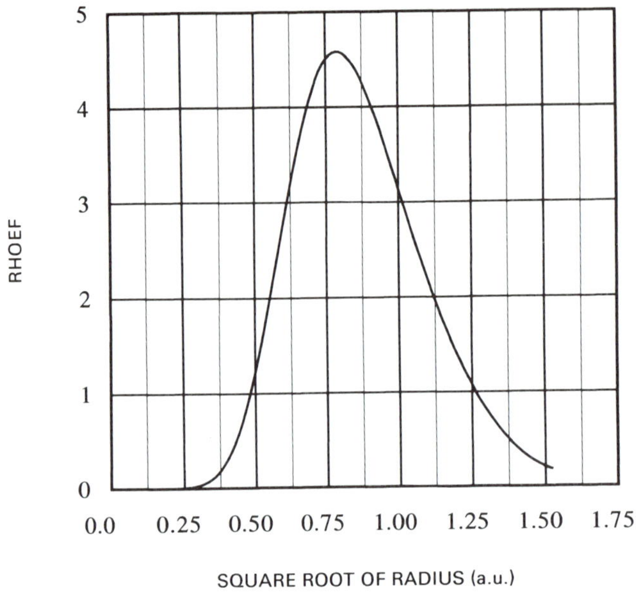

RHOEF=$r^2 \partial\rho(E,r)/\partial E$ (at $E=E_f$)

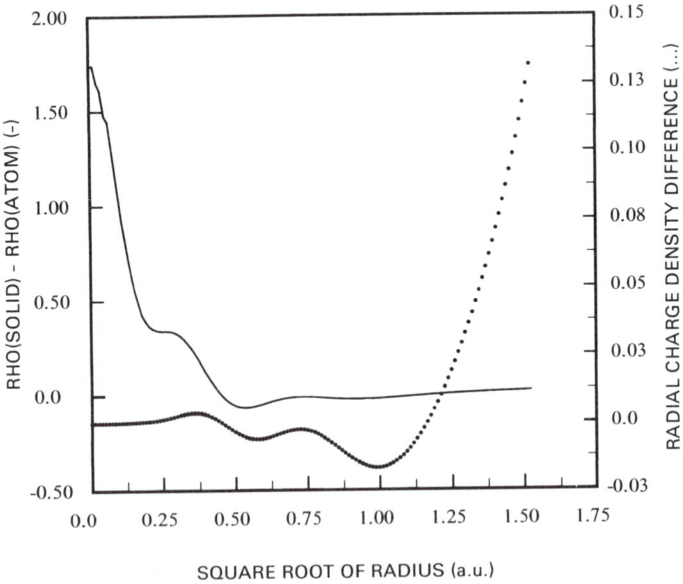

Moruzzi
Janak
Williams

Z=28

Nickel

r	2z(r)	$r^2\rho$	r	2z(r)	$r^2\rho$
1.413436E - 04	5.597035E+01	2.950006E - 04	5.971768E - 01	1.256901E+01	1.610909E+00
5.653745E - 04	5.588126E+01	4.609250E - 03	6.156928E - 01	1.208915E+01	1.620014E+00
1.272093E - 03	5.573228E+01	2.242909E - 02	6.344915E - 01	1.162174E+01	1.621864E+00
2.261498E - 03	5.552257E+01	6.706744E - 02	6.535729E - 01	1.116706E+01	1.616475E+00
3.533591E - 03	5.525160E+01	1.524884E - 01	6.729370E - 01	1.072536E+01	1.604045E+00
5.088370E - 03	5.491925E+01	2.898641E - 01	6.925837E - 01	1.029682E+01	1.584926E+00
6.925836E - 03	5.452612E+01	4.845974E - 01	7.125132E - 01	9.881527E+00	1.559587E+00
9.045992E - 03	5.407368E+01	7.344177E - 01	7.327253E - 01	9.479542E+00	1.528589E+00
1.144883E - 02	5.356451E+01	1.028922E+00	7.532201E - 01	9.090821E+00	1.492552E+00
1.413436E - 02	5.300214E+01	1.350650E+00	7.739977E - 01	8.715270E+00	1.452136E+00
1.710258E - 02	5.239063E+01	1.677393E+00	7.950579E - 01	8.352741E+00	1.408015E+00
2.035348E - 02	5.173482E+01	1.985346E+00	8.164008E - 01	8.003038E+00	1.360862E+00
2.388707E - 02	5.103969E+01	2.252449E+00	8.380263E - 01	7.665897E+00	1.311330E+00
2.770335E - 02	5.031009E+01	2.461358E+00	8.599346E - 01	7.341038E+00	1.260043E+00
3.180232E - 02	4.955049E+01	2.601580E+00	8.821256E - 01	7.028101E+00	1.207590E+00
3.618397E - 02	4.876486E+01	2.670507E+00	9.045992E - 01	6.726766E+00	1.154509E+00
4.084831E - 02	4.795668E+01	2.673262E+00	9.273555E - 01	6.436639E+00	1.101293E+00
4.579533E - 02	4.712886E+01	2.621537E+00	9.503945E - 01	6.157350E+00	1.048379E+00
5.102505E - 02	4.628403E+01	2.531674E+00	9.737162E - 01	5.888485E+00	9.961565E - 01
5.653745E - 02	4.542464E+01	2.422353E+00	9.973206E - 01	5.629637E+00	9.449583E - 01
6.233254E - 02	4.455331E+01	2.312253E+00	1.021207E+00	5.380624E+00	8.950667E - 01
6.841028E - 02	4.367268E+01	2.217997E+00	1.045377E+00	5.140589E+00	8.467221E - 01
7.477075E - 02	4.278554E+01	2.152566E+00	1.069829E+00	4.909495E+00	8.001114E - 01
8.141392E - 02	4.189445E+01	2.124352E+00	1.094564E+00	4.686724E+00	7.553866E - 01
8.833975E - 02	4.100171E+01	2.136815E+00	1.119582E+00	4.472154E+00	7.126576E - 01
9.554827E - 02	4.010899E+01	2.188708E+00	1.144883E+00	4.265192E+00	6.720021E - 01
1.030395E - 01	3.921754E+01	2.274756E+00	1.170466E+00	4.065753E+00	6.334678E - 01
1.108134E - 01	3.832828E+01	2.386606E+00	1.196332E+00	3.873524E+00	5.970747E - 01
1.188700E - 01	3.744203E+01	2.513945E+00	1.222481E+00	3.687963E+00	5.628222E - 01
1.272092E - 01	3.655986E+01	2.645635E+00	1.248912E+00	3.509040E+00	5.306868E - 01
1.358312E - 01	3.568297E+01	2.770738E+00	1.275626E+00	3.336250E+00	5.006323E - 01
1.447359E - 01	3.481276E+01	2.879394E+00	1.302623E+00	3.169602E+00	4.726065E - 01
1.539232E - 01	3.395064E+01	2.963469E+00	1.329902E+00	3.008873E+00	4.465478E - 01
1.633932E - 01	3.309802E+01	3.016983E+00	1.357464E+00	2.853613E+00	4.223866E - 01
1.731459E - 01	3.225621E+01	3.036309E+00	1.385308E+00	2.703877E+00	4.000455E - 01
1.831813E - 01	3.142621E+01	3.020179E+00	1.413436E+00	2.559249E+00	3.794451E - 01
1.934994E - 01	3.060875E+01	2.969506E+00	1.441846E+00	2.419818E+00	3.605015E - 01
2.041002E - 01	2.980426E+01	2.887114E+00	1.470538E+00	2.285433E+00	3.431301E - 01
2.149836E - 01	2.901292E+01	2.777343E+00	1.499514E+00	2.155722E+00	3.272477E - 01
2.261498E - 01	2.823457E+01	2.645643E+00	1.528772E+00	2.030560E+00	3.127705E - 01
2.375986E - 01	2.746895E+01	2.498139E+00	1.558313E+00	1.910077E+00	2.996169E - 01
2.493301E - 01	2.671561E+01	2.341214E+00	1.588137E+00	1.794179E+00	2.877082E - 01
2.613443E - 01	2.597414E+01	2.181144E+00	1.618242E+00	1.682769E+00	2.769686E - 01
2.736412E - 01	2.524416E+01	2.023777E+00	1.648632E+00	1.575269E+00	2.673268E - 01
2.862208E - 01	2.452544E+01	1.874290E+00	1.679303E+00	1.472338E+00	2.587132E - 01
2.990831E - 01	2.381786E+01	1.737004E+00	1.710258E+00	1.373410E+00	2.510645E - 01
3.122280E - 01	2.312152E+01	1.615290E+00	1.741494E+00	1.278670E+00	2.443200E - 01
3.256557E - 01	2.243666E+01	1.511501E+00	1.773014E+00	1.187811E+00	2.384245E - 01
3.393660E - 01	2.176364E+01	1.427017E+00	1.804816E+00	1.101027E+00	2.333260E - 01
3.533590E - 01	2.110274E+01	1.362277E+00	1.836902E+00	1.018022E+00	2.289776E - 01
3.676347E - 01	2.045424E+01	1.316892E+00	1.869269E+00	9.390003E - 01	2.253360E - 01
3.821931E - 01	1.981815E+01	1.289762E+00	1.901919E+00	8.639306E - 01	2.223619E - 01
3.970342E - 01	1.919437E+01	1.279222E+00	1.934853E+00	7.925358E - 01	2.200217E - 01
4.121580E - 01	1.858258E+01	1.283170E+00	1.968068E+00	7.250425E - 01	2.182831E - 01
4.275644E - 01	1.798239E+01	1.299219E+00	2.001567E+00	6.611882E - 01	2.171198E - 01
4.432536E - 01	1.739334E+01	1.324828E+00	2.035348E+00	6.012133E - 01	2.165079E - 01
4.592254E - 01	1.681505E+01	1.357416E+00	2.069411E+00	5.451180E - 01	2.164270E - 01
4.754799E - 01	1.624722E+01	1.394464E+00	2.103758E+00	4.926643E - 01	2.168613E - 01
4.920171E - 01	1.568970E+01	1.433604E+00	2.138388E+00	4.438687E - 01	2.177970E - 01
5.088370E - 01	1.514251E+01	1.472678E+00	2.173299E+00	3.992459E - 01	2.192234E - 01
5.259396E - 01	1.460579E+01	1.509790E+00	2.208493E+00	3.583390E - 01	2.211340E - 01
5.433249E - 01	1.407972E+01	1.543329E+00	2.243971E+00	3.211881E - 01	2.235242E - 01
5.609928E - 01	1.356466E+01	1.571985E+00	2.279731E+00	2.880920E - 01	2.263921E - 01
5.789435E - 01	1.306096E+01	1.594751E+00	2.315774E+00	2.588591E - 01	2.297394E - 01

Z=28

Copper

Copper

Atomic Number	29
Lattice	fcc
Lattice constant a_o	6.76 a.u.
Pressure at a_o	4.94 Kbar
Total energy (solid)	-3275.768 Ry
Total energy (atom)	-3275.464 Ry
Zero point energy	0.003 Ry
Cohesive energy	0.301 Ry
Bulk modulus	1.55 Mbar
Fermi energy	0.628 Ry
Interstitial electrons (per atom)	0.647
χ/χ_o	1.12
$N(E_f)$	0.29 states/eV-atom
ρ_o	1.657×10^4 electrons/Bohr^3
$(d\rho(r,E)/dE)_{E=E_f,r=0}$	6.2372 electrons/Bohr^3-Ry

Symmetry point	Band energies in Rydbergs (relative to muffin tin zero)					
Γ	-0.064	0.390	0.390	0.390	0.451	0.451
X	0.241	0.284	0.494	0.509	0.509	0.748
L	0.242	0.386	0.386	0.497	0.497	0.544
W	0.297	0.354	0.354	0.448	0.509	
K	0.286	0.311	0.413	0.461	0.494	

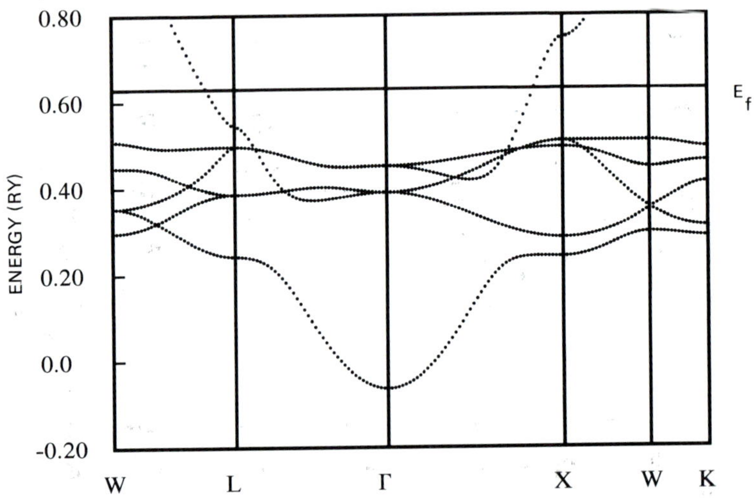

Z=29

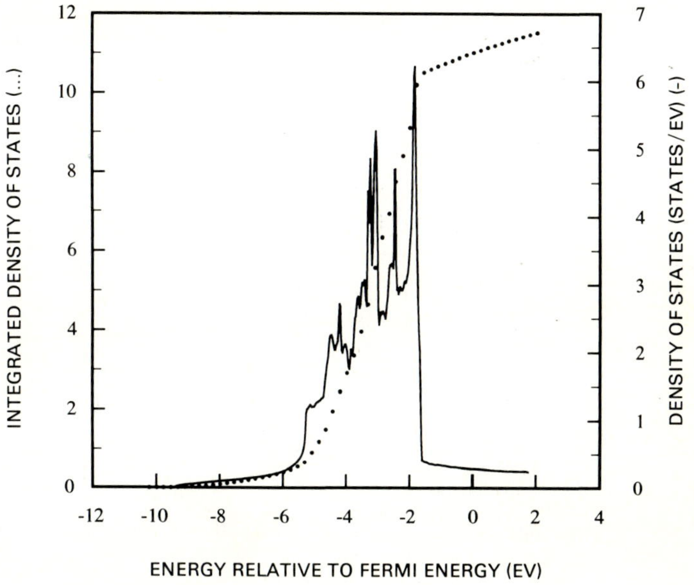

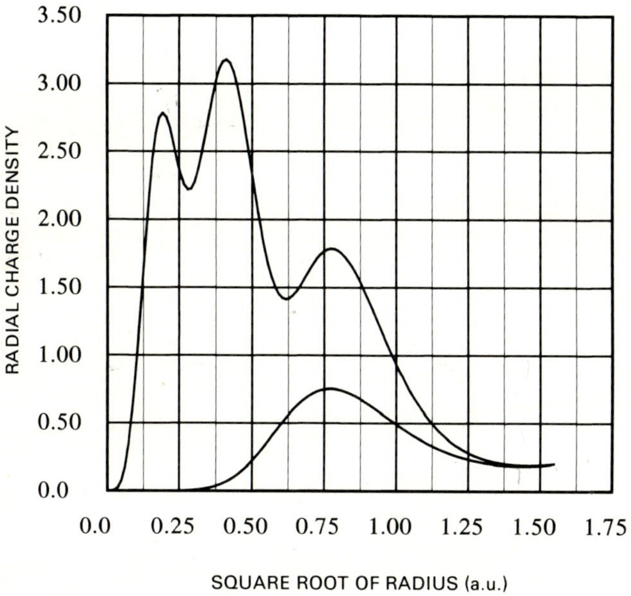

Moruzzi
Janak
Williams

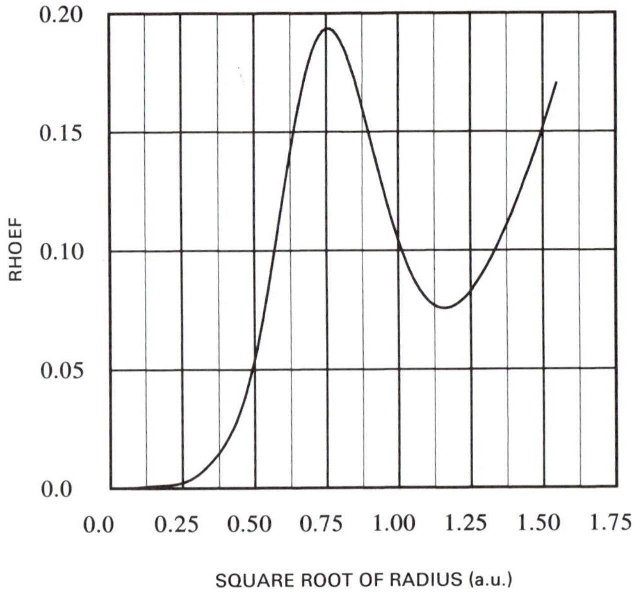

RHOEF=$r^2\partial\rho(E,r)/\partial E$ (at $E=E_f$)

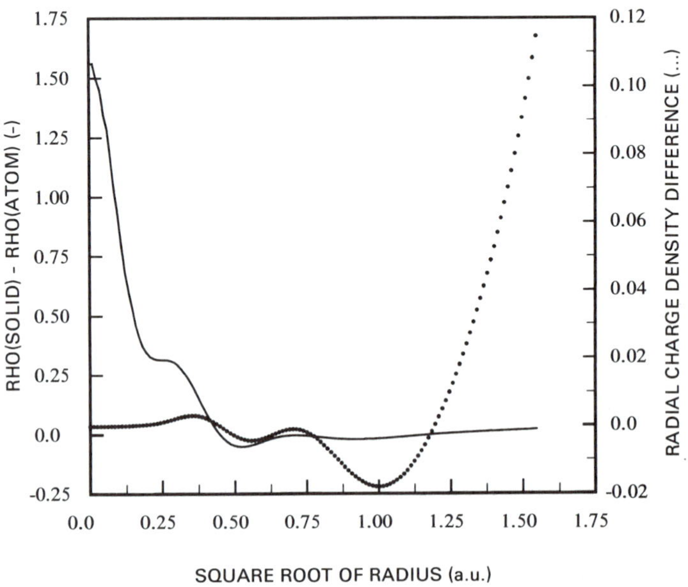

Moruzzi
Janak
Williams

Z=29

Copper

r	2z(r)	$r^2\rho$	r	2z(r)	$r^2\rho$
1.458753E - 04	5.796773E+01	3.496481E - 04	6.163230E - 01	1.201586E+01	1.781885E+00
5.835011E - 04	5.787076E+01	5.454164E - 03	6.354327E - 01	1.153001E+01	1.770859E+00
1.312877E - 03	5.770851E+01	2.646826E - 02	6.548340E - 01	1.105925E+01	1.751857E+00
2.334004E - 03	5.748013E+01	7.884395E - 02	6.745272E - 01	1.060370E+01	1.725391E+00
3.646882E - 03	5.718506E+01	1.783871E - 01	6.945121E - 01	1.016347E+01	1.692102E+00
5.251508E - 03	5.682329E+01	3.370717E - 01	7.147888E - 01	9.738505E+00	1.652712E+00
7.147886E - 03	5.639558E+01	5.595560E - 01	7.353572E - 01	9.328745E+00	1.608001E+00
9.336017E - 03	5.590395E+01	8.411624E - 01	7.562174E - 01	8.933986E+00	1.558778E+00
1.181589E - 02	5.535141E+01	1.167730E+00	7.773693E - 01	8.554024E+00	1.505848E+00
1.458753E - 02	5.474217E+01	1.517365E+00	7.988129E - 01	8.188542E+00	1.450002E+00
1.765091E - 02	5.408101E+01	1.863626E+00	8.205484E - 01	7.837202E+00	1.391995E+00
2.100604E - 02	5.337337E+01	2.179527E+00	8.425755E - 01	7.499621E+00	1.332534E+00
2.465292E - 02	5.262476E+01	2.441565E+00	8.648944E - 01	7.175350E+00	1.272271E+00
2.859155E - 02	5.184036E+01	2.633001E+00	8.875051E - 01	6.863964E+00	1.211797E+00
3.282193E - 02	5.102498E+01	2.745929E+00	9.104075E - 01	6.564997E+00	1.151639E+00
3.734407E - 02	5.018271E+01	2.781872E+00	9.336017E - 01	6.277933E+00	1.092259E+00
4.215795E - 02	4.931706E+01	2.750947E+00	9.570876E - 01	6.002316E+00	1.034054E+00
4.726359E - 02	4.843109E+01	2.669906E+00	9.808652E - 01	5.737646E+00	9.773587E - 01
5.266097E - 02	4.752765E+01	2.559437E+00	1.004934E+00	5.483534E+00	9.224474E - 01
5.835010E - 02	4.660951E+01	2.441273E+00	1.029295E+00	5.239458E+00	8.695419E - 01
6.433094E - 02	4.567970E+01	2.335487E+00	1.053948E+00	5.004637E+00	8.188142E - 01
7.060361E - 02	4.474120E+01	2.258328E+00	1.078893E+00	4.779140E+00	7.703860E - 01
7.716799E - 02	4.379706E+01	2.220820E+00	1.104130E+00	4.562024E+00	7.243436E - 01
8.402413E - 02	4.284988E+01	2.228149E+00	1.129658E+00	4.353394E+00	6.807317E - 01
9.117204E - 02	4.190163E+01	2.279792E+00	1.155478E+00	4.152606E+00	6.395690E - 01
9.861165E - 02	4.095380E+01	2.370253E+00	1.181589E+00	3.959291E+00	6.008453E - 01
1.063430E - 01	4.000746E+01	2.490210E+00	1.207993E+00	3.773108E+00	5.645261E - 01
1.143662E - 01	3.906358E+01	2.627880E+00	1.234688E+00	3.593733E+00	5.305615E - 01
1.226811E - 01	3.812335E+01	2.770428E+00	1.261675E+00	3.420859E+00	4.988847E - 01
1.312877E - 01	3.718816E+01	2.905242E+00	1.288954E+00	3.254210E+00	4.694186E - 01
1.401861E - 01	3.625949E+01	3.021024E+00	1.316524E+00	3.093778E+00	4.420733E - 01
1.493763E - 01	3.533916E+01	3.108586E+00	1.344386E+00	2.938809E+00	4.167583E - 01
1.588581E - 01	3.442856E+01	3.161372E+00	1.372540E+00	2.789598E+00	3.933746E - 01
1.686318E - 01	3.352917E+01	3.175661E+00	1.400986E+00	2.645415E+00	3.718231E - 01
1.786972E - 01	3.264211E+01	3.150546E+00	1.429723E+00	2.506593E+00	3.520020E - 01
1.890543E - 01	3.176816E+01	3.087683E+00	1.458753E+00	2.372437E+00	3.338128E - 01
1.997032E - 01	3.090775E+01	2.990917E+00	1.488073E+00	2.243303E+00	3.171566E - 01
2.106438E - 01	3.006100E+01	2.865772E+00	1.517686E+00	2.118788E+00	3.019381E - 01
2.218763E - 01	2.922778E+01	2.718928E+00	1.547590E+00	1.998751E+00	2.880657E - 01
2.334004E - 01	2.840779E+01	2.557680E+00	1.577786E+00	1.883072E+00	2.754515E - 01
2.452163E - 01	2.760062E+01	2.389431E+00	1.608274E+00	1.771634E+00	2.640117E - 01
2.573239E - 01	2.680592E+01	2.221275E+00	1.639054E+00	1.664331E+00	2.536681E - 01
2.697234E - 01	2.602341E+01	2.059609E+00	1.670125E+00	1.561325E+00	2.443450E - 01
2.824145E - 01	2.525301E+01	1.909891E+00	1.701488E+00	1.462015E+00	2.359753E - 01
2.953974E - 01	2.449480E+01	1.776452E+00	1.733144E+00	1.366573E+00	2.284940E - 01
3.086720E - 01	2.374904E+01	1.662419E+00	1.765090E+00	1.275191E+00	2.218414E - 01
3.222384E - 01	2.301608E+01	1.569700E+00	1.797329E+00	1.187284E+00	2.159641E - 01
3.360966E - 01	2.229631E+01	1.499045E+00	1.829859E+00	1.103309E+00	2.108115E - 01
3.502465E - 01	2.159000E+01	1.450172E+00	1.862680E+00	1.022960E+00	2.063390E - 01
3.646882E - 01	2.089723E+01	1.421894E+00	1.895795E+00	9.459358E - 01	2.025064E - 01
3.794215E - 01	2.021793E+01	1.412308E+00	1.929200E+00	8.727099E - 01	1.992764E - 01
3.944467E - 01	1.955177E+01	1.418971E+00	1.962897E+00	8.029981E - 01	1.966172E - 01
4.097636E - 01	1.889835E+01	1.439075E+00	1.996886E+00	7.367796E - 01	1.945001E - 01
4.253722E - 01	1.825726E+01	1.469623E+00	2.031167E+00	6.740353E - 01	1.929008E - 01
4.412727E - 01	1.762817E+01	1.507574E+00	2.065740E+00	6.147618E - 01	1.917974E - 01
4.574648E - 01	1.701089E+01	1.549973E+00	2.100603E+00	5.592108E - 01	1.911726E - 01
4.739487E - 01	1.640533E+01	1.594061E+00	2.135759E+00	5.068808E - 01	1.910118E - 01
4.907244E - 01	1.581160E+01	1.637351E+00	2.171207E+00	4.580389E - 01	1.913038E - 01
5.077918E - 01	1.522996E+01	1.677674E+00	2.206946E+00	4.129712E - 01	1.920394E - 01
5.251510E - 01	1.466071E+01	1.713233E+00	2.242978E+00	3.711888E - 01	1.932139E - 01
5.428019E - 01	1.410430E+01	1.742594E+00	2.279301E+00	3.332452E - 01	1.948236E - 01
5.607445E - 01	1.356118E+01	1.764695E+00	2.315915E+00	2.989336E - 01	1.968682E - 01
5.789789E - 01	1.303180E+01	1.778824E+00	2.352821E+00	2.683100E - 01	1.993505E - 01
5.975051E - 01	1.251656E+01	1.784589E+00	2.390020E+00	2.411752E - 01	2.022759E - 01

Z=29

Zinc

Atomic Number	30
Lattice	fcc
Lattice constant a_o	7.25 a.u.
Pressure at a_o	15.8 Kbar
Total energy (solid)	-3553.139 Ry
Total energy (atom)	-3553.007 Ry
Zero point energy	0.002 Ry
Cohesive energy	0.129 Ry
Bulk modulus	0.82 Mbar
Fermi energy	0.610 Ry
Interstitial electrons (per atom)	0.632
χ/χ_o	1.18
$N(E_f)$	0.30 states/eV-atom
ρ_o	1.838×10^4 electrons/Bohr^3
$(d\rho(r,E)/dE)_{E=E_f,r=0}$	15.567 electrons/Bohr^3-Ry

Symmetry point	Band energies in Rydbergs (relative to muffin tin zero)						
Γ	-0.189	0.006	0.006	0.006	0.036	0.036	
X	-0.054	-0.044	0.058	0.066	0.066	0.526	
L	-0.051	0.004	0.004	0.060	0.060	0.365	0.522
W	-0.026	-0.002	-0.002	0.038	0.066		
K	-0.035	-0.021	0.030	0.043	0.058	0.663	0.731

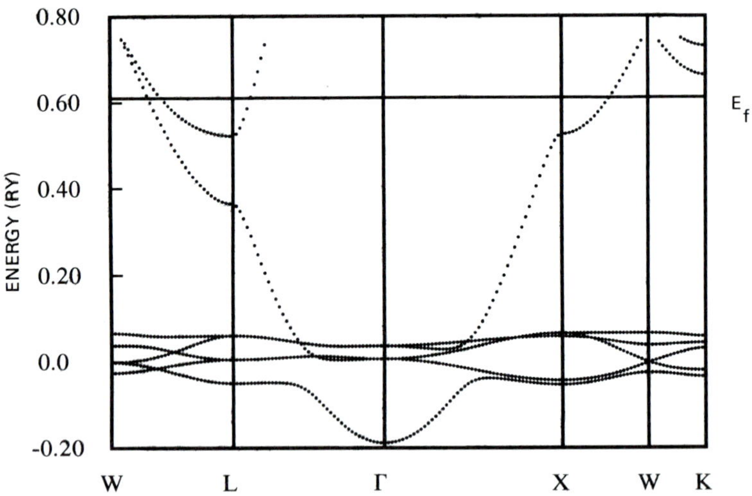

Z=30

Moruzzi
Janak
Williams

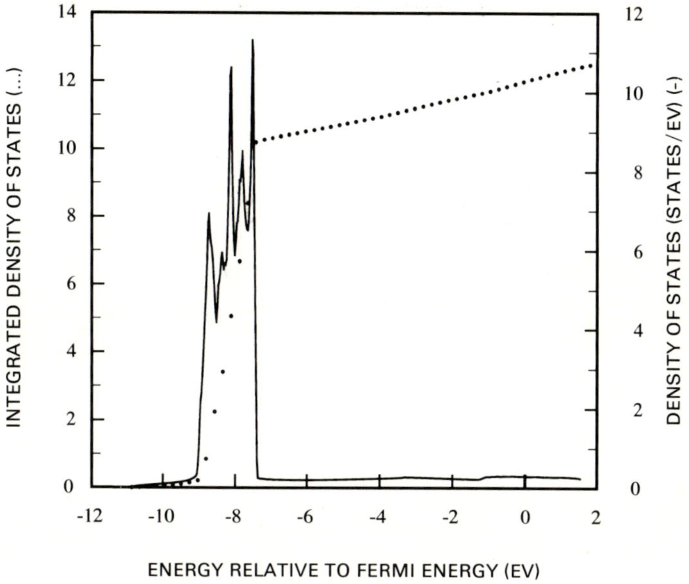

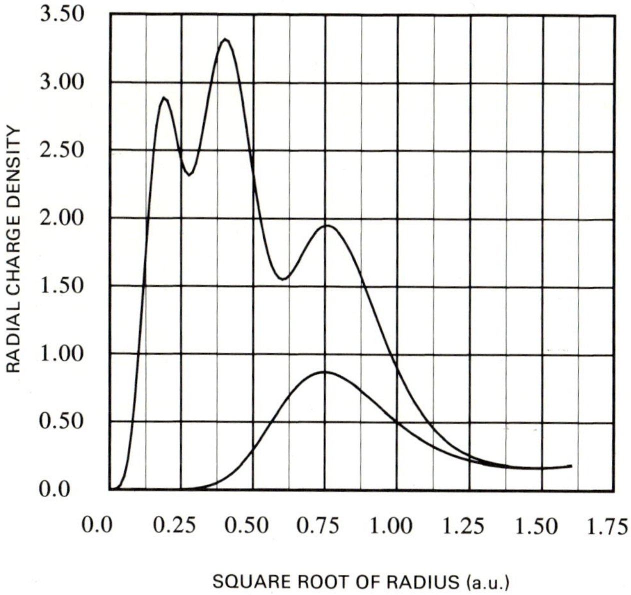

Moruzzi
Janak
Williams

Z=30

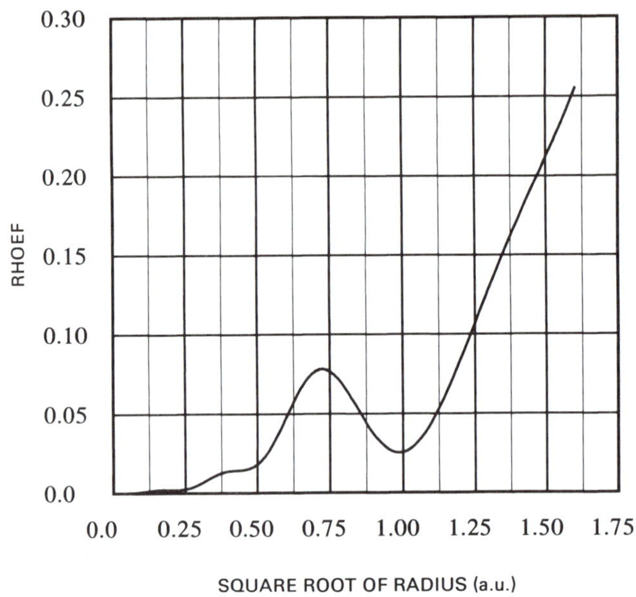

RHOEF=$r^2\partial\rho(E,r)/\partial E$ (at $E=E_f$)

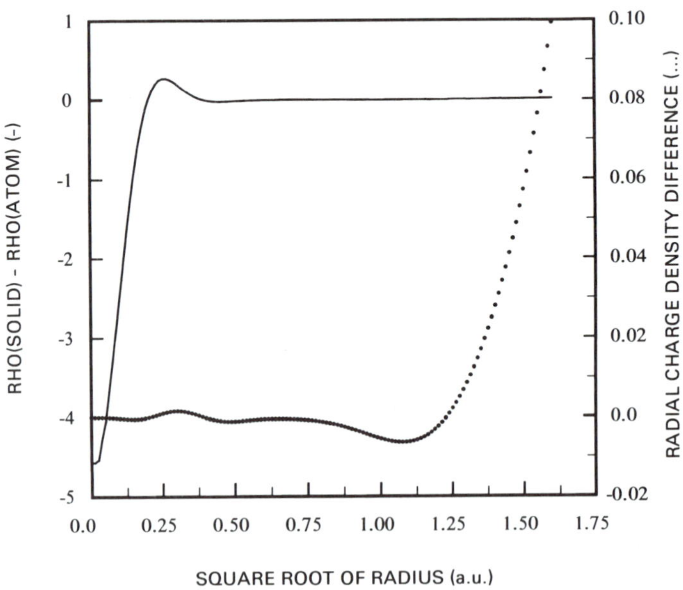

Moruzzi
Janak
Williams

Z=30

Zinc

r	2z(r)	$r^2\rho$	r	2z(r)	$r^2\rho$
1.564490E-04	5.996361E+01	4.457624E-04	6.609972E-01	1.100894E+01	1.849457E+00
6.257962E-04	5.985419E+01	6.934155E-03	6.814920E-01	1.054218E+01	1.804025E+00
1.408041E-03	5.967104E+01	3.349502E-02	7.022997E-01	1.009333E+01	1.752019E+00
2.503185E-03	5.941318E+01	9.913111E-02	7.234203E-01	9.662190E+00	1.694477E+00
3.911223E-03	5.908008E+01	2.224289E-01	7.448539E-01	9.248422E+00	1.632438E+00
5.632162E-03	5.867189E+01	4.160442E-01	7.666003E-01	8.851652E+00	1.566917E+00
7.665999E-03	5.818996E+01	6.824347E-01	7.886596E-01	8.471397E+00	1.498884E+00
1.001274E-02	5.763702E+01	1.011858E+00	8.110318E-01	8.107168E+00	1.429240E+00
1.267237E-02	5.701718E+01	1.383084E+00	8.337169E-01	7.758388E+00	1.358812E+00
1.564490E-02	5.633585E+01	1.766599E+00	8.567149E-01	7.424462E+00	1.288342E+00
1.893033E-02	5.559889E+01	2.129478E+00	8.800258E-01	7.104802E+00	1.218484E+00
2.252866E-02	5.481282E+01	2.440919E+00	9.036496E-01	6.798768E+00	1.149801E+00
2.643989E-02	5.398395E+01	2.677163E+00	9.275863E-01	6.505744E+00	1.082774E+00
3.066401E-02	5.311806E+01	2.824966E+00	9.518359E-01	6.225095E+00	1.017797E+00
3.520103E-02	5.222017E+01	2.883076E+00	9.763984E-01	5.956223E+00	9.551879E-01
4.005095E-02	5.129457E+01	2.861606E+00	1.001273E+00	5.698699E+00	8.951902E-01
4.521377E-02	5.034500E+01	2.779707E+00	1.026462E+00	5.451537E+00	8.379884E-01
5.068948E-02	4.937477E+01	2.662077E+00	1.051963E+00	5.214437E+00	7.837031E-01
5.647810E-02	4.838724E+01	2.535055E+00	1.077777E+00	4.986868E+00	7.324055E-01
6.257957E-02	4.738593E+01	2.422957E+00	1.103904E+00	4.768328E+00	6.841223E-01
6.899399E-02	4.637440E+01	2.345163E+00	1.130343E+00	4.558341E+00	6.388427E-01
7.572132E-02	4.535620E+01	2.314265E+00	1.157097E+00	4.356193E+00	5.965262E-01
8.276153E-02	4.433424E+01	2.335347E+00	1.184162E+00	4.162008E+00	5.571025E-01
9.011459E-02	4.331081E+01	2.406306E+00	1.211541E+00	3.974859E+00	5.204843E-01
9.778064E-02	4.228740E+01	2.518993E+00	1.239232E+00	3.794923E+00	4.865643E-01
1.057595E-01	4.126534E+01	2.660851E+00	1.267237E+00	3.621330E+00	4.552283E-01
1.140513E-01	4.024577E+01	2.816824E+00	1.295554E+00	3.454041E+00	4.263491E-01
1.226560E-01	3.923038E+01	2.971216E+00	1.324184E+00	3.292772E+00	3.997962E-01
1.315736E-01	3.822078E+01	3.109328E+00	1.353127E+00	3.137261E+00	3.754361E-01
1.408041E-01	3.721902E+01	3.218755E+00	1.382383E+00	2.987267E+00	3.531352E-01
1.503475E-01	3.622701E+01	3.290253E+00	1.411952E+00	2.842566E+00	3.327610E-01
1.602038E-01	3.524643E+01	3.318195E+00	1.441833E+00	2.702954E+00	3.141832E-01
1.703730E-01	3.427884E+01	3.300632E+00	1.472029E+00	2.567973E+00	2.972779E-01
1.808550E-01	3.332523E+01	3.239031E+00	1.502536E+00	2.437987E+00	2.819229E-01
1.916500E-01	3.238623E+01	3.137788E+00	1.533357E+00	2.312293E+00	2.680050E-01
2.027579E-01	3.146210E+01	3.003556E+00	1.564490E+00	2.191015E+00	2.554144E-01
2.141787E-01	3.055275E+01	2.844521E+00	1.595936E+00	2.074275E+00	2.440486E-01
2.259124E-01	2.965791E+01	2.669662E+00	1.627695E+00	1.961411E+00	2.338131E-01
2.379590E-01	2.877722E+01	2.488058E+00	1.659767E+00	1.852564E+00	2.246181E-01
2.503184E-01	2.791040E+01	2.308298E+00	1.692152E+00	1.747628E+00	2.163813E-01
2.629908E-01	2.705722E+01	2.138003E+00	1.724850E+00	1.646496E+00	2.090270E-01
2.759761E-01	2.621773E+01	1.983480E+00	1.757861E+00	1.548800E+00	2.024857E-01
2.892742E-01	2.539218E+01	1.849524E+00	1.791184E+00	1.454992E+00	1.966928E-01
3.028853E-01	2.458098E+01	1.739331E+00	1.824821E+00	1.364455E+00	1.915914E-01
3.168093E-01	2.378459E+01	1.654529E+00	1.858770E+00	1.277648E+00	1.871282E-01
3.310461E-01	2.300339E+01	1.595301E+00	1.893033E+00	1.193965E+00	1.832570E-01
3.455959E-01	2.223758E+01	1.560561E+00	1.927608E+00	1.113624E+00	1.799351E-01
3.604586E-01	2.148712E+01	1.548194E+00	1.962496E+00	1.036835E+00	1.771241E-01
3.756341E-01	2.075174E+01	1.555299E+00	1.997697E+00	9.630194E-01	1.747920E-01
3.911226E-01	2.003108E+01	1.578441E+00	2.033211E+00	8.924122E-01	1.729087E-01
4.069239E-01	1.932474E+01	1.613892E+00	2.069038E+00	8.247132E-01	1.714493E-01
4.230382E-01	1.863235E+01	1.657854E+00	2.105178E+00	7.604484E-01	1.703912E-01
4.394653E-01	1.795380E+01	1.706628E+00	2.141630E+00	6.993312E-01	1.697162E-01
4.562054E-01	1.728911E+01	1.756777E+00	2.178396E+00	6.410910E-01	1.694091E-01
4.732583E-01	1.663847E+01	1.805230E+00	2.215474E+00	5.862766E-01	1.694570E-01
4.906241E-01	1.600229E+01	1.849357E+00	2.252866E+00	5.343548E-01	1.698505E-01
5.083029E-01	1.538102E+01	1.887012E+00	2.290570E+00	4.856216E-01	1.705824E-01
5.262945E-01	1.477522E+01	1.916553E+00	2.328587E+00	4.403754E-01	1.716477E-01
5.445991E-01	1.418545E+01	1.936810E+00	2.366917E+00	3.981157E-01	1.730446E-01
5.632165E-01	1.361226E+01	1.947079E+00	2.405560E+00	3.591551E-01	1.747730E-01
5.821468E-01	1.305614E+01	1.947060E+00	2.444516E+00	3.232766E-01	1.768349E-01
6.013901E-01	1.251750E+01	1.936813E+00	2.483785E+00	2.910842E-01	1.792343E-01
6.209462E-01	1.199662E+01	1.916701E+00	2.523366E+00	2.623740E-01	1.819776E-01
6.408153E-01	1.149374E+01	1.887321E+00	2.563261E+00	2.369533E-01	1.850733E-01

Z=30

Gallium

Gallium

Atomic Number	31
Lattice	fcc
Lattice constant a_o	7.83 a.u.
Pressure at a_o	6.92 Kbar
Total energy (solid)	-3843.796 Ry
Total energy (atom)	-3843.556 Ry
Zero point energy	0.003 Ry
Cohesive energy	0.237 Ry
Bulk modulus	0.44 Mbar
Fermi energy	0.516 Ry
Interstitial electrons (per atom)	0.631
χ/χ_o	1.26
$N(E_f)$	0.41 states/eV-atom
ρ_o	2.033×10^4 electrons/Bohr^3
$(d\rho(r,E)/dE)_{E=E_f,r=0}$	19.618 electrons/Bohr^3-Ry

Symmetry point	Band energies in Rydbergs (relative to muffin tin zero)								
Γ	-0.633	-0.633	-0.633	-0.623	-0.623	-0.324			
X	-0.646	-0.643	-0.615	-0.613	-0.613	0.315	0.358		
L	-0.639	-0.633	-0.633	-0.615	-0.615	0.158	0.195		
W	-0.636	-0.631	-0.631	-0.621	-0.613	0.437	0.494	0.494	0.600
K	-0.641	-0.634	-0.620	-0.620	-0.615	0.377	0.407	0.482	

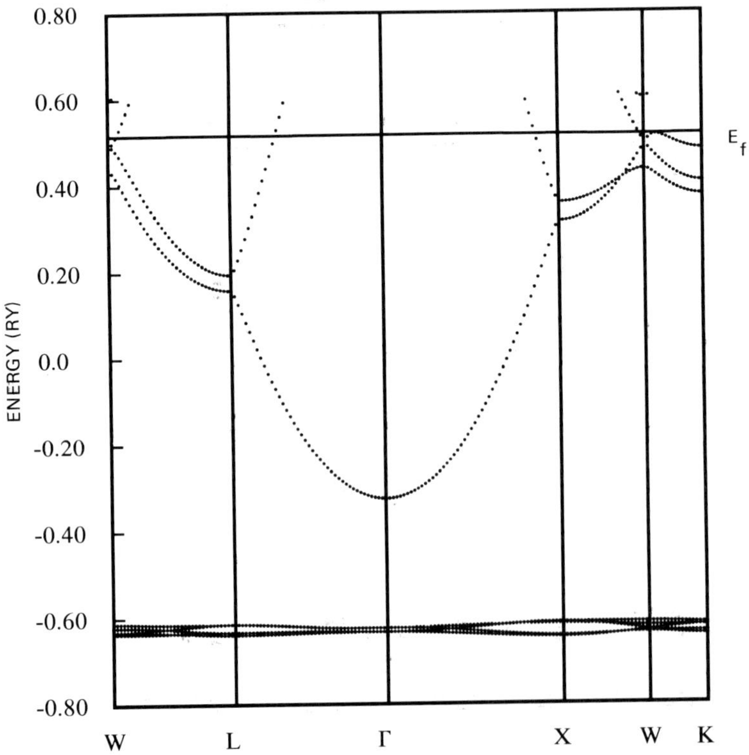

Moruzzi
Janak
Williams

Z=31

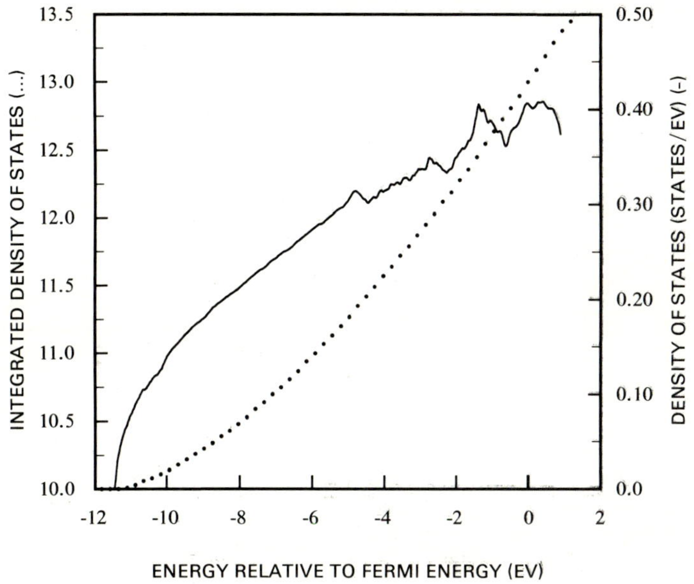

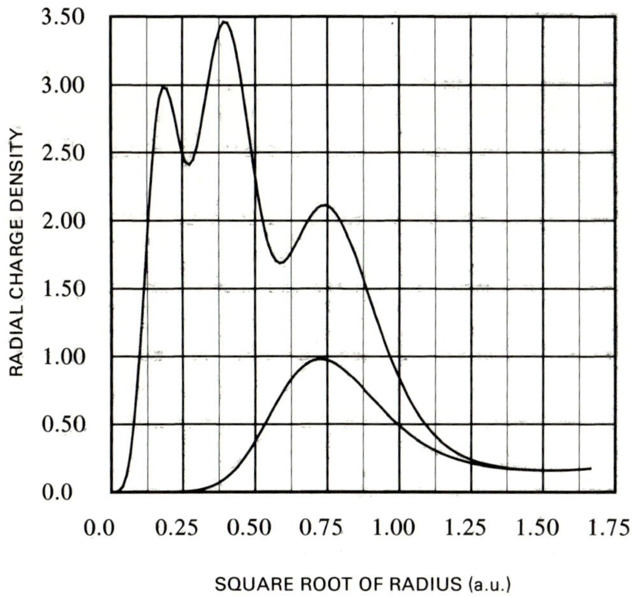

Moruzzi
Janak
Williams

Z=31

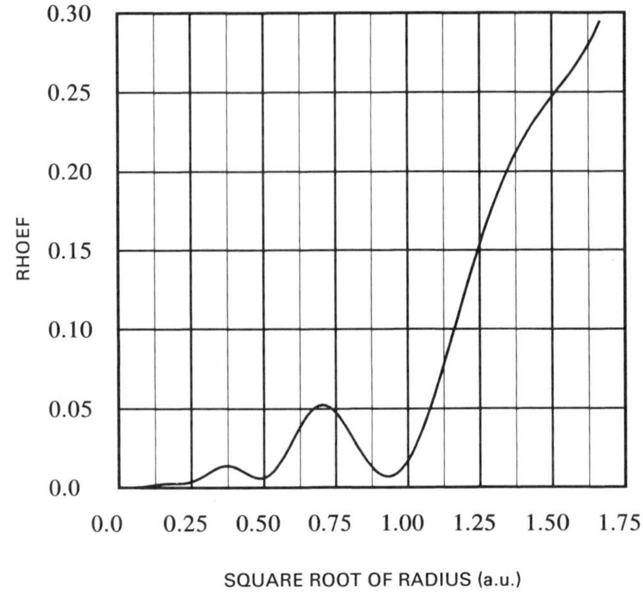

$RHOEF=r^2 \partial\rho(E,r)/\partial E$ (at $E=E_f$)

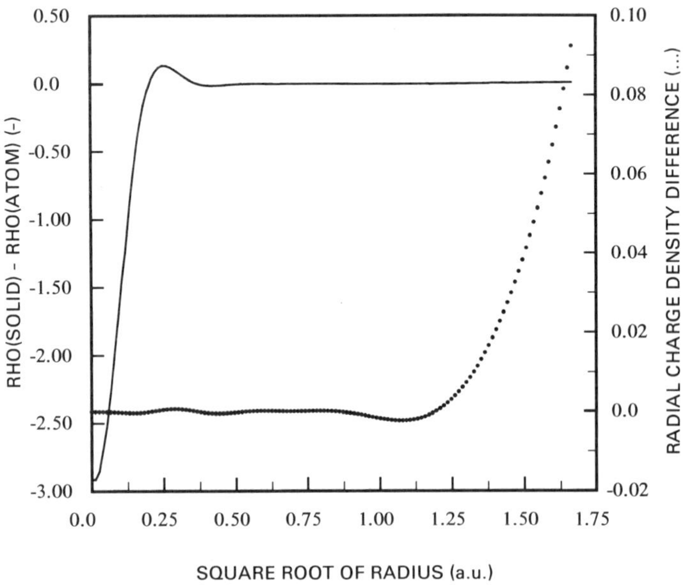

Moruzzi
Janak
Williams

Z=31

Gallium

r	2z(r)	$r^2\rho$	r	2z(r)	$r^2\rho$
1.689650E - 04	6.195874E+01	5.743571E - 04	7.138770E - 01	1.013288E+01	1.750987E+00
6.758599E - 04	6.183467E+01	8.905414E - 03	7.360114E - 01	9.702355E+00	1.673397E+00
1.520685E - 03	6.162704E+01	4.278356E - 02	7.584837E - 01	9.290959E+00	1.592958E+00
2.703439E - 03	6.133459E+01	1.256608E - 01	7.812940E - 01	8.897976E+00	1.510848E+00
4.224122E - 03	6.095689E+01	2.792112E - 01	8.044422E - 01	8.522694E+00	1.428137E+00
6.082736E - 03	6.049442E+01	5.160581E - 01	8.279283E - 01	8.164317E+00	1.345775E+00
8.279283E - 03	5.994931E+01	8.346676E - 01	8.517523E - 01	7.822076E+00	1.264586E+00
1.081376E - 02	5.932536E+01	1.217752E+00	8.759143E - 01	7.495146E+00	1.185274E+00
1.368616E - 02	5.862808E+01	1.634553E+00	9.004143E - 01	7.182731E+00	1.108416E+00
1.689649E - 02	5.786433E+01	2.046318E+00	9.252521E - 01	6.884066E+00	1.034478E+00
2.044476E - 02	5.704141E+01	2.413525E+00	9.504279E - 01	6.598385E+00	9.638197E - 01
2.433095E - 02	5.616695E+01	2.703126E+00	9.759416E - 01	6.324937E+00	8.967034E - 01
2.855508E - 02	5.524805E+01	2.894213E+00	1.001793E+00	6.063124E+00	8.333048E - 01
3.311713E - 02	5.429094E+01	2.980985E+00	1.027983E+00	5.812019E+00	7.737280E - 01
3.801712E - 02	5.330090E+01	2.972740E+00	1.054510E+00	5.571259E+00	7.180036E - 01
4.325503E - 02	5.228247E+01	2.891191E+00	1.081375E+00	5.340261E+00	6.661138E - 01
4.883087E - 02	5.123969E+01	2.765920E+00	1.108579E+00	5.118196E+00	6.179925E - 01
5.474465E - 02	5.017651E+01	2.629091E+00	1.136120E+00	4.904846E+00	5.735336E - 01
6.099635E - 02	4.909717E+01	2.510393E+00	1.164000E+00	4.699462E+00	5.326062E - 01
6.758595E - 02	4.800594E+01	2.433064E+00	1.192216E+00	4.502177E+00	4.950522E - 01
7.451349E - 02	4.690689E+01	2.411435E+00	1.220772E+00	4.311757E+00	4.607018E - 01
8.177900E - 02	4.580344E+01	2.450091E+00	1.249664E+00	4.128682E+00	4.293698E - 01
8.938241E - 02	4.469803E+01	2.544541E+00	1.278895E+00	3.952055E+00	4.008681E - 01
9.732378E - 02	4.359238E+01	2.682948E+00	1.308464E+00	3.781853E+00	3.750047E - 01
1.056031E - 01	4.248799E+01	2.848536E+00	1.338371E+00	3.617509E+00	3.515896E - 01
1.142203E - 01	4.138657E+01	3.022215E+00	1.368616E+00	3.459040E+00	3.304355E - 01
1.231754E - 01	4.029005E+01	3.185051E+00	1.399199E+00	3.306200E+00	3.113607E - 01
1.324685E - 01	3.920071E+01	3.320319E+00	1.430119E+00	3.158757E+00	2.941913E - 01
1.420995E - 01	3.812093E+01	3.414996E+00	1.461377E+00	3.016207E+00	2.787626E - 01
1.520684E - 01	3.705293E+01	3.460626E+00	1.492974E+00	2.878344E+00	2.649177E - 01
1.623753E - 01	3.599852E+01	3.453580E+00	1.524908E+00	2.745531E+00	2.525097E - 01
1.730201E - 01	3.495915E+01	3.394837E+00	1.557180E+00	2.617020E+00	2.414029E - 01
1.840028E - 01	3.393570E+01	3.289345E+00	1.589791E+00	2.492627E+00	2.314714E - 01
1.953235E - 01	3.292857E+01	3.145142E+00	1.622739E+00	2.372746E+00	2.225981E - 01
2.069821E - 01	3.193776E+01	2.972338E+00	1.656025E+00	2.256648E+00	2.146780E - 01
2.189786E - 01	3.096307E+01	2.782086E+00	1.689650E+00	2.144179E+00	2.076142E - 01
2.313130E - 01	3.000417E+01	2.585615E+00	1.723611E+00	2.035759E+00	2.013186E - 01
2.439854E - 01	2.906079E+01	2.393412E+00	1.757911E+00	1.930690E+00	1.957136E - 01
2.569957E - 01	2.813289E+01	2.214576E+00	1.792549E+00	1.828843E+00	1.907280E - 01
2.703439E - 01	2.722070E+01	2.056390E+00	1.827524E+00	1.730672E+00	1.862985E - 01
2.840301E - 01	2.632465E+01	1.924071E+00	1.862838E+00	1.635505E+00	1.823704E - 01
2.980542E - 01	2.544528E+01	1.820715E+00	1.898490E+00	1.543249E+00	1.788945E - 01
3.124162E - 01	2.458313E+01	1.747394E+00	1.934480E+00	1.454108E+00	1.758274E - 01
3.271161E - 01	2.373856E+01	1.703367E+00	1.970807E+00	1.368011E+00	1.731324E - 01
3.421540E - 01	2.291164E+01	1.686389E+00	2.007472E+00	1.284900E+00	1.707774E - 01
3.575298E - 01	2.210216E+01	1.693046E+00	2.044476E+00	1.204444E+00	1.687360E - 01
3.732436E - 01	2.130977E+01	1.719122E+00	2.081817E+00	1.126889E+00	1.669850E - 01
3.892953E - 01	2.053401E+01	1.759941E+00	2.119496E+00	1.051930E+00	1.655064E - 01
4.056848E - 01	1.977466E+01	1.810684E+00	2.157514E+00	9.798360E - 01	1.642849E - 01
4.224124E - 01	1.903156E+01	1.866666E+00	2.195868E+00	9.105992E - 01	1.633093E - 01
4.394779E - 01	1.830484E+01	1.923536E+00	2.234561E+00	8.442285E - 01	1.625713E - 01
4.568812E - 01	1.759486E+01	1.977453E+00	2.273592E+00	7.804574E - 01	1.620659E - 01
4.746225E - 01	1.690216E+01	2.025188E+00	2.312961E+00	7.195938E - 01	1.617901E - 01
4.927018E - 01	1.622736E+01	2.064166E+00	2.352668E+00	6.613857E - 01	1.617439E - 01
5.111190E - 01	1.557117E+01	2.092488E+00	2.392713E+00	6.061655E - 01	1.619292E - 01
5.298741E - 01	1.493427E+01	2.108909E+00	2.433095E+00	5.539823E - 01	1.623502E - 01
5.489671E - 01	1.431726E+01	2.112778E+00	2.473816E+00	5.046133E - 01	1.630138E - 01
5.683981E - 01	1.372071E+01	2.103977E+00	2.514874E+00	4.584088E - 01	1.639274E - 01
5.881670E - 01	1.314500E+01	2.082835E+00	2.556271E+00	4.154494E - 01	1.651013E - 01
6.082739E - 01	1.259038E+01	2.050046E+00	2.598005E+00	3.755285E - 01	1.665478E - 01
6.287186E - 01	1.205701E+01	2.006583E+00	2.640077E+00	3.393086E - 01	1.682798E - 01
6.495013E - 01	1.154481E+01	1.953619E+00	2.682487E+00	3.060356E - 01	1.703131E - 01
6.706219E - 01	1.105362E+01	1.892454E+00	2.725235E+00	2.766657E - 01	1.726643E - 01
6.920805E - 01	1.058311E+01	1.824452E+00	2.768322E+00	2.504572E - 01	1.753521E - 01

Z=31

Rubidium

Atomic Number	37
Lattice	bcc
Lattice constant a_o	10.21 a.u.
Pressure at a_o	0.56 Kbar
Total energy (solid)	-5872.525 Ry
Total energy (atom)	-5872.477 Ry
Zero point energy	0.001 Ry
Cohesive energy	0.047 Ry
Bulk modulus	0.03 Mbar
Fermi energy	0.172 Ry
Interstitial electrons (per atom)	0.397
χ/χ_o	2.12
$N(E_f)$	0.90 states/eV-atom
ρ_o	3.496×10^4 electrons/Bohr^3
$(d\rho(r,E)/dE)_{E=E_f,r=0}$	19.394 electrons/Bohr^3-Ry

Symmetry point	Band energies in Rydbergs (relative to muffin tin zero)		
Γ	0.014		
H	0.247	0.247	
N	0.184	0.238	0.352
P	0.267	0.267	0.267

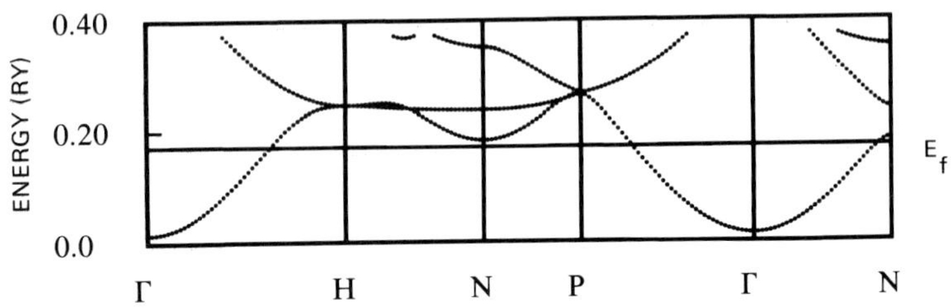

Z=37

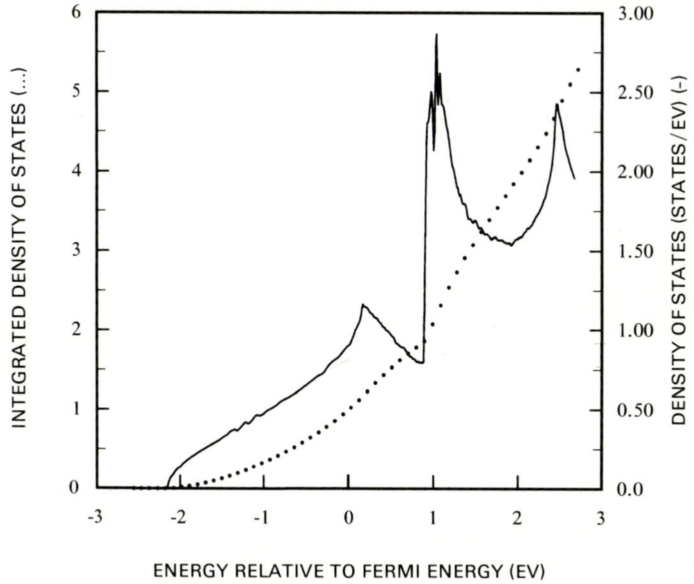

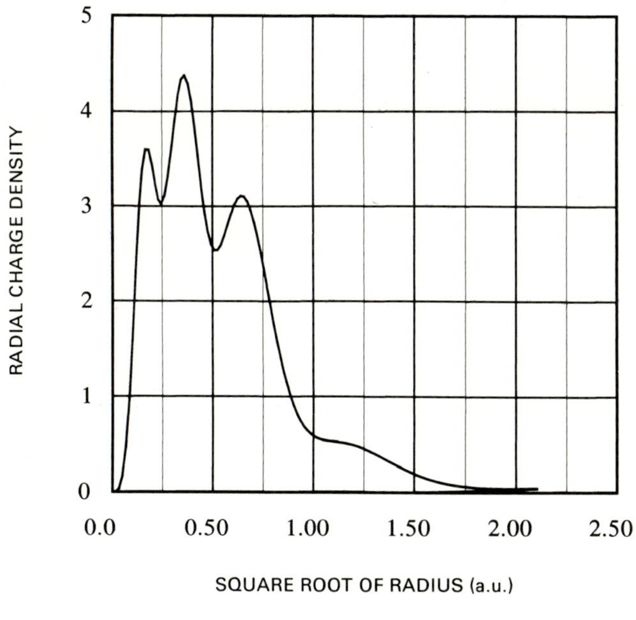

Moruzzi
Janak
Williams

Z=37

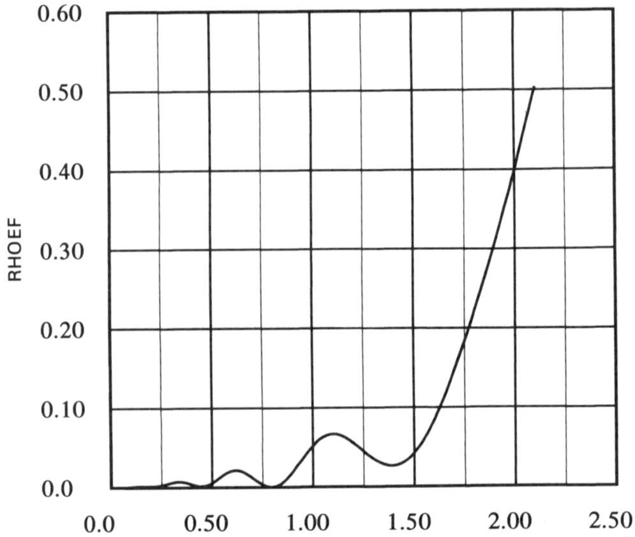

RHOEF=$r^2 \partial\rho(E,r)/\partial E$ (at $E=E_f$)

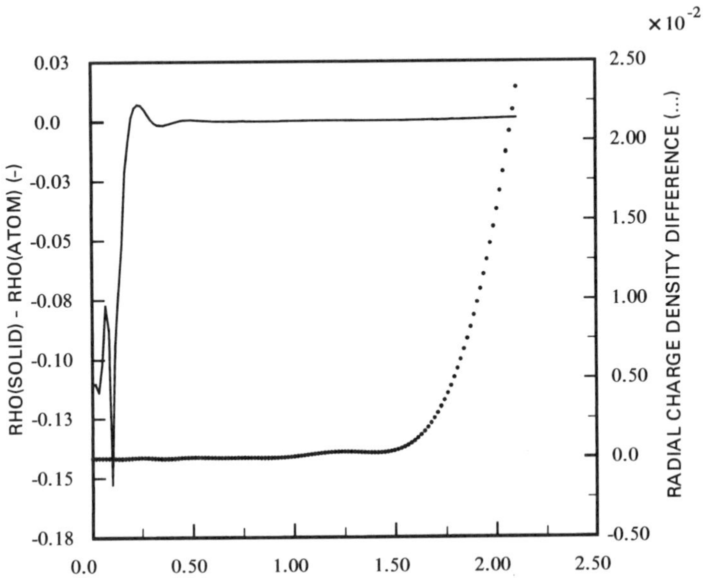

Moruzzi
Janak
Williams

Z=37

Rubidium

r	2z(r)	$r^2\rho$	r	2z(r)	$r^2\rho$
2.698400E-04	7.391490E+01	2.495349E-03	1.140074E+00	7.305051E+00	5.424978E-01
1.079360E-03	7.365842E+01	3.760438E-02	1.175423E+00	7.014509E+00	5.376792E-01
2.428560E-03	7.322836E+01	1.722907E-01	1.211311E+00	6.734630E+00	5.336615E-01
4.317440E-03	7.262289E+01	4.735662E-01	1.247740E+00	6.464749E+00	5.297188E-01
6.745998E-03	7.184503E+01	9.664117E-01	1.284708E+00	6.204338E+00	5.252804E-01
9.714238E-03	7.090369E+01	1.610495E+00	1.322215E+00	5.953324E+00	5.199197E-01
1.322216E-02	6.981386E+01	2.307021E+00	1.360263E+00	5.710640E+00	5.133377E-01
1.726976E-02	6.859462E+01	2.931734E+00	1.398850E+00	5.476358E+00	5.053481E-01
2.185704E-02	6.726596E+01	3.378294E+00	1.437977E+00	5.250606E+00	4.958602E-01
2.698400E-02	6.584647E+01	3.593190E+00	1.477643E+00	5.032801E+00	4.848615E-01
3.265064E-02	6.435153E+01	3.589242E+00	1.517850E+00	4.822378E+00	4.724041E-01
3.885696E-02	6.279376E+01	3.435440E+00	1.558596E+00	4.619867E+00	4.585900E-01
4.560296E-02	6.118457E+01	3.230093E+00	1.599881E+00	4.424703E+00	4.435576E-01
5.288864E-02	5.953566E+01	3.069090E+00	1.641706E+00	4.236673E+00	4.274708E-01
6.071400E-02	5.785887E+01	3.020091E+00	1.684071E+00	4.055923E+00	4.105096E-01
6.907904E-02	5.616435E+01	3.109170E+00	1.726975E+00	3.881496E+00	3.928600E-01
7.798374E-02	5.445921E+01	3.320909E+00	1.770420E+00	3.713516E+00	3.747104E-01
8.742815E-02	5.274847E+01	3.608856E+00	1.814404E+00	3.552095E+00	3.562444E-01
9.741223E-02	5.103729E+01	3.911311E+00	1.858928E+00	3.396251E+00	3.376355E-01
1.079360E-01	4.933189E+01	4.167388E+00	1.903991E+00	3.246449E+00	3.190466E-01
1.189994E-01	4.763907E+01	4.329695E+00	1.949594E+00	3.102058E+00	3.006253E-01
1.306025E-01	4.596530E+01	4.371744E+00	1.995736E+00	2.962828E+00	2.825044E-01
1.427453E-01	4.431561E+01	4.289867E+00	2.042418E+00	2.828494E+00	2.647997E-01
1.554278E-01	4.269331E+01	4.100513E+00	2.089641E+00	2.698805E+00	2.476115E-01
1.686500E-01	4.110022E+01	3.834558E+00	2.137403E+00	2.573521E+00	2.310231E-01
1.824118E-01	3.953699E+01	3.530230E+00	2.185703E+00	2.452773E+00	2.151034E-01
1.967133E-01	3.800426E+01	3.226216E+00	2.234545E+00	2.335619E+00	1.999052E-01
2.115545E-01	3.650313E+01	2.955981E+00	2.283925E+00	2.222580E+00	1.854690E-01
2.269354E-01	3.503539E+01	2.743946E+00	2.333846E+00	2.112751E+00	1.718218E-01
2.428560E-01	3.360313E+01	2.603699E+00	2.384306E+00	2.006682E+00	1.589804E-01
2.593162E-01	3.220818E+01	2.538047E+00	2.435306E+00	1.903864E+00	1.469510E-01
2.763162E-01	3.085138E+01	2.540492E+00	2.486845E+00	1.804168E+00	1.357310E-01
2.938557E-01	2.953291E+01	2.597608E+00	2.538924E+00	1.707480E+00	1.253107E-01
3.119350E-01	2.825252E+01	2.691787E+00	2.591543E+00	1.613715E+00	1.156738E-01
3.305539E-01	2.701042E+01	2.803911E+00	2.644701E+00	1.523161E+00	1.067988E-01
3.497126E-01	2.580746E+01	2.915656E+00	2.698400E+00	1.435040E+00	9.865952E-02
3.694109E-01	2.464528E+01	3.011198E+00	2.752638E+00	1.349687E+00	9.122705E-02
3.896489E-01	2.352588E+01	3.078307E+00	2.807415E+00	1.267443E+00	8.446968E-02
4.104266E-01	2.245129E+01	3.108816E+00	2.862732E+00	1.187944E+00	7.835382E-02
4.317440E-01	2.142329E+01	3.098597E+00	2.918589E+00	1.111202E+00	7.284480E-02
4.536010E-01	2.044308E+01	3.047174E+00	2.974985E+00	1.037238E+00	6.790733E-02
4.759977E-01	1.951109E+01	2.957055E+00	3.031921E+00	9.660797E-01	6.350619E-02
4.989341E-01	1.862708E+01	2.833013E+00	3.089397E+00	8.977577E-01	5.960650E-02
5.224102E-01	1.778995E+01	2.681301E+00	3.147413E+00	8.323021E-01	5.617399E-02
5.464260E-01	1.699803E+01	2.508944E+00	3.205969E+00	7.697408E-01	5.317550E-02
5.709814E-01	1.624913E+01	2.323148E+00	3.265063E+00	7.104532E-01	5.057905E-02
5.960765E-01	1.554068E+01	2.130805E+00	3.324698E+00	6.537234E-01	4.835393E-02
6.217114E-01	1.486990E+01	1.938157E+00	3.384872E+00	6.002673E-01	4.647111E-02
6.478858E-01	1.423401E+01	1.750559E+00	3.445586E+00	5.497014E-01	4.490308E-02
6.746000E-01	1.363025E+01	1.572356E+00	3.506841E+00	5.016262E-01	4.362393E-02
7.018538E-01	1.305618E+01	1.406850E+00	3.568634E+00	4.567078E-01	4.260955E-02
7.296473E-01	1.250952E+01	1.256332E+00	3.630966E+00	4.148792E-01	4.183744E-02
7.579805E-01	1.198838E+01	1.122158E+00	3.693839E+00	3.753307E-01	4.128665E-02
7.868534E-01	1.149117E+01	1.004865E+00	3.757252E+00	3.383326E-01	4.093794E-02
8.162660E-01	1.101663E+01	9.043013E-01	3.821204E+00	3.037833E-01	4.077354E-02
8.462182E-01	1.056372E+01	8.197619E-01	3.885695E+00	2.719407E-01	4.077726E-02
8.767102E-01	1.013152E+01	7.501249E-01	3.950727E+00	2.419777E-01	4.093414E-02
9.077417E-01	9.719226E+00	6.939756E-01	4.016298E+00	2.145092E-01	4.123065E-02
9.393131E-01	9.325873E+00	6.497180E-01	4.082409E+00	1.890783E-01	4.165448E-02
9.714240E-01	8.950561E+00	6.156719E-01	4.149059E+00	1.655917E-01	4.219442E-02
1.004074E+00	8.592370E+00	5.901517E-01	4.216249E+00	1.439641E-01	4.284032E-02
1.037265E+00	8.249612E+00	5.715275E-01	4.283979E+00	1.241189E-01	4.358309E-02
1.070994E+00	7.921791E+00	5.582763E-01	4.352249E+00	1.059917E-01	4.441441E-02
1.105265E+00	7.607045E+00	5.490095E-01	4.421059E+00	8.952570E-02	4.532683E-02

Z=37

Moruzzi
Janak
Williams

Strontium

Atomic Number	38
Lattice	fcc
Lattice constant a_o	10.88 a.u.
Pressure at a_o	-0.41 Kbar
Total energy (solid)	-6258.822 Ry
Total energy (atom)	-6258.683 Ry
Zero point energy	0.001 Ry
Cohesive energy	0.138 Ry
Bulk modulus	0.15 Mbar
Fermi energy	0.302 Ry
Interstitial electrons (per atom)	0.634
χ/χ_o	1.15
$N(E_f)$	0.31 states/eV-atom
ρ_o	3.794×10^4 electrons/Bohr^3
$(d\rho(r,E)/dE)_{E=E_f,r=0}$	1.4924 electrons/Bohr^3-Ry

Symmetry point	Band energies in Rydbergs (relative to muffin tin zero)		
Γ	0.038		
X	0.236	0.321	
L	0.229	0.330	
W	0.290	0.364	0.364
K	0.280	0.326	

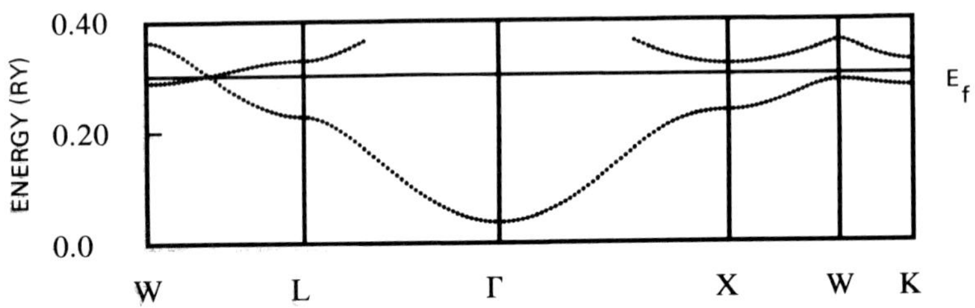

Z=38

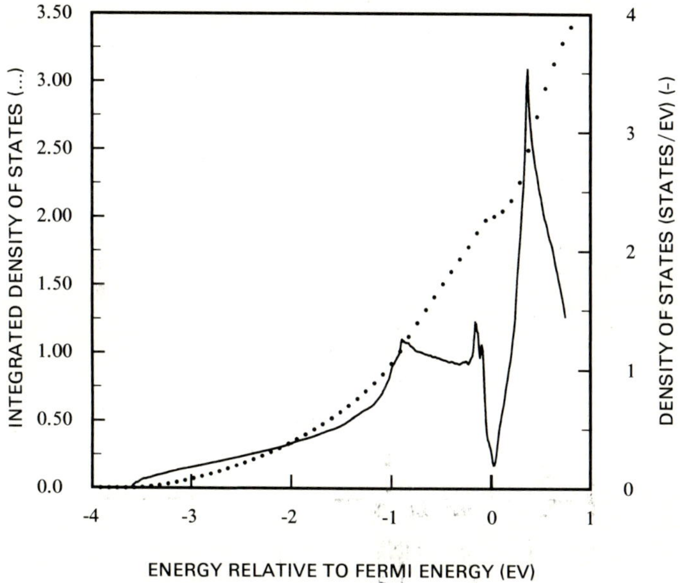

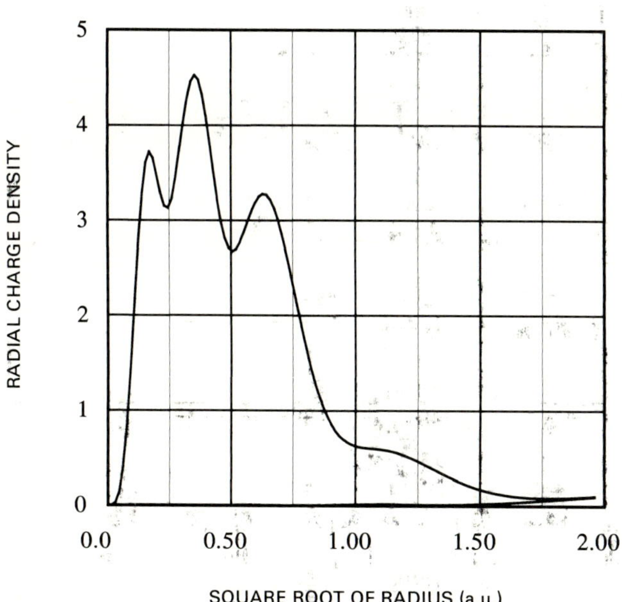

Moruzzi
Janak
Williams

Z=38

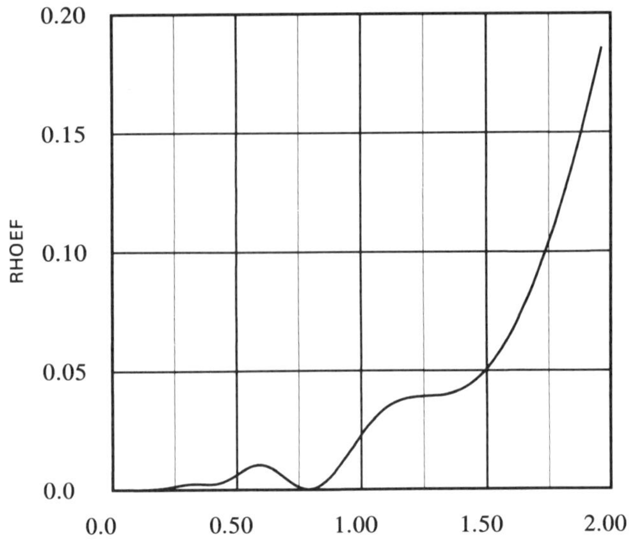

RHOEF=$r^2\partial\rho(E,r)/\partial E$ (at $E=E_f$)

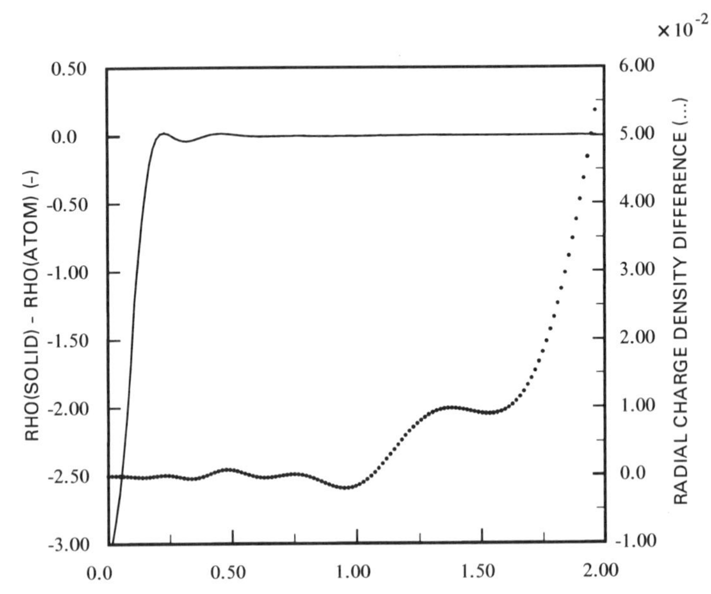

Z=38

Strontium

r	2z(r)	$r^2\rho$	r	2z(r)	$r^2\rho$
2.347815E-04	7.592308E+01	2.054265E-03	9.919517E-01	9.076505E+00	6.306277E-01
9.391259E-04	7.569142E+01	3.115527E-02	1.022708E+00	8.739604E+00	6.197423E-01
2.113033E-03	7.530296E+01	1.442660E-01	1.053933E+00	8.415267E+00	6.123070E-01
3.756504E-03	7.475606E+01	4.024618E-01	1.085629E+00	8.101718E+00	6.071255E-01
5.869534E-03	7.405232E+01	8.370684E-01	1.117794E+00	7.799266E+00	6.031699E-01
8.452132E-03	7.319797E+01	1.427515E+00	1.150429E+00	7.506475E+00	5.995804E-01
1.150429E-02	7.220419E+01	2.100738E+00	1.183533E+00	7.223153E+00	5.956621E-01
1.502601E-02	7.108624E+01	2.751916E+00	1.217107E+00	6.948464E+00	5.908759E-01
1.901730E-02	6.986136E+01	3.277422E+00	1.251150E+00	6.682412E+00	5.848258E-01
2.347815E-02	6.854572E+01	3.606491E+00	1.285663E+00	6.424698E+00	5.772445E-01
2.840856E-02	6.715398E+01	3.720370E+00	1.320645E+00	6.174692E+00	5.679767E-01
3.380853E-02	6.569809E+01	3.653573E+00	1.356097E+00	5.932553E+00	5.569641E-01
3.967807E-02	6.418825E+01	3.479270E+00	1.392019E+00	5.697707E+00	5.442275E-01
4.601717E-02	6.263416E+01	3.285604E+00	1.428410E+00	5.470718E+00	5.298530E-01
5.282583E-02	6.104596E+01	3.151468E+00	1.465271E+00	5.250271E+00	5.139751E-01
6.010406E-02	5.943359E+01	3.128574E+00	1.502601E+00	5.037303E+00	4.967670E-01
6.785184E-02	5.780533E+01	3.233306E+00	1.540401E+00	4.830864E+00	4.784254E-01
7.606918E-02	5.616695E+01	3.448385E+00	1.578671E+00	4.630747E+00	4.591622E-01
8.475608E-02	5.452275E+01	3.731791E+00	1.617409E+00	4.437491E+00	4.391969E-01
9.391254E-02	5.287720E+01	4.029314E+00	1.656617E+00	4.250124E+00	4.187464E-01
1.035386E-01	5.123557E+01	4.287229E+00	1.696296E+00	4.068420E+00	3.980212E-01
1.136342E-01	4.960370E+01	4.462485E+00	1.736444E+00	3.892528E+00	3.772220E-01
1.241994E-01	4.798729E+01	4.528991E+00	1.777061E+00	3.722219E+00	3.565339E-01
1.352341E-01	4.639101E+01	4.479762E+00	1.818148E+00	3.556897E+00	3.361252E-01
1.467384E-01	4.481808E+01	4.325417E+00	1.859704E+00	3.396722E+00	3.161463E-01
1.587123E-01	4.327043E+01	4.090119E+00	1.901730E+00	3.241486E+00	2.967282E-01
1.711557E-01	4.174919E+01	3.806147E+00	1.944225E+00	3.090613E+00	2.779822E-01
1.840686E-01	4.025484E+01	3.508218E+00	1.987190E+00	2.944307E+00	2.600015E-01
1.974512E-01	3.878824E+01	3.228477E+00	2.030624E+00	2.802401E+00	2.428609E-01
2.113033E-01	3.735083E+01	2.992740E+00	2.074529E+00	2.664373E+00	2.266175E-01
2.256250E-01	3.594441E+01	2.818233E+00	2.118902E+00	2.530477E+00	2.113132E-01
2.404162E-01	3.457083E+01	2.712852E+00	2.163746E+00	2.400231E+00	1.969750E-01
2.556770E-01	3.323145E+01	2.675718E+00	2.209059E+00	2.273932E+00	1.836172E-01
2.714074E-01	3.192679E+01	2.698667E+00	2.254841E+00	2.151519E+00	1.712423E-01
2.876073E-01	3.065694E+01	2.768308E+00	2.301093E+00	2.032565E+00	1.598427E-01
3.042768E-01	2.942180E+01	2.868266E+00	2.347815E+00	1.917435E+00	1.494025E-01
3.214158E-01	2.822179E+01	2.981302E+00	2.395005E+00	1.806120E+00	1.398983E-01
3.390244E-01	2.705777E+01	3.091126E+00	2.442666E+00	1.698254E+00	1.313007E-01
3.571026E-01	2.593126E+01	3.183730E+00	2.490796E+00	1.594230E+00	1.235761E-01
3.756503E-01	2.484398E+01	3.248250E+00	2.539396E+00	1.493704E+00	1.166868E-01
3.946676E-01	2.379770E+01	3.277349E+00	2.588466E+00	1.397081E+00	1.105928E-01
4.141545E-01	2.279390E+01	3.267214E+00	2.638004E+00	1.304384E+00	1.052527E-01
4.341109E-01	2.183354E+01	3.217224E+00	2.688013E+00	1.215264E+00	1.006234E-01
4.545369E-01	2.091708E+01	3.129480E+00	2.738491E+00	1.130108E+00	9.666234E-02
4.754325E-01	2.004427E+01	3.008149E+00	2.789438E+00	1.048911E+00	9.332687E-02
4.967976E-01	1.921423E+01	2.858848E+00	2.840856E+00	9.712843E-01	9.057522E-02
5.186322E-01	1.842555E+01	2.688029E+00	2.892742E+00	8.975666E-01	8.836693E-02
5.409365E-01	1.767638E+01	2.502442E+00	2.945098E+00	8.277065E-01	8.666289E-02
5.637103E-01	1.696451E+01	2.308687E+00	2.997924E+00	7.612590E-01	8.542585E-02
5.869536E-01	1.628755E+01	2.112875E+00	3.051220E+00	6.981480E-01	8.462042E-02
6.106666E-01	1.564307E+01	1.920397E+00	3.104984E+00	6.390252E-01	8.421320E-02
6.348491E-01	1.502870E+01	1.735760E+00	3.159219E+00	5.830428E-01	8.417302E-02
6.595011E-01	1.444218E+01	1.562539E+00	3.213923E+00	5.301001E-01	8.447051E-02
6.846228E-01	1.388154E+01	1.403352E+00	3.269096E+00	4.808452E-01	8.507890E-02
7.102139E-01	1.334507E+01	1.259926E+00	3.324740E+00	4.340521E-01	8.597302E-02
7.362747E-01	1.283130E+01	1.133161E+00	3.380853E+00	3.907533E-01	8.712995E-02
7.628050E-01	1.233908E+01	1.023247E+00	3.437435E+00	3.501174E-01	8.852851E-02
7.898049E-01	1.186750E+01	9.297817E-01	3.494487E+00	3.124443E-01	9.014958E-02
8.172743E-01	1.141574E+01	8.518851E-01	3.552009E+00	2.772999E-01	9.197557E-02
8.452133E-01	1.098306E+01	7.883274E-01	3.610000E+00	2.450107E-01	9.399045E-02
8.736218E-01	1.056873E+01	7.376339E-01	3.668460E+00	2.155398E-01	9.617972E-02
9.025000E-01	1.017187E+01	6.981851E-01	3.727390E+00	1.884890E-01	9.853041E-02
9.318476E-01	9.791589E+00	6.683024E-01	3.786790E+00	1.642218E-01	1.010303E-01
9.616649E-01	9.426815E+00	6.463171E-01	3.846660E+00	1.423677E-01	1.036688E-01

Z=38

Yttrium

Atomic Number	39
Lattice	fcc
Lattice constant a_o	9.23 a.u.
Pressure at a_o	-0.38 Kbar
Total energy (solid)	-6659.183 Ry
Total energy (atom)	-6658.819 Ry
Zero point energy	0.002 Ry
Cohesive energy	0.362 Ry
Bulk modulus	0.33 Mbar
Fermi energy	0.460 Ry
Interstitial electrons (per atom)	0.843
χ/χ_o	1.88
$N(E_f)$	1.41 states/eV-atom
ρ_o	4.108×10^4 electrons/Bohr^3
$(d\rho(r,E)/dE)_{E=E_f,r=0}$	4.4856 electrons/Bohr^3-Ry

Symmetry point	Band energies in Rydbergs (relative to muffin tin zero)			
Γ	0.134	0.593	0.593	0.593
X	0.299	0.384		
L	0.325	0.539	0.572	0.572
W	0.378	0.493	0.493	
K	0.361	0.428	0.571	

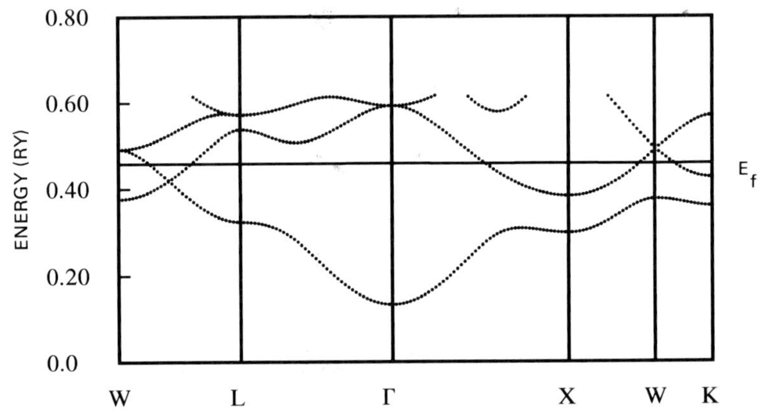

Z=39

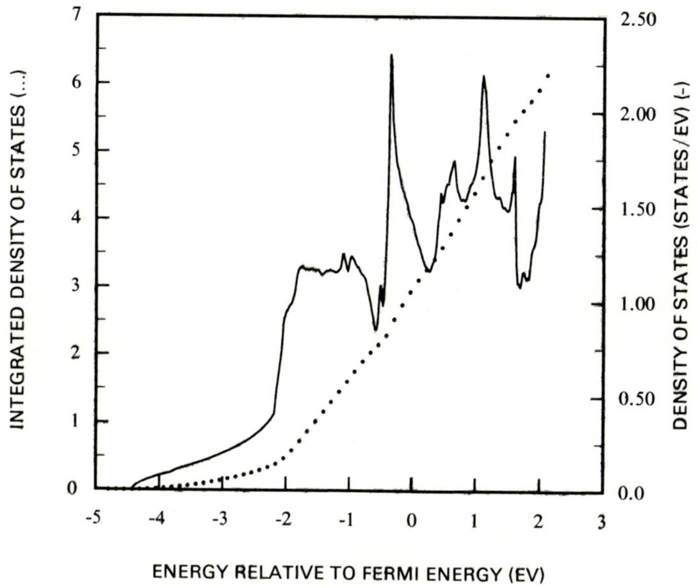

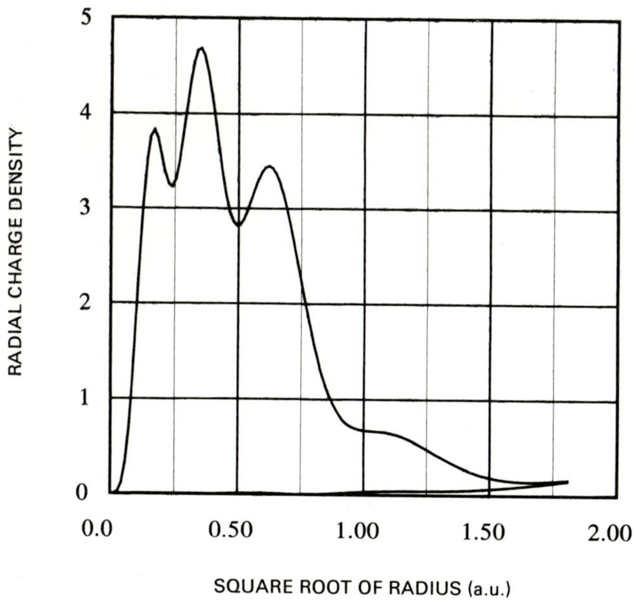

Moruzzi
Janak
Williams

Z=39

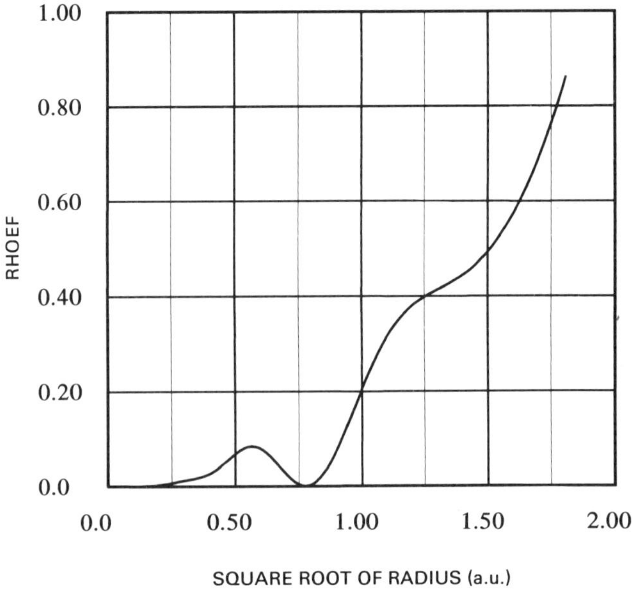

RHOEF=$r^2 \partial\rho(E,r)/\partial E$ (at $E=E_f$)

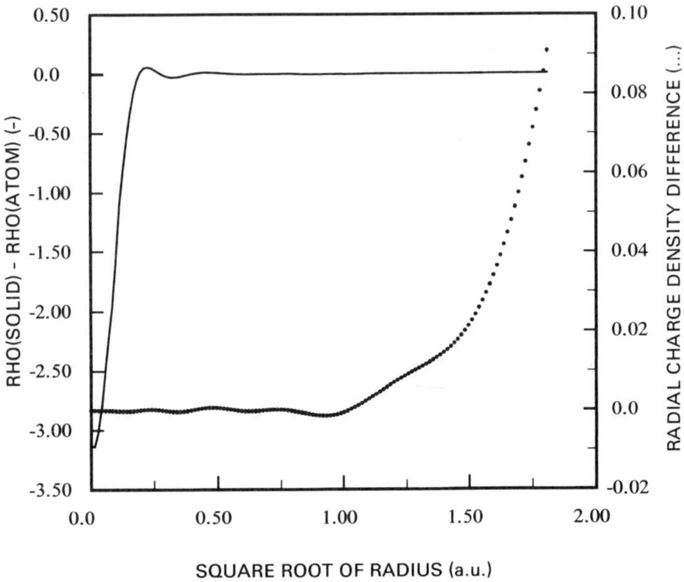

Moruzzi
Janak
Williams

Z=39

Yttrium

r	2z(r)	$r^2\rho$	r	2z(r)	$r^2\rho$
1.991758E - 04	7.793225E+01	1.604522E - 03	8.415179E - 01	1.128065E+01	7.822018E - 01
7.967034E - 04	7.772832E+01	2.450340E - 02	8.676100E - 01	1.088091E+01	7.475270E - 01
1.792582E - 03	7.738652E+01	1.147798E - 01	8.941004E - 01	1.049587E+01	7.216337E - 01
3.186814E - 03	7.690530E+01	3.254073E - 01	9.209891E - 01	1.012462E+01	7.029655E - 01
4.979394E - 03	7.628519E+01	6.909394E - 01	9.482762E - 01	9.766218E+00	6.900318E - 01
7.170327E - 03	7.553003E+01	1.208300E+00	9.759616E - 01	9.419689E+00	6.814439E - 01
9.759616E - 03	7.464752E+01	1.831240E+00	1.004045E+00	9.084555E+00	6.759402E - 01
1.274725E - 02	7.364885E+01	2.480330E+00	1.032527E+00	8.759174E+00	6.724019E - 01
1.613324E - 02	7.254733E+01	3.064435E+00	1.061408E+00	8.443233E+00	6.698631E - 01
1.991758E - 02	7.135712E+01	3.505747E+00	1.090687E+00	8.136552E+00	6.675125E - 01
2.410027E - 02	7.009094E+01	3.759857E+00	1.120363E+00	7.838661E+00	6.646898E - 01
2.868132E - 02	6.876012E+01	3.825132E+00	1.150439E+00	7.548412E+00	6.608784E - 01
3.366072E - 02	6.737399E+01	3.739759E+00	1.180913E+00	7.266282E+00	6.556956E - 01
3.903846E - 02	6.594075E+01	3.568836E+00	1.211785E+00	6.991642E+00	6.488793E - 01
4.481456E - 02	6.446838E+01	3.386477E+00	1.243056E+00	6.724299E+00	6.402755E - 01
5.098902E - 02	6.296523E+01	3.258242E+00	1.274725E+00	6.464476E+00	6.298220E - 01
5.756182E - 02	6.143948E+01	3.228227E+00	1.306792E+00	6.211636E+00	6.175346E - 01
6.453294E - 02	5.989807E+01	3.312856E+00	1.339258E+00	5.965638E+00	6.034933E - 01
7.190245E - 02	5.834608E+01	3.501537E+00	1.372122E+00	5.726740E+00	5.878284E - 01
7.967031E - 02	5.678717E+01	3.762547E+00	1.405384E+00	5.494411E+00	5.707084E - 01
8.783650E - 02	5.522491E+01	4.051966E+00	1.439045E+00	5.268511E+00	5.523283E - 01
9.640110E - 02	5.366339E+01	4.323235E+00	1.473104E+00	5.048889E+00	5.329009E - 01
1.053640E - 01	5.210739E+01	4.535500E+00	1.507562E+00	4.835779E+00	5.126482E - 01
1.147252E - 01	5.056175E+01	4.659541E+00	1.542418E+00	4.628629E+00	4.917923E - 01
1.244849E - 01	4.903078E+01	4.680798E+00	1.577671E+00	4.427670E+00	4.705518E - 01
1.346428E - 01	4.751801E+01	4.599654E+00	1.613324E+00	4.231948E+00	4.491351E - 01
1.451992E - 01	4.602568E+01	4.429466E+00	1.649375E+00	4.042078E+00	4.277392E - 01
1.561539E - 01	4.455539E+01	4.193097E+00	1.685823E+00	3.857903E+00	4.065449E - 01
1.675069E - 01	4.310782E+01	3.918770E+00	1.722672E+00	3.678478E+00	3.857145E - 01
1.792582E - 01	4.168375E+01	3.635848E+00	1.759917E+00	3.504440E+00	3.653938E - 01
1.914080E - 01	4.028386E+01	3.371241E+00	1.797562E+00	3.335259E+00	3.457083E - 01
2.039561E - 01	3.890950E+01	3.146644E+00	1.835604E+00	3.171192E+00	3.267654E - 01
2.169025E - 01	3.756224E+01	2.976894E+00	1.874045E+00	3.011352E+00	3.086533E - 01
2.302473E - 01	3.624367E+01	2.869354E+00	1.912885E+00	2.856417E+00	2.914435E - 01
2.439904E - 01	3.495500E+01	2.824239E+00	1.952122E+00	2.706299E+00	2.751911E - 01
2.581319E - 01	3.369687E+01	2.835647E+00	1.991758E+00	2.560159E+00	2.599343E - 01
2.726717E - 01	3.246947E+01	2.893035E+00	2.031793E+00	2.418721E+00	2.456996E - 01
2.876099E - 01	3.127280E+01	2.982901E+00	2.072225E+00	2.281943E+00	2.324996E - 01
3.029464E - 01	3.010704E+01	3.090466E+00	2.113056E+00	2.149031E+00	2.203358E - 01
3.186813E - 01	2.897287E+01	3.201118E+00	2.154285E+00	2.020745E+00	2.092012E - 01
3.348145E - 01	2.787128E+01	3.301654E+00	2.195913E+00	1.896690E+00	1.990795E - 01
3.513461E - 01	2.680365E+01	3.381099E+00	2.237939E+00	1.777267E+00	1.899481E - 01
3.682761E - 01	2.577142E+01	3.431207E+00	2.280364E+00	1.661704E+00	1.817784E - 01
3.856044E - 01	2.477602E+01	3.446671E+00	2.323187E+00	1.550780E+00	1.745377E - 01
4.033310E - 01	2.381848E+01	3.425008E+00	2.366407E+00	1.444510E+00	1.681895E - 01
4.214560E - 01	2.289949E+01	3.366301E+00	2.410028E+00	1.342114E+00	1.626951E - 01
4.399794E - 01	2.201923E+01	3.272779E+00	2.454045E+00	1.244362E+00	1.580138E - 01
4.589011E - 01	2.117741E+01	3.148334E+00	2.498462E+00	1.150858E+00	1.541044E - 01
4.782212E - 01	2.037317E+01	2.998034E+00	2.543276E+00	1.061959E+00	1.509251E - 01
4.979396E - 01	1.960529E+01	2.827631E+00	2.588489E+00	9.772452E - 01	1.484342E - 01
5.180563E - 01	1.887219E+01	2.643153E+00	2.634100E+00	8.966725E - 01	1.465913E - 01
5.385715E - 01	1.817201E+01	2.450535E+00	2.680110E+00	8.201876E - 01	1.453565E - 01
5.594849E - 01	1.750278E+01	2.255342E+00	2.726518E+00	7.481250E - 01	1.446915E - 01
5.807967E - 01	1.686241E+01	2.062553E+00	2.773324E+00	6.800305E - 01	1.445594E - 01
6.025069E - 01	1.624892E+01	1.876438E+00	2.820529E+00	6.158372E - 01	1.449250E - 01
6.246154E - 01	1.566049E+01	1.700469E+00	2.868132E+00	5.558717E - 01	1.457552E - 01
6.471223E - 01	1.509533E+01	1.537299E+00	2.916133E+00	4.996840E - 01	1.470186E - 01
6.700275E - 01	1.455207E+01	1.388792E+00	2.964533E+00	4.472209E - 01	1.486858E - 01
6.933311E - 01	1.402938E+01	1.256072E+00	3.013330E+00	3.988147E - 01	1.507288E - 01
7.170330E - 01	1.352633E+01	1.139590E+00	3.062528E+00	3.536537E - 01	1.531222E - 01
7.411333E - 01	1.304211E+01	1.039235E+00	3.112123E+00	3.124843E - 01	1.558416E - 01
7.656319E - 01	1.257603E+01	9.544152E - 01	3.162115E+00	2.752821E - 01	1.588647E - 01
7.905289E - 01	1.212751E+01	8.841616E - 01	3.212506E+00	2.416517E - 01	1.621706E - 01
8.158243E - 01	1.169592E+01	8.272339E - 01	3.263297E+00	2.112048E - 01	1.657398E - 01

Z=39

Zirconium

Atomic Number	40
Lattice	fcc
Lattice constant a_o	8.32 a.u.
Pressure at a_o	-2.5 Kbar
Total energy (solid)	-7073.818 Ry
Total energy (atom)	-7073.319 Ry
Zero point energy	0.002 Ry
Cohesive energy	0.497 Ry
Bulk modulus	0.94 Mbar
Fermi energy	0.601 Ry
Interstitial electrons (per atom)	0.991
χ/χ_o	1.64
$N(E_f)$	1.28 states/eV-atom
ρ_o	4.439×10^4 electrons/Bohr^3
$(d\rho(r,E)/dE)_{E=E_f,r=0}$	8.2697 electrons/Bohr^3-Ry

Symmetry point	Band energies in Rydbergs (relative to muffin tin zero)			
Γ	0.224	0.665	0.665	0.665
X	0.337	0.420		
L	0.389	0.643	0.643	0.713
W	0.435	0.579	0.579	
K	0.410	0.494	0.697	

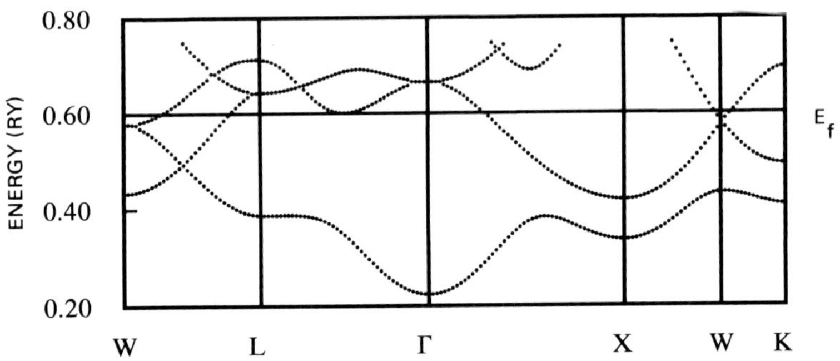

Z=40

Moruzzi
Janak
Williams

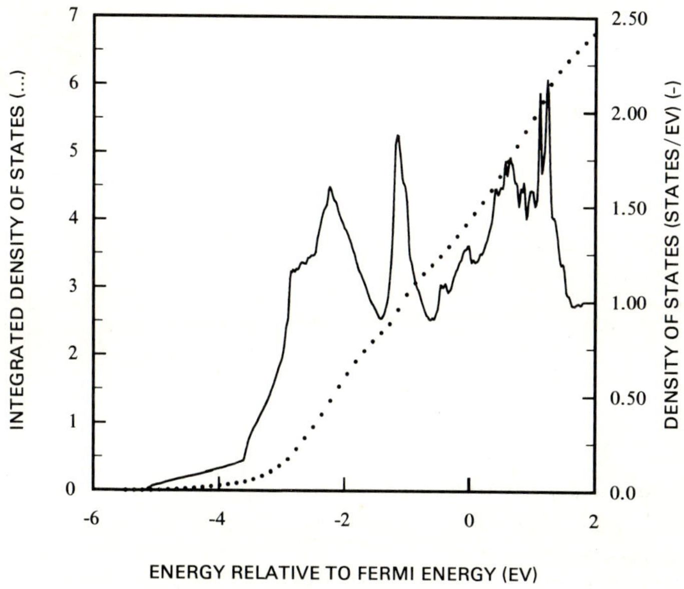

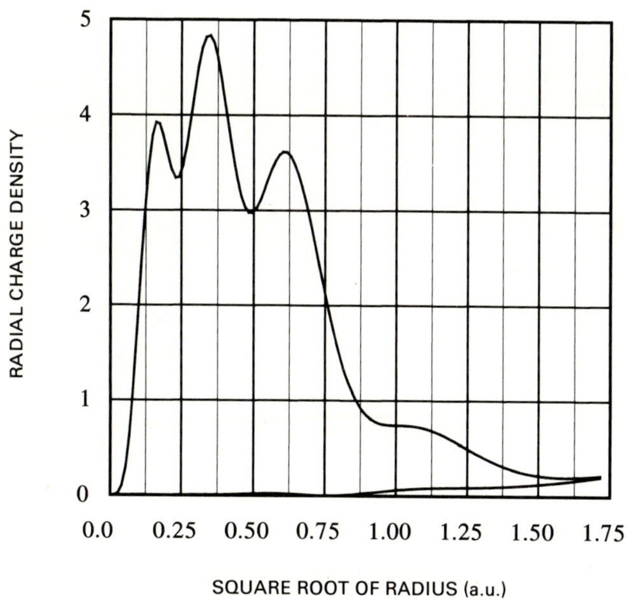

Moruzzi
Janak
Williams

Z=40

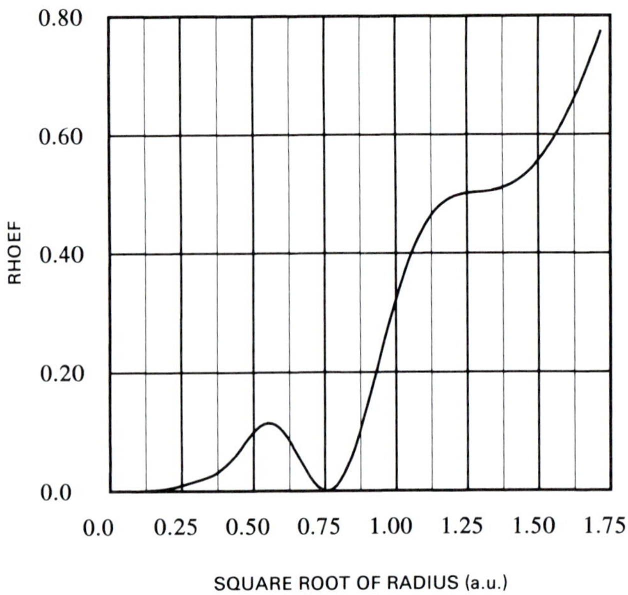

RHOEF=$r^2\partial\rho(E,r)/\partial E$ (at $E=E_f$)

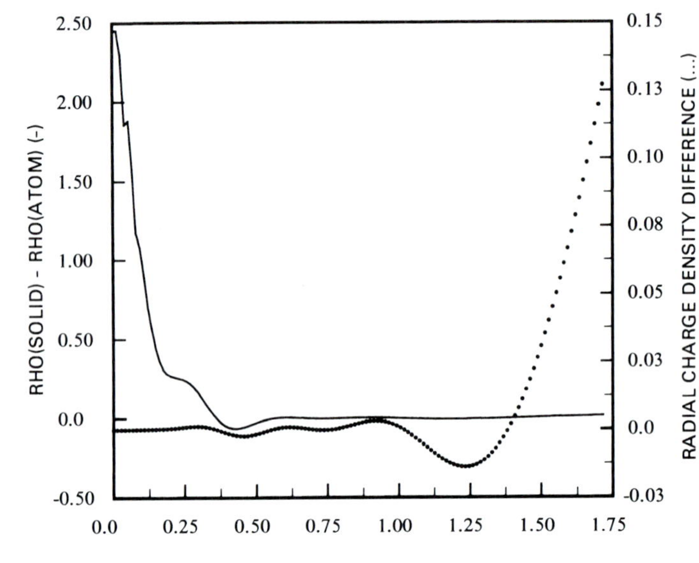

Moruzzi
Janak
Williams

Z=40

Zirconium

r	2z(r)	$r^2\rho$	r	2z(r)	$r^2\rho$
1.795388E - 04	7.993665E+01	1.410404E - 03	7.585514E - 01	1.296741E+01	9.377079E - 01
7.181552E - 04	7.974591E+01	2.161482E - 02	7.820709E - 01	1.252061E+01	8.822722E - 01
1.615849E - 03	7.942662E+01	1.018440E - 01	8.059496E - 01	1.208931E+01	8.388678E - 01
2.872621E - 03	7.897696E+01	2.911087E - 01	8.301874E - 01	1.167286E+01	8.059376E - 01
4.488468E - 03	7.839719E+01	6.246443E - 01	8.547842E - 01	1.127043E+01	7.818841E - 01
6.463394E - 03	7.769019E+01	1.106429E+00	8.797401E - 01	1.088116E+01	7.651276E - 01
8.797400E - 03	7.686208E+01	1.702216E+00	9.050550E - 01	1.050414E+01	7.541518E - 01
1.149048E - 02	7.592210E+01	2.345336E+00	9.307291E - 01	1.013852E+01	7.475398E - 01
1.454264E - 02	7.488181E+01	2.953115E+00	9.567622E - 01	9.783477E+00	7.439994E - 01
1.795388E - 02	7.375354E+01	3.447997E+00	9.831544E - 01	9.438272E+00	7.423819E - 01
2.172419E - 02	7.254953E+01	3.776672E+00	1.009905E+00	9.102670E+00	7.416903E - 01
2.585359E - 02	7.128024E+01	3.921700E+00	1.037016E+00	8.775148E+00	7.410842E - 01
3.034205E - 02	6.995482E+01	3.903108E+00	1.064485E+00	8.456559E+00	7.398757E - 01
3.518960E - 02	6.858095E+01	3.770697E+00	1.092314E+00	8.145374E+00	7.375247E - 01
4.039623E - 02	6.716582E+01	3.590264E+00	1.120502E+00	7.842150E+00	7.336282E - 01
4.596193E - 02	6.571672E+01	3.427987E+00	1.149048E+00	7.546689E+00	7.279083E - 01
5.188671E - 02	6.424118E+01	3.336886E+00	1.177954E+00	7.258442E+00	7.201982E - 01
5.817057E - 02	6.274623E+01	3.348099E+00	1.207218E+00	6.977685E+00	7.104285E - 01
6.481349E - 02	6.123758E+01	3.467852E+00	1.236842E+00	6.703907E+00	6.986120E - 01
7.181549E - 02	5.971925E+01	3.679714E+00	1.266826E+00	6.437011E+00	6.848291E - 01
7.917660E - 02	5.819449E+01	3.950636E+00	1.297168E+00	6.177300E+00	6.692151E - 01
8.689672E - 02	5.666658E+01	4.238889E+00	1.327868E+00	5.924675E+00	6.519468E - 01
9.497601E - 02	5.513939E+01	4.502131E+00	1.358929E+00	5.678624E+00	6.332301E - 01
1.034143E - 01	5.361725E+01	4.704193E+00	1.390347E+00	5.439438E+00	6.132909E - 01
1.122117E - 01	5.210452E+01	4.819793E+00	1.422127E+00	5.206187E+00	5.923653E - 01
1.213682E - 01	5.060513E+01	4.836835E+00	1.454264E+00	4.979956E+00	5.706936E - 01
1.308838E - 01	4.912201E+01	4.756457E+00	1.486760E+00	4.759803E+00	5.485108E - 01
1.407584E - 01	4.765753E+01	4.591271E+00	1.519616E+00	4.545595E+00	5.260445E - 01
1.509921E - 01	4.621297E+01	4.362354E+00	1.552831E+00	4.337594E+00	5.035093E - 01
1.615849E - 01	4.478915E+01	4.095672E+00	1.586404E+00	4.135666E+00	4.811041E - 01
1.725367E - 01	4.338692E+01	3.818452E+00	1.620337E+00	3.938876E+00	4.590089E - 01
1.838477E - 01	4.200688E+01	3.555980E+00	1.654629E+00	3.748298E+00	4.373867E - 01
1.955177E - 01	4.065034E+01	3.329148E+00	1.689281E+00	3.562612E+00	4.163786E - 01
2.075468E - 01	3.931870E+01	3.152840E+00	1.724290E+00	3.382913E+00	3.961076E - 01
2.199350E - 01	3.801347E+01	3.035258E+00	1.759659E+00	3.208296E+00	3.766759E - 01
2.326822E - 01	3.673593E+01	2.978043E+00	1.795387E+00	3.038673E+00	3.581675E - 01
2.457886E - 01	3.548677E+01	2.977012E+00	1.831474E+00	2.874374E+00	3.406486E - 01
2.592540E - 01	3.426633E+01	3.023387E+00	1.867921E+00	2.714931E+00	3.241690E - 01
2.730785E - 01	3.307468E+01	3.105218E+00	1.904727E+00	2.560292E+00	3.087636E - 01
2.872620E - 01	3.191202E+01	3.208861E+00	1.941891E+00	2.411223E+00	2.944540E - 01
3.018047E - 01	3.077875E+01	3.320361E+00	1.979415E+00	2.266489E+00	2.812484E - 01
3.167064E - 01	2.967569E+01	3.426620E+00	2.017298E+00	2.126872E+00	2.691458E - 01
3.319672E - 01	2.860406E+01	3.516275E+00	2.055539E+00	1.992360E+00	2.581354E - 01
3.475870E - 01	2.756508E+01	3.580298E+00	2.094140E+00	1.862540E+00	2.481986E - 01
3.635660E - 01	2.656006E+01	3.612288E+00	2.133100E+00	1.737807E+00	2.393109E - 01
3.799040E - 01	2.559012E+01	3.608530E+00	2.172419E+00	1.617755E+00	2.314413E - 01
3.966011E - 01	2.465602E+01	3.567827E+00	2.212097E+00	1.502377E+00	2.245566E - 01
4.136574E - 01	2.375812E+01	3.491211E+00	2.252134E+00	1.392059E+00	2.186189E - 01
4.310726E - 01	2.289639E+01	3.381541E+00	2.292530E+00	1.286793E+00	2.135891E - 01
4.488469E - 01	2.207028E+01	3.243056E+00	2.333285E+00	1.186159E+00	2.094263E - 01
4.669804E - 01	2.127885E+01	3.080945E+00	2.374400E+00	1.090129E+00	2.060894E - 01
4.854729E - 01	2.052086E+01	2.900920E+00	2.415874E+00	9.990780E - 01	2.035367E - 01
5.043244E - 01	1.979471E+01	2.708840E+00	2.457706E+00	9.125640E - 01	2.017274E - 01
5.235351E - 01	1.909866E+01	2.510418E+00	2.499898E+00	8.309545E - 01	2.006212E - 01
5.431048E - 01	1.843085E+01	2.310951E+00	2.542448E+00	7.542141E - 01	2.001790E - 01
5.630336E - 01	1.778943E+01	2.115169E+00	2.585359E+00	6.814886E - 01	2.003642E - 01
5.833215E - 01	1.717264E+01	1.927092E+00	2.628627E+00	6.139581E - 01	2.011397E - 01
6.039685E - 01	1.657877E+01	1.749979E+00	2.672255E+00	5.507779E - 01	2.024723E - 01
6.249745E - 01	1.600642E+01	1.586312E+00	2.716242E+00	4.919158E - 01	2.043294E - 01
6.463397E - 01	1.545427E+01	1.437805E+00	2.760588E+00	4.377546E - 01	2.066804E - 01
6.680638E - 01	1.492135E+01	1.305474E+00	2.805293E+00	3.878777E - 01	2.094973E - 01
6.901471E - 01	1.440681E+01	1.189692E+00	2.850357E+00	3.426808E - 01	2.127529E - 01
7.125894E - 01	1.391000E+01	1.090287E+00	2.895781E+00	3.017602E - 01	2.164226E - 01
7.353909E - 01	1.343034E+01	1.006624E+00	2.941564E+00	2.651362E - 01	2.204829E - 01

Z=40

Niobium

Atomic Number	41
Lattice	bcc
Lattice constant a_o	6.20 a.u.
Pressure at a_o	-49.4 Kbar
Total energy (solid)	-7502.913 Ry
Total energy (atom)	-7502.361 Ry
Zero point energy	0.002 Ry
Cohesive energy	0.551 Ry
Bulk modulus	1.68 Mbar
Fermi energy	0.678 Ry
Interstitial electrons (per atom)	1.315
χ/χ_o	1.72.
$N(E_f)$	1.40 states/eV-atom
ρ_o	4.787×10^4 electrons/Bohr^3
$(d\rho(r,E)/dE)_{E=E_f,r=0}$	12.747 electrons/Bohr^3-Ry

Symmetry point	Band energies in Rydbergs (relative to muffin tin zero)			
Γ	0.291	0.701	0.701	0.701
H	0.376	0.376		
N	0.405	0.550		
P	0.601	0.601	0.601	

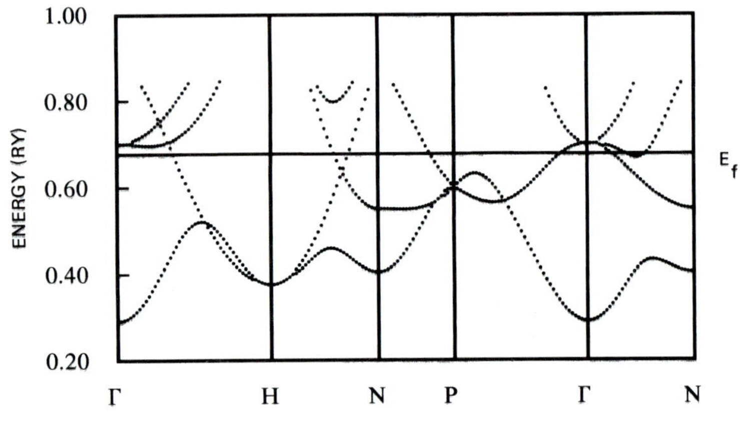

Z=41

Moruzzi
Janak
Williams

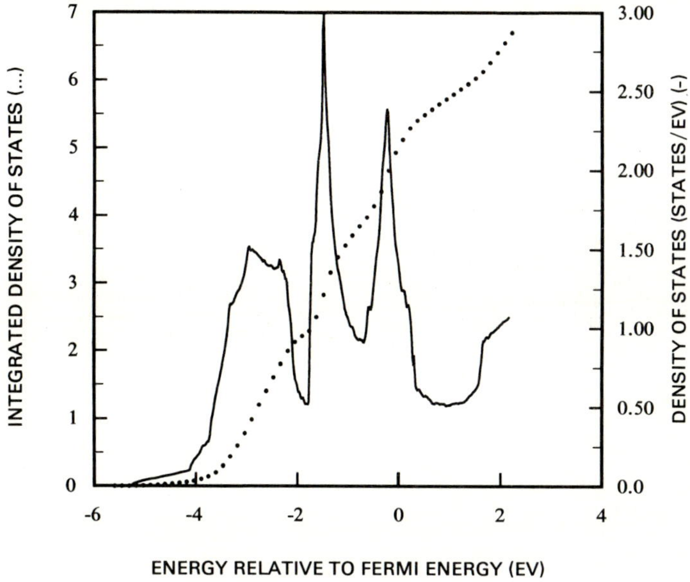

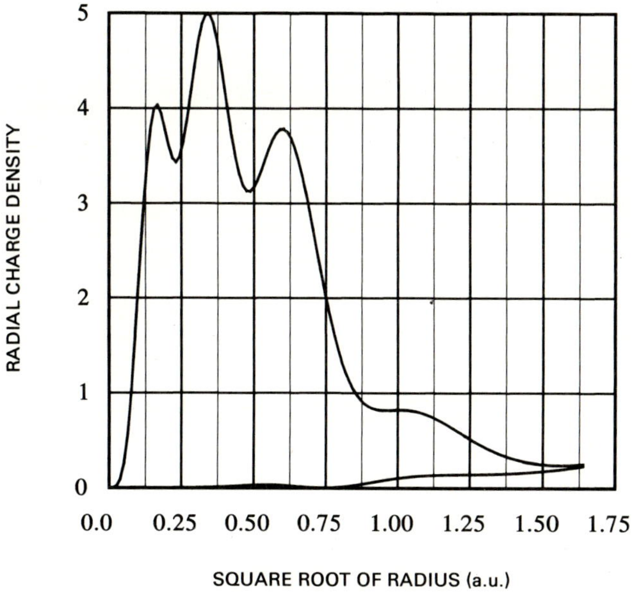

Moruzzi
Janak
Williams

Z=41

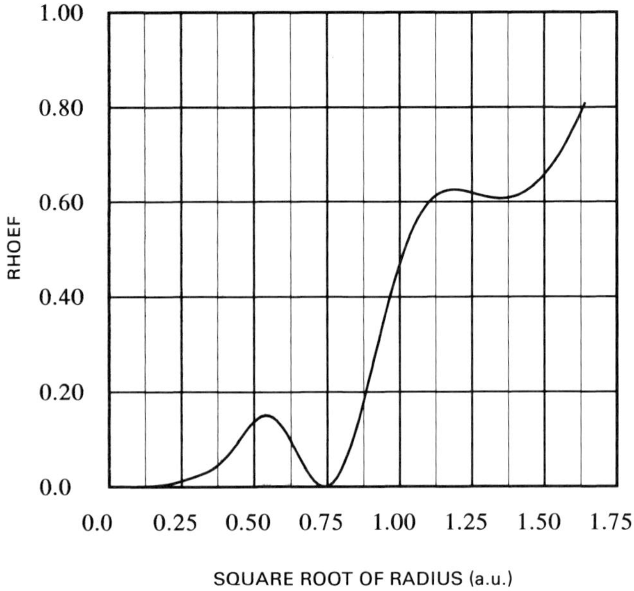

RHOEF=$r^2\partial\rho(E,r)/\partial E$ (at $E=E_f$)

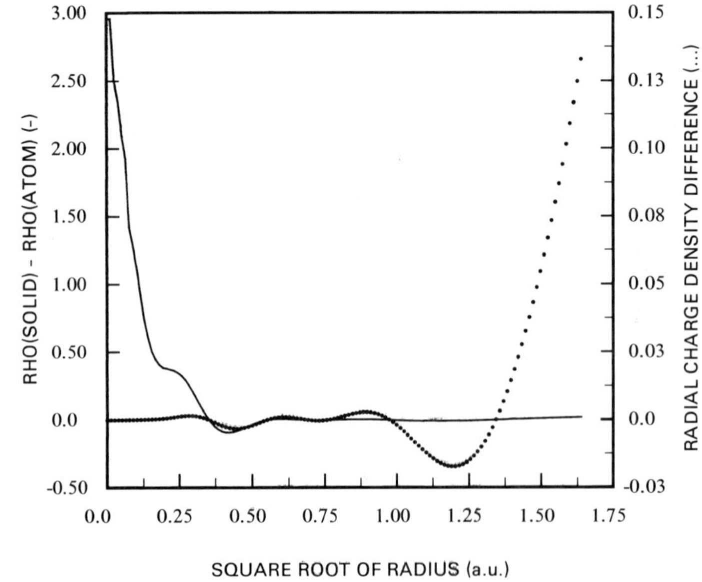

Moruzzi
Janak
Williams

Z=41

Niobium

r	2z(r)	$r^2\rho$	r	2z(r)	$r^2\rho$
1.638597E - 04	8.194008E+01	1.268158E - 03	6.923074E - 01	1.464109E+01	1.117633E+00
6.554390E - 04	8.175978E+01	1.948895E - 02	7.137730E - 01	1.415141E+01	1.039589E+00
1.474738E - 03	8.145787E+01	9.225351E - 02	7.355664E - 01	1.367758E+01	9.762229E - 01
2.621756E - 03	8.103282E+01	2.654087E - 01	7.576874E - 01	1.321915E+01	9.261724E - 01
4.096493E - 03	8.048459E+01	5.742509E - 01	7.801362E - 01	1.277549E+01	8.879144E - 01
5.898949E - 03	7.981537E+01	1.027514E+00	8.029127E - 01	1.234593E+01	8.598377E - 01
8.029126E - 03	7.903017E+01	1.599706E+00	8.260170E - 01	1.192962E+01	8.403077E - 01
1.048702E - 02	7.813680E+01	2.234214E+00	8.494489E - 01	1.152576E+01	8.277202E - 01
1.327264E - 02	7.714534E+01	2.855999E+00	8.732086E - 01	1.113349E+01	8.205448E - 01
1.638597E - 02	7.606694E+01	3.389614E+00	8.972960E - 01	1.075202E+01	8.173577E - 01
1.982703E - 02	7.491286E+01	3.776937E+00	9.217110E - 01	1.038060E+01	8.168672E - 01
2.359580E - 02	7.369312E+01	3.989714E+00	9.464539E - 01	1.001860E+01	8.179288E - 01
2.769230E - 02	7.241656E+01	4.034028E+00	9.715244E - 01	9.665500E+00	8.195540E - 01
3.211651E - 02	7.109065E+01	3.946331E+00	9.969227E - 01	9.320886E+00	8.209130E - 01
3.686844E - 02	6.972200E+01	3.783192E+00	1.022648E+00	8.984766E+00	8.213310E - 01
4.194809E - 02	6.831721E+01	3.607828E+00	1.048702E+00	8.655902E+00	8.202835E - 01
4.735547E - 02	6.688307E+01	3.477074E+00	1.075084E+00	8.335020E+00	8.173822E - 01
5.309056E - 02	6.542635E+01	3.431424E+00	1.101792E+00	8.021985E+00	8.123662E - 01
5.915337E - 02	6.395300E+01	3.489717E+00	1.128829E+00	7.716277E+00	8.050870E - 01
6.554389E - 02	6.246759E+01	3.648606E+00	1.156194E+00	7.417813E+00	7.954924E - 01
7.226211E - 02	6.097348E+01	3.886061E+00	1.183887E+00	7.126529E+00	7.836140E - 01
7.930809E - 02	5.947350E+01	4.167501E+00	1.211906E+00	6.842771E+00	7.695511E - 01
8.668178E - 02	5.797087E+01	4.452954E+00	1.240254E+00	6.566057E+00	7.534572E - 01
9.438318E - 02	5.646935E+01	4.703904E+00	1.268929E+00	6.296729E+00	7.355272E - 01
1.024123E - 01	5.497292E+01	4.888809E+00	1.297933E+00	6.034290E+00	7.159854E - 01
1.107692E - 01	5.348549E+01	4.986723E+00	1.327264E+00	5.778646E+00	6.950755E - 01
1.194537E - 01	5.201060E+01	4.988857E+00	1.356922E+00	5.530116E+00	6.730516E - 01
1.284660E - 01	5.055090E+01	4.898293E+00	1.386909E+00	5.288170E+00	6.501698E - 01
1.378060E - 01	4.910841E+01	4.728197E+00	1.417222E+00	5.053109E+00	6.266820E - 01
1.474738E - 01	4.768431E+01	4.499063E+00	1.447865E+00	4.823980E+00	6.028304E - 01
1.574692E - 01	4.627953E+01	4.235532E+00	1.478834E+00	4.601487E+00	5.788445E - 01
1.677924E - 01	4.489470E+01	3.963162E+00	1.510131E+00	4.385510E+00	5.549374E - 01
1.784433E - 01	4.353084E+01	3.705632E+00	1.541756E+00	4.175504E+00	5.313024E - 01
1.894218E - 01	4.218890E+01	3.482558E+00	1.573709E+00	3.971347E+00	5.081132E - 01
2.007282E - 01	4.087035E+01	3.308065E+00	1.605989E+00	3.773339E+00	4.855238E - 01
2.123622E - 01	3.957660E+01	3.190149E+00	1.638597E+00	3.580958E+00	4.636661E - 01
2.243240E - 01	3.830882E+01	3.130732E+00	1.671533E+00	3.394096E+00	4.426525E - 01
2.366135E - 01	3.706783E+01	3.126296E+00	1.704797E+00	3.212664E+00	4.225758E - 01
2.492306E - 01	3.585396E+01	3.168935E+00	1.738387E+00	3.037416E+00	4.035100E - 01
2.621756E - 01	3.466733E+01	3.247604E+00	1.772306E+00	2.867023E+00	3.855122E - 01
2.754482E - 01	3.350813E+01	3.349476E+00	1.806553E+00	2.702257E+00	3.686233E - 01
2.890486E - 01	3.237679E+01	3.461193E+00	1.841127E+00	2.542650E+00	3.528699E - 01
3.029767E - 01	3.127399E+01	3.569972E+00	1.876030E+00	2.388157E+00	3.382666E - 01
3.172324E - 01	3.020078E+01	3.664484E+00	1.911260E+00	2.239154E+00	3.248158E - 01
3.318160E - 01	2.915828E+01	3.735451E+00	1.946817E+00	2.095199E+00	3.125104E - 01
3.467272E - 01	2.814769E+01	3.776000E+00	1.982702E+00	1.956681E+00	3.013355E - 01
3.619661E - 01	2.717007E+01	3.781785E+00	2.018915E+00	1.823155E+00	2.912689E - 01
3.775328E - 01	2.622618E+01	3.750905E+00	2.055456E+00	1.694614E+00	2.822826E - 01
3.934272E - 01	2.531650E+01	3.683667E+00	2.092324E+00	1.571446E+00	2.743445E - 01
4.096493E - 01	2.444102E+01	3.582263E+00	2.129520E+00	1.453219E+00	2.674189E - 01
4.261992E - 01	2.359940E+01	3.450369E+00	2.167045E+00	1.339911E+00	2.614679E - 01
4.430767E - 01	2.279089E+01	3.292747E+00	2.204896E+00	1.231917E+00	2.564512E - 01
4.602820E - 01	2.201442E+01	3.114823E+00	2.243075E+00	1.128798E+00	2.523292E - 01
4.778150E - 01	2.126855E+01	2.922339E+00	2.281583E+00	1.030527E+00	2.490606E - 01
4.956757E - 01	2.055174E+01	2.721030E+00	2.320417E+00	9.374948E - 01	2.466050E - 01
5.138642E - 01	1.986230E+01	2.516351E+00	2.359580E+00	8.492557E - 01	2.449229E - 01
5.323803E - 01	1.919853E+01	2.313285E+00	2.399070E+00	7.661974E - 01	2.439758E - 01
5.512242E - 01	1.855870E+01	2.116196E+00	2.438888E+00	6.878852E - 01	2.437267E - 01
5.703958E - 01	1.794125E+01	1.928734E+00	2.479033E+00	6.142913E - 01	2.441403E - 01
5.898951E - 01	1.734477E+01	1.753793E+00	2.519507E+00	5.453985E - 01	2.451833E - 01
6.097221E - 01	1.676804E+01	1.593513E+00	2.560308E+00	4.816096E - 01	2.468234E - 01
6.298769E - 01	1.621002E+01	1.449306E+00	2.601437E+00	4.229219E - 01	2.490324E - 01
6.503593E - 01	1.566995E+01	1.321913E+00	2.642894E+00	3.685002E - 01	2.517821E - 01
6.711695E - 01	1.514711E+01	1.211479E+00	2.684678E+00	3.191853E - 01	2.550468E - 01

Z=41

Molybdenum

Atomic Number	42
Lattice	bcc
Lattice constant a_o	5.89 a.u.
Pressure at a_o	30.8 Kbar
Total energy (solid)	-7946.585 Ry
Total energy (atom)	-7946.087 Ry
Zero point energy	0.003 Ry
Cohesive energy	0.495 Ry
Bulk modulus	2.51 Mbar
Fermi energy	0.805 Ry
Interstitial electrons (per atom)	1.398
χ/χ_o	1.23
$N(E_f)$	0.65 states/eV-atom
ρ_o	5.153×10^4 electrons/Bohr^3
$(d\rho(r,E)/dE)_{E=E_f,r=0}$	2.8234 electrons/Bohr^3-Ry

Symmetry point	Band energies in Rydbergs (relative to muffin tin zero)					
Γ	0.327	0.701	0.701	0.701	0.904	0.904
H	0.369	0.369				
N	0.408	0.545	0.921			
P	0.614	0.614	0.614			

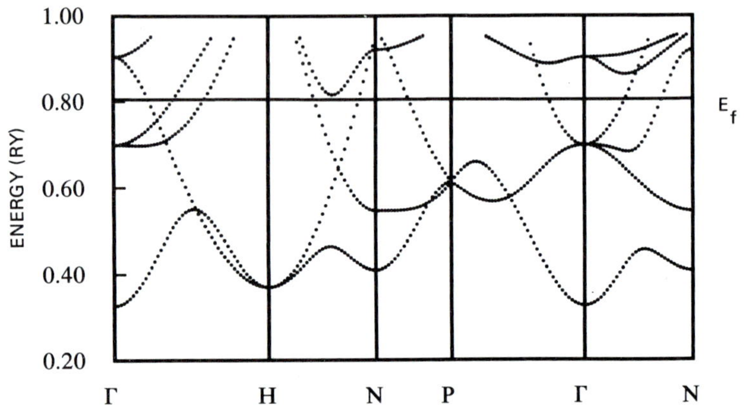

Z=42

Moruzzi
Janak
Williams

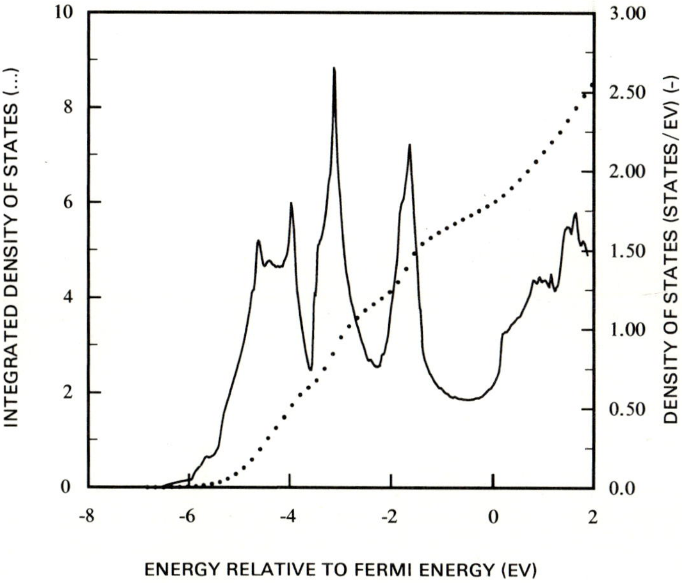

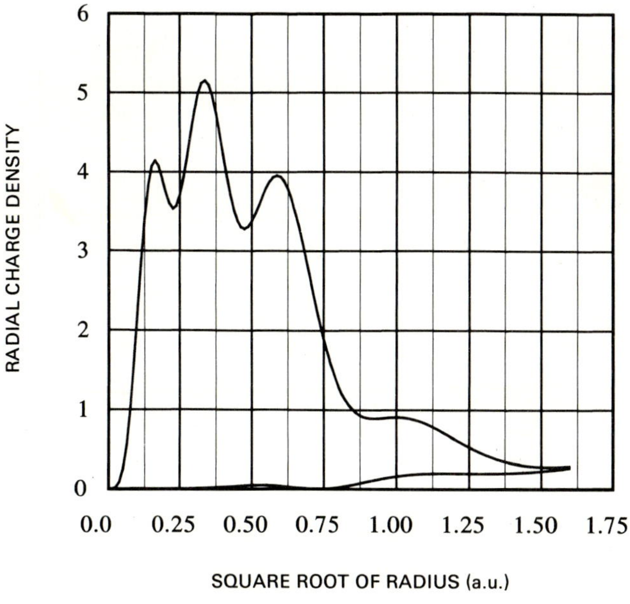

Moruzzi
Janak
Williams

Z=42

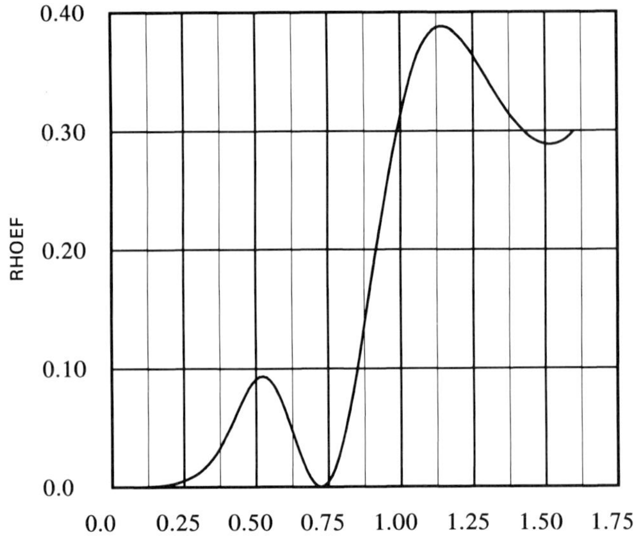

SQUARE ROOT OF RADIUS (a.u.)

$RHOEF = r^2 \partial\rho(E,r)/\partial E$ (at $E=E_f$)

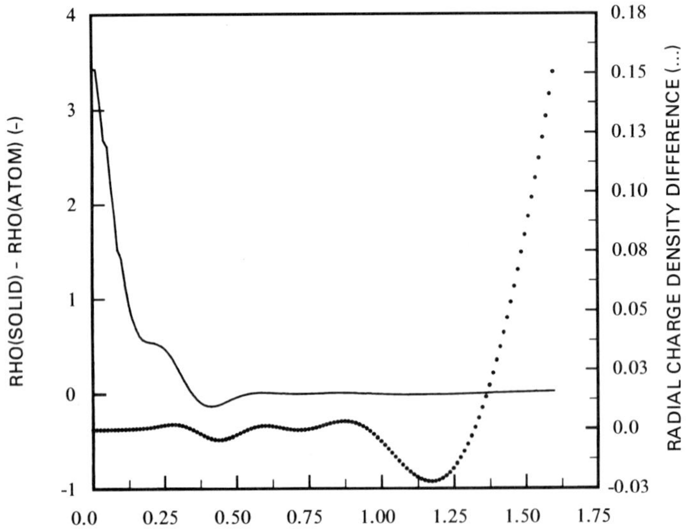

SQUARE ROOT OF RADIUS (a.u.)

Moruzzi
Janak
Williams

Molybdenum

r	2z(r)	$r^2\rho$	r	2z(r)	$r^2\rho$
1.556668E - 04	8.394106E+01	1.232463E - 03	6.576921E - 01	1.576884E+01	1.216542E+00
6.226671E - 04	8.376370E+01	1.896089E - 02	6.780845E - 01	1.524816E+01	1.131065E+00
1.401001E - 03	8.346678E+01	8.991551E - 02	6.987882E - 01	1.474356E+01	1.061678E+00
2.490669E - 03	8.304878E+01	2.593352E - 01	7.198032E - 01	1.425464E+01	1.007011E+00
3.891669E - 03	8.250961E+01	5.629277E - 01	7.411295E - 01	1.378076E+01	9.654827E - 01
5.604003E - 03	8.185118E+01	1.011230E+00	7.627672E - 01	1.332131E+01	9.353843E - 01
7.627670E - 03	8.107813E+01	1.581656E+00	7.847162E - 01	1.287548E+01	9.149550E - 01
9.962674E - 03	8.019777E+01	2.220725E+00	8.069766E - 01	1.244248E+01	9.024424E - 01
1.260901E - 02	7.921959E+01	2.855569E+00	8.295482E - 01	1.202143E+01	8.961557E - 01
1.556668E - 02	7.815425E+01	3.410962E+00	8.524313E - 01	1.161149E+01	8.945067E - 01
1.883568E - 02	7.701274E+01	3.826659E+00	8.756256E - 01	1.121199E+01	8.960398E - 01
2.241601E - 02	7.580475E+01	4.070291E+00	8.991313E - 01	1.082222E+01	8.994545E - 01
2.630768E - 02	7.453908E+01	4.142779E+00	9.229484E - 01	1.044168E+01	9.036173E - 01
3.051069E - 02	7.322311E+01	4.075736E+00	9.470767E - 01	1.006998E+01	9.075674E - 01
3.502502E - 02	7.186334E+01	3.922359E+00	9.715164E - 01	9.706772E+00	9.105158E - 01
3.985070E - 02	7.046613E+01	3.744802E+00	9.962674E - 01	9.351843E+00	9.118399E - 01
4.498770E - 02	6.903799E+01	3.601313E+00	1.021329E+00	9.005378E+00	9.110724E - 01
5.043603E - 02	6.758554E+01	3.535880E+00	1.046703E+00	8.666222E+00	9.078913E - 01
5.619571E - 02	6.611481E+01	3.572090E+00	1.072388E+00	8.335613E+00	9.021006E - 01
6.226671E - 02	6.463051E+01	3.711588E+00	1.098384E+00	8.012644E+00	8.936203E - 01
6.864899E - 02	6.313618E+01	3.936664E+00	1.124692E+00	7.697268E+00	8.824643E - 01
7.534271E - 02	6.163460E+01	4.215639E+00	1.151311E+00	7.389878E+00	8.687279E - 01
8.234769E - 02	6.012886E+01	4.509672E+00	1.178242E+00	7.090002E+00	8.525712E - 01
8.966404E - 02	5.862238E+01	4.779601E+00	1.205483E+00	6.798014E+00	8.342035E - 01
9.729171E - 02	5.711906E+01	4.991704E+00	1.233036E+00	6.513851E+00	8.138718E - 01
1.052307E - 01	5.562283E+01	5.121807E+00	1.260900E+00	6.237005E+00	7.918463E - 01
1.134810E - 01	5.413724E+01	5.157408E+00	1.289076E+00	5.967820E+00	7.684119E - 01
1.220427E - 01	5.266521E+01	5.097934E+00	1.317563E+00	5.705761E+00	7.438575E - 01
1.309158E - 01	5.120889E+01	4.953433E+00	1.346361E+00	5.451146E+00	7.184700E - 01
1.401001E - 01	4.976985E+01	4.742172E+00	1.375471E+00	5.203427E+00	6.925266E - 01
1.495957E - 01	4.834908E+01	4.487589E+00	1.404892E+00	4.962907E+00	6.662910E - 01
1.594028E - 01	4.694727E+01	4.215140E+00	1.434625E+00	4.729020E+00	6.400094E - 01
1.695211E - 01	4.556544E+01	3.949373E+00	1.464668E+00	4.502070E+00	6.139067E - 01
1.799508E - 01	4.420467E+01	3.711528E+00	1.495024E+00	4.281493E+00	5.881866E - 01
1.906918E - 01	4.286612E+01	3.517858E+00	1.525690E+00	4.067595E+00	5.630301E - 01
2.017441E - 01	4.155130E+01	3.378693E+00	1.556667E+00	3.860248E+00	5.385937E - 01
2.131078E - 01	4.026152E+01	3.298237E+00	1.587956E+00	3.658904E+00	5.150121E - 01
2.247828E - 01	3.899773E+01	3.274981E+00	1.619556E+00	3.463884E+00	4.923978E - 01
2.367691E - 01	3.776047E+01	3.302574E+00	1.651468E+00	3.274666E+00	4.708416E - 01
2.490668E - 01	3.654994E+01	3.371015E+00	1.683691E+00	3.091577E+00	4.504150E - 01
2.616758E - 01	3.536638E+01	3.467937E+00	1.716226E+00	2.914111E+00	4.311721E - 01
2.745962E - 01	3.421010E+01	3.579935E+00	1.749071E+00	2.742626E+00	4.131500E - 01
2.878278E - 01	3.308168E+01	3.693727E+00	1.782228E+00	2.576630E+00	3.963711E - 01
3.013709E - 01	3.198203E+01	3.797127E+00	1.815697E+00	2.416493E+00	3.808445E - 01
3.152252E - 01	3.091222E+01	3.879780E+00	1.849477E+00	2.261744E+00	3.665689E - 01
3.293909E - 01	2.987340E+01	3.933638E+00	1.883568E+00	2.112764E+00	3.535325E - 01
3.438679E - 01	2.886665E+01	3.953170E+00	1.917970E+00	1.969522E+00	3.417159E - 01
3.586562E - 01	2.789282E+01	3.935385E+00	1.952683E+00	1.831559E+00	3.310930E - 01
3.737559E - 01	2.695245E+01	3.879663E+00	1.987709E+00	1.698836E+00	3.216318E - 01
3.891670E - 01	2.604573E+01	3.787471E+00	2.023045E+00	1.572200E+00	3.132967E - 01
4.048893E - 01	2.517242E+01	3.661984E+00	2.058693E+00	1.450320E+00	3.060493E - 01
4.209229E - 01	2.433192E+01	3.507689E+00	2.094652E+00	1.334042E+00	2.998485E - 01
4.372680E - 01	2.352322E+01	3.329947E+00	2.130922E+00	1.223337E+00	2.946522E - 01
4.539243E - 01	2.274510E+01	3.134629E+00	2.167504E+00	1.117750E+00	2.904186E - 01
4.708920E - 01	2.199602E+01	2.927738E+00	2.204397E+00	1.017690E+00	2.871048E - 01
4.881710E - 01	2.127443E+01	2.715121E+00	2.241601E+00	9.231291E - 01	2.846696E - 01
5.057613E - 01	2.057857E+01	2.502228E+00	2.279117E+00	8.336187E - 01	2.830731E - 01
5.236630E - 01	1.990680E+01	2.293921E+00	2.316944E+00	7.491403E - 01	2.822764E - 01
5.418761E - 01	1.925752E+01	2.094370E+00	2.355083E+00	6.701118E - 01	2.822423E - 01
5.604004E - 01	1.862939E+01	1.906963E+00	2.393532E+00	5.965187E - 01	2.829360E - 01
5.792361E - 01	1.802106E+01	1.734291E+00	2.432293E+00	5.279312E - 01	2.843253E - 01
5.983831E - 01	1.743159E+01	1.578169E+00	2.471365E+00	4.647769E - 01	2.863796E - 01
6.178414E - 01	1.686011E+01	1.439668E+00	2.510749E+00	4.066431E - 01	2.890713E - 01
6.376112E - 01	1.630602E+01	1.319185E+00	2.550445E+00	3.535492E - 01	2.923746E - 01

Z=42

Moruzzi
Janak
Williams

Technetium

Atomic Number	43
Lattice	fcc
Lattice constant a_o	7.28 a.u.
Pressure at a_o	-0.76 Kbar
Total energy (solid)	-8404.992 Ry
Total energy (atom)	-8404.435 Ry
Zero point energy	0.003 Ry
Cohesive energy	0.554 Ry
Bulk modulus	2.93 Mbar
Fermi energy	0.743 Ry
Interstitial electrons (per atom)	1.074
χ/χ_o	1.57
N(E)	1.22 states/eV-atom
ρ_o	5.537×10^4 electrons/Bohr^3
$(d\rho(r,E)/dE)_{E=E_f,r=0}$	5.6760 electrons/Bohr^3-Ry

Symmetry point	Band energies in Rydbergs (relative to muffin tin zero)					
Γ	0.262	0.612	0.612	0.612	0.796	0.796
X	0.296	0.358				
L	0.363	0.594	0.594	0.859		
W	0.406	0.552	0.552	0.763		
K	0.373	0.458	0.704	0.814		

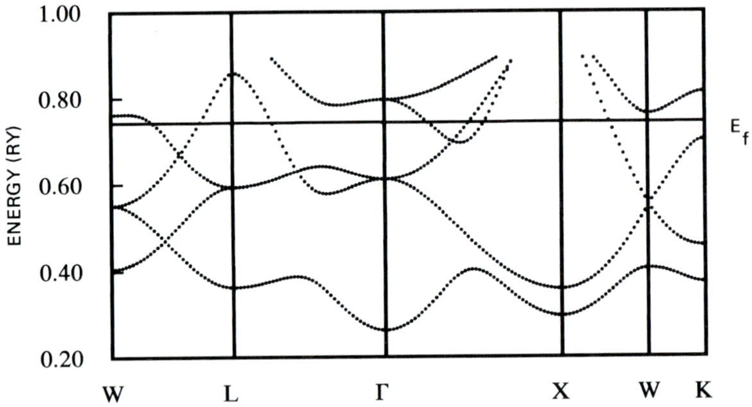

Z=43

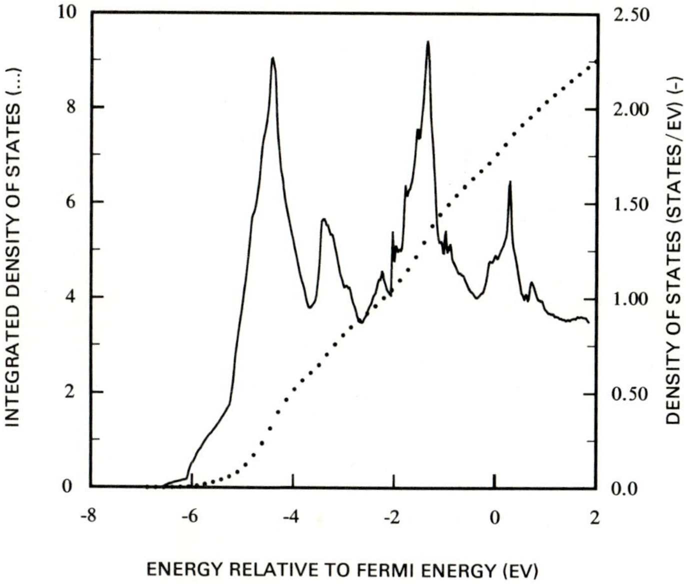

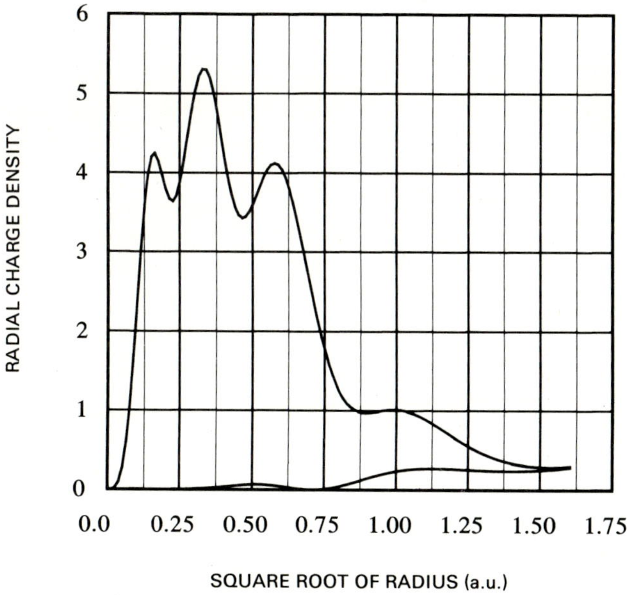

Moruzzi
Janak
Williams

Z=43

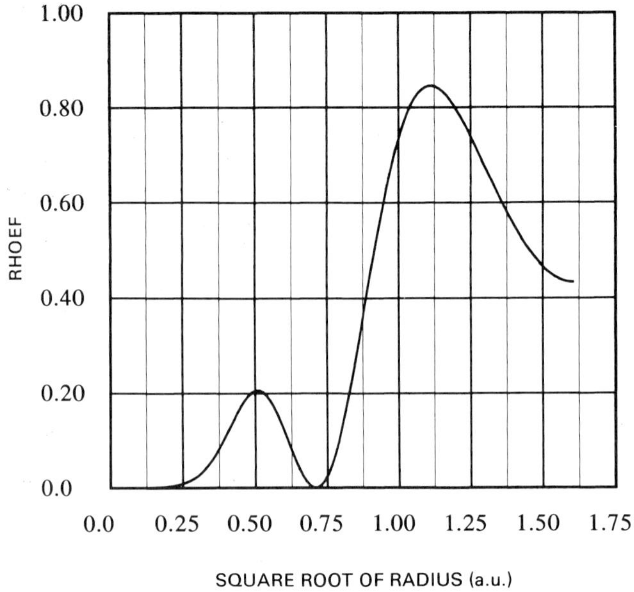

RHOEF=$r^2 \partial \rho(E,r)/\partial E$ (at $E=E_f$)

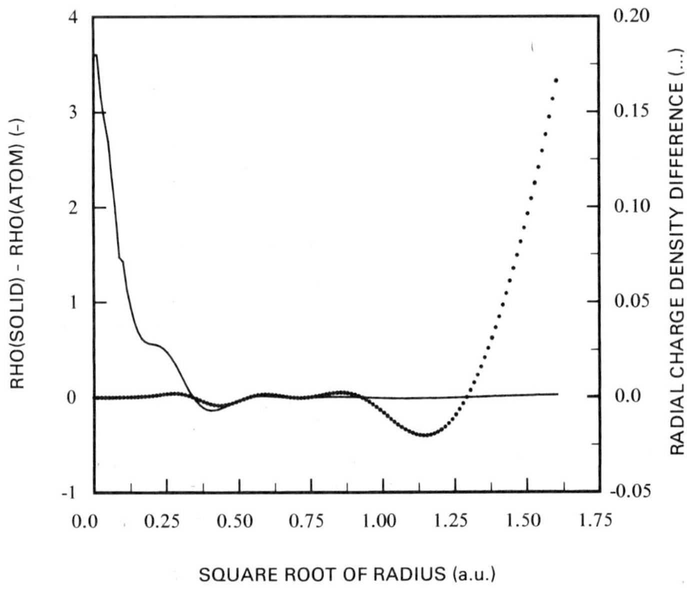

Moruzzi
Janak
Williams

Z=43

Technetium

r	2z(r)	r²ρ	r	2z(r)	r²ρ
1.570964E-04	8.593846E+01	1.348197E-03	6.637323E-01	1.591171E+01	1.155112E+00
6.283857E-04	8.575328E+01	2.071440E-02	6.843120E-01	1.537765E+01	1.092733E+00
1.413868E-03	8.544327E+01	9.801817E-02	7.052058E-01	1.485993E+01	1.045868E+00
2.513543E-03	8.500684E+01	2.818474E-01	7.264138E-01	1.435783E+01	1.012571E+00
3.927410E-03	8.444400E+01	6.094136E-01	7.479361E-01	1.387057E+01	9.907998E-01
5.655471E-03	8.375697E+01	1.089551E+00	7.697725E-01	1.339728E+01	9.784900E-01
7.697724E-03	8.295090E+01	1.694674E+00	7.919230E-01	1.293701E+01	9.736310E-01
1.005417E-02	8.203378E+01	2.364293E+00	8.143879E-01	1.248885E+01	9.743197E-01
1.272481E-02	8.101578E+01	3.018682E+00	8.371668E-01	1.205207E+01	9.788027E-01
1.570964E-02	7.990819E+01	3.578164E+00	8.602600E-01	1.162588E+01	9.855074E-01
1.900867E-02	7.872238E+01	3.981964E+00	8.836673E-01	1.120982E+01	9.930603E-01
2.262188E-02	7.746873E+01	4.201379E+00	9.073889E-01	1.080342E+01	1.000297E+00
2.654929E-02	7.615585E+01	4.244247E+00	9.314247E-01	1.040641E+01	1.006261E+00
3.079090E-02	7.479143E+01	4.150459E+00	9.557746E-01	1.001858E+01	1.010201E+00
3.534669E-02	7.338225E+01	3.980731E+00	9.804388E-01	9.639767E+00	1.011554E+00
4.021668E-02	7.193506E+01	3.802159E+00	1.005417E+00	9.270044E+00	1.009934E+00
4.540087E-02	7.045683E+01	3.674357E+00	1.030709E+00	8.909204E+00	1.005116E+00
5.089924E-02	6.895447E+01	3.638977E+00	1.056315E+00	8.557484E+00	9.970057E-01
5.671181E-02	6.743398E+01	3.714309E+00	1.082237E+00	8.213982E+00	9.856322E-01
6.283855E-02	6.589990E+01	3.895003E+00	1.108472E+00	7.880019E+00	9.711170E-01
6.927949E-02	6.435565E+01	4.156102E+00	1.135021E+00	7.554685E+00	9.536625E-01
7.603467E-02	6.280429E+01	4.459778E+00	1.161884E+00	7.238383E+00	9.335274E-01
8.310395E-02	6.124919E+01	4.763104E+00	1.189062E+00	6.930609E+00	9.110138E-01
9.048754E-02	5.969424E+01	5.025402E+00	1.216555E+00	6.631286E+00	8.864517E-01
9.818524E-02	5.814371E+01	5.214049E+00	1.244360E+00	6.341205E+00	8.601841E-01
1.061972E-01	5.660161E+01	5.308191E+00	1.272481E+00	6.058918E+00	8.325593E-01
1.145233E-01	5.507149E+01	5.300228E+00	1.300915E+00	5.785627E+00	8.039179E-01
1.231636E-01	5.355612E+01	5.195250E+00	1.329663E+00	5.520297E+00	7.745896E-01
1.321180E-01	5.205750E+01	5.008929E+00	1.358727E+00	5.262775E+00	7.448837E-01
1.413867E-01	5.057703E+01	4.764405E+00	1.388103E+00	5.013339E+00	7.150880E-01
1.509696E-01	4.911569E+01	4.488711E+00	1.417795E+00	4.771378E+00	6.854637E-01
1.608667E-01	4.767439E+01	4.209291E+00	1.447801E+00	4.537169E+00	6.562456E-01
1.710780E-01	4.625436E+01	3.950992E+00	1.478120E+00	4.310540E+00	6.276392E-01
1.816034E-01	4.485695E+01	3.733779E+00	1.508754E+00	4.090883E+00	5.998234E-01
1.924431E-01	4.348360E+01	3.571340E+00	1.539701E+00	3.878483E+00	5.729476E-01
2.035969E-01	4.213583E+01	3.470537E+00	1.570964E+00	3.672743E+00	5.471372E-01
2.150650E-01	4.081485E+01	3.431626E+00	1.602540E+00	3.473958E+00	5.224915E-01
2.268472E-01	3.952138E+01	3.449095E+00	1.634431E+00	3.281552E+00	4.990878E-01
2.389436E-01	3.825580E+01	3.512927E+00	1.666636E+00	3.095841E+00	4.769820E-01
2.513543E-01	3.701831E+01	3.610053E+00	1.699155E+00	2.916270E+00	4.562124E-01
2.640790E-01	3.580933E+01	3.725866E+00	1.731988E+00	2.743171E+00	4.367996E-01
2.771181E-01	3.462938E+01	3.845627E+00	1.765135E+00	2.576448E+00	4.187506E-01
2.904713E-01	3.347945E+01	3.955621E+00	1.798596E+00	2.415573E+00	4.020598E-01
3.041387E-01	3.236063E+01	4.044083E+00	1.832373E+00	2.260468E+00	3.867112E-01
3.181202E-01	3.127414E+01	4.101780E+00	1.866462E+00	2.111948E+00	3.726798E-01
3.324160E-01	3.022112E+01	4.122335E+00	1.900867E+00	1.968619E+00	3.599346E-01
3.470260E-01	2.920241E+01	4.102253E+00	1.935585E+00	1.831310E+00	3.484387E-01
3.619502E-01	2.821860E+01	4.040762E+00	1.970617E+00	1.699975E+00	3.381506E-01
3.771885E-01	2.726982E+01	3.939499E+00	2.005963E+00	1.574556E+00	3.290268E-01
3.927410E-01	2.635579E+01	3.802066E+00	2.041625E+00	1.454127E+00	3.210225E-01
4.086078E-01	2.547577E+01	3.633564E+00	2.077600E+00	1.339979E+00	3.140910E-01
4.247887E-01	2.462866E+01	3.440105E+00	2.113889E+00	1.231188E+00	3.081868E-01
4.412838E-01	2.381306E+01	3.228354E+00	2.150493E+00	1.127715E+00	3.032649E-01
4.580932E-01	2.302734E+01	3.005098E+00	2.187410E+00	1.029985E+00	2.992810E-01
4.752166E-01	2.226974E+01	2.776916E+00	2.224642E+00	9.379628E-01	2.961939E-01
4.926544E-01	2.153842E+01	2.549883E+00	2.262188E+00	8.511941E-01	2.939637E-01
5.104063E-01	2.083167E+01	2.329376E+00	2.300048E+00	7.701037E-01	2.925534E-01
5.284724E-01	2.014783E+01	2.119925E+00	2.338223E+00	6.942378E-01	2.919292E-01
5.468526E-01	1.948550E+01	1.925147E+00	2.376711E+00	6.240447E-01	2.920595E-01
5.655471E-01	1.884348E+01	1.747717E+00	2.415514E+00	5.590845E-01	2.929166E-01
5.845557E-01	1.822083E+01	1.589408E+00	2.454631E+00	4.998187E-01	2.944756E-01
6.038786E-01	1.761679E+01	1.451142E+00	2.494062E+00	4.458302E-01	2.967151E-01
6.235157E-01	1.703087E+01	1.333084E+00	2.533808E+00	3.975988E-01	2.996166E-01
6.434669E-01	1.646262E+01	1.234748E+00	2.573868E+00	3.547260E-01	3.031651E-01

Z=43

Ruthenium

Atomic Number	44
Lattice	fcc
Lattice constant a_o	7.20 a.u.
Pressure at a_o	-8.3 Kbar
Total energy (solid)	-8878.368 Ry
Total energy (atom)	-8877.807 Ry
Zero point energy	0.004 Ry
Cohesive energy	0.557 Ry
Bulk modulus	2.89 Mbar
Fermi energy	0.714 Ry
Interstitial electrons (per atom)	1.014
χ/χ_o	1.53
$N(E_f)$	1.13 states/eV-atom
ρ_o	5.940×10^4 electrons/Bohr^3
$(d\rho(r,E)/dE)_{E=E_f,r=0}$	4.5348 electrons/Bohr^3-Ry

Symmetry point	Band energies in Rydbergs (relative to muffin tin zero)					
Γ	0.212	0.534	0.534	0.534	0.699	0.699
X	0.249	0.302	0.826			
L	0.307	0.519	0.519	0.810	0.832	0.832
W	0.351	0.481	0.481	0.677		
K	0.320	0.395	0.624	0.719	0.819	

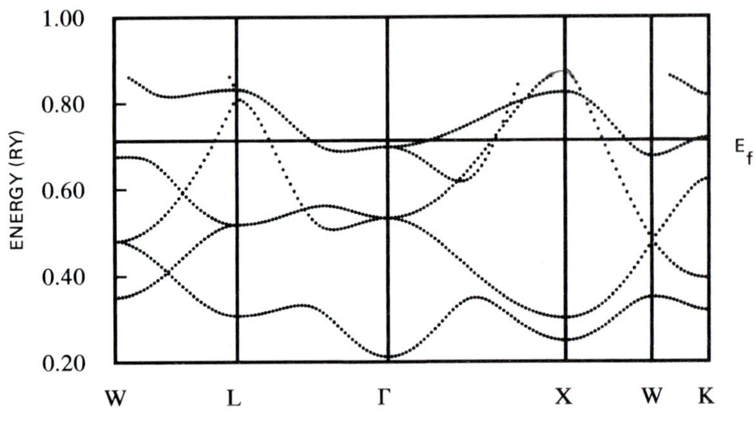

Z=44

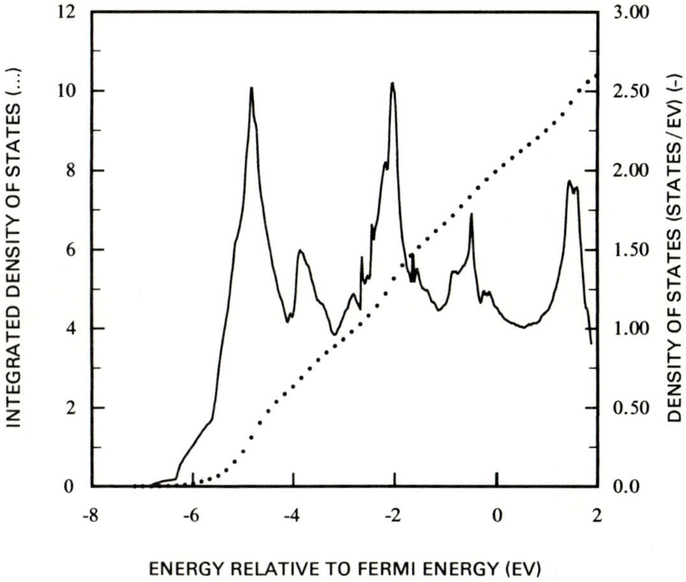
12
10
8
6
4
2
0
3.00
2.50
2.00
1.50
1.00
0.50
0.0
-8
-6
-4
-2
0
2
INTEGRATED DENSITY OF STATES (....)
DENSITY OF STATES (STATES/EV) (-)
ENERGY RELATIVE TO FERMI ENERGY (EV)

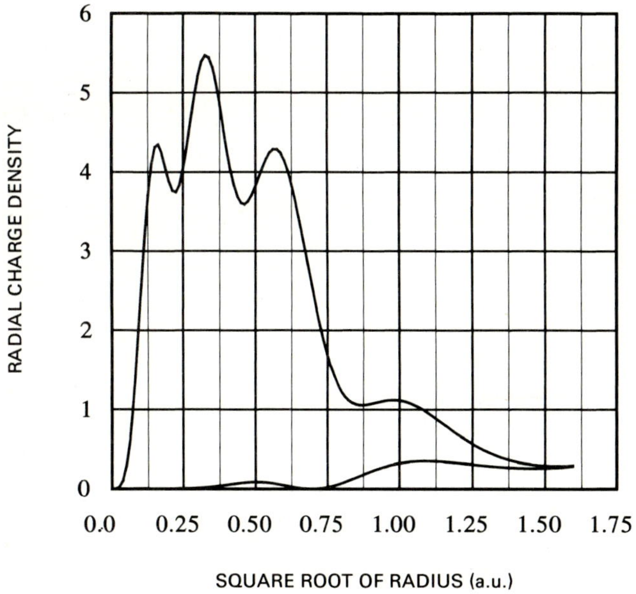
6
5
4
3
2
1
0
0.0
0.25
0.50
0.75
1.00
1.25
1.50
1.75
RADIAL CHARGE DENSITY
SQUARE ROOT OF RADIUS (a.u.)

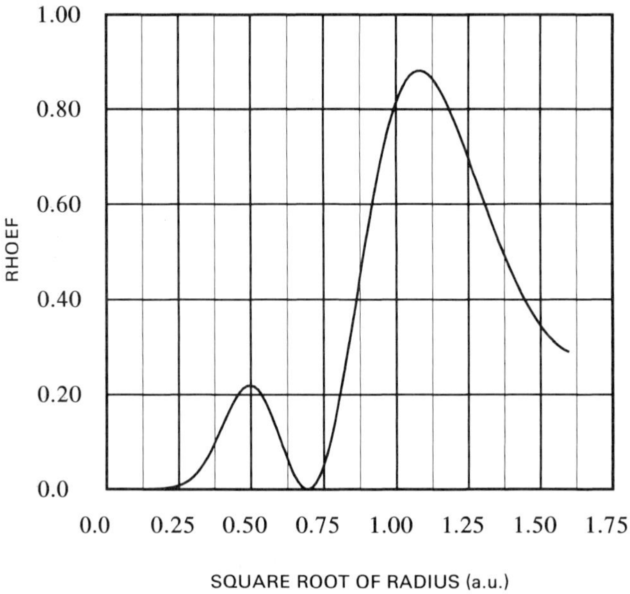

RHOEF=$r^2\partial\rho(E,r)/\partial E$ (at $E=E_f$)

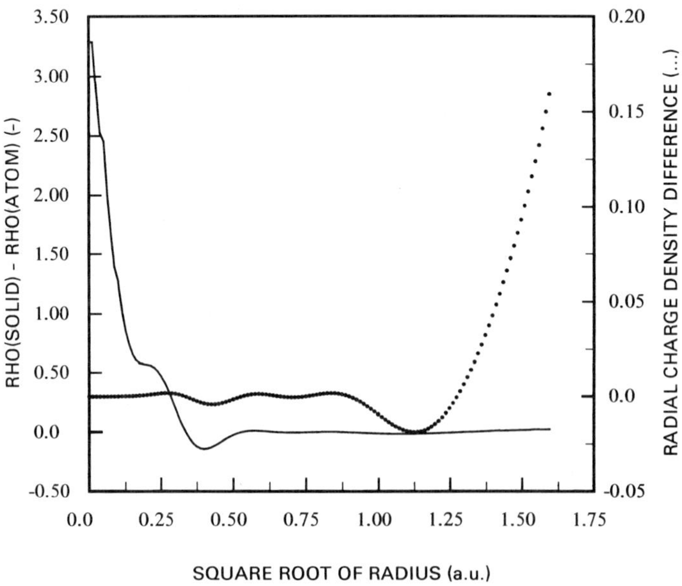

Z=44

Ruthenium

r	2z(r)	r²ρ	r	2z(r)	r²ρ
1.553701E-04	8.793709E+01	1.414427E-03	6.564386E-01	1.636133E+01	1.169405E+00
6.214804E-04	8.774779E+01	2.172141E-02	6.767921E-01	1.580716E+01	1.120930E+00
1.398331E-03	8.743077E+01	1.026998E-01	6.974564E-01	1.526940E+01	1.087596E+00
2.485922E-03	8.698456E+01	2.949747E-01	7.184313E-01	1.474723E+01	1.067107E+00
3.884252E-03	8.640918E+01	6.368692E-01	7.397170E-01	1.423968E+01	1.057134E+00
5.593322E-03	8.570697E+01	1.136617E+00	7.613134E-01	1.374579E+01	1.055391E+00
7.613134E-03	8.488329E+01	1.764204E+00	7.832206E-01	1.326463E+01	1.059713E+00
9.943686E-03	8.394633E+01	2.455448E+00	8.054386E-01	1.279541E+01	1.068103E+00
1.258498E-02	8.290657E+01	3.126797E+00	8.279672E-01	1.233748E+01	1.078765E+00
1.553701E-02	8.177563E+01	3.695780E+00	8.508067E-01	1.189025E+01	1.090138E+00
1.879978E-02	8.056491E+01	4.100700E+00	8.739568E-01	1.145338E+01	1.100897E+00
2.237329E-02	7.928516E+01	4.314107E+00	8.974177E-01	1.102659E+01	1.109961E+00
2.625754E-02	7.794485E+01	4.346849E+00	9.211893E-01	1.060973E+01	1.116482E+00
3.045254E-02	7.655186E+01	4.242751E+00	9.452717E-01	1.020273E+01	1.119836E+00
3.495827E-02	7.511311E+01	4.066348E+00	9.696648E-01	9.805580E+00	1.119599E+00
3.977475E-02	7.363554E+01	3.887503E+00	9.943686E-01	9.418255E+00	1.115531E+00
4.490196E-02	7.212633E+01	3.766969E+00	1.019382E+00	9.041154E+00	1.107547E+00
5.033991E-02	7.059247E+01	3.745776E+00	1.044708E+00	8.673387E+00	1.095698E+00
5.608860E-02	6.903996E+01	3.840078E+00	1.070344E+00	8.315911E+00	1.080145E+00
6.214804E-02	6.747333E+01	4.041442E+00	1.096291E+00	7.968261E+00	1.061133E+00
6.851816E-02	6.589601E+01	4.321532E+00	1.122548E+00	7.630412E+00	1.038975E+00
7.519913E-02	6.431111E+01	4.639526E+00	1.149117E+00	7.302302E+00	1.014029E+00
8.219075E-02	6.272231E+01	4.950365E+00	1.175996E+00	6.983854E+00	9.866816E-01
8.949316E-02	6.113367E+01	5.212420E+00	1.203186E+00	6.674958E+00	9.573333E-01
9.710628E-02	5.954955E+01	5.393346E+00	1.230686E+00	6.375937E+00	9.263868E-01
1.050302E-01	5.797409E+01	5.473660E+00	1.258497E+00	6.086181E+00	8.942357E-01
1.132648E-01	5.641078E+01	5.447944E+00	1.286619E+00	5.805513E+00	8.612569E-01
1.218101E-01	5.486246E+01	5.323922E+00	1.315052E+00	5.533750E+00	8.278036E-01
1.306662E-01	5.333110E+01	5.119944E+00	1.343796E+00	5.270689E+00	7.942022E-01
1.398330E-01	5.181796E+01	4.861514E+00	1.372849E+00	5.016581E+00	7.607465E-01
1.493106E-01	5.032431E+01	4.577442E+00	1.402215E+00	4.770298E+00	7.277015E-01
1.590990E-01	4.885115E+01	4.296195E+00	1.431890E+00	4.532546E+00	6.952967E-01
1.691980E-01	4.739973E+01	4.042814E+00	1.461877E+00	4.302650E+00	6.637309E-01
1.796078E-01	4.597174E+01	3.836683E+00	1.492174E+00	4.080408E+00	6.331706E-01
1.903284E-01	4.456868E+01	3.690243E+00	1.522782E+00	3.865614E+00	6.037533E-01
2.013596E-01	4.319202E+01	3.608644E+00	1.553700E+00	3.658535E+00	5.755878E-01
2.127016E-01	4.184291E+01	3.590170E+00	1.584929E+00	3.458528E+00	5.487585E-01
2.243544E-01	4.052199E+01	3.627296E+00	1.616470E+00	3.264959E+00	5.233265E-01
2.363179E-01	3.922963E+01	3.708144E+00	1.648321E+00	3.078579E+00	4.993309E-01
2.485921E-01	3.796613E+01	3.818078E+00	1.680483E+00	2.898779E+00	4.767951E-01
2.611771E-01	3.673198E+01	3.941343E+00	1.712955E+00	2.725877E+00	4.557245E-01
2.740728E-01	3.552800E+01	4.062508E+00	1.745738E+00	2.558826E+00	4.361133E-01
2.872793E-01	3.435516E+01	4.167654E+00	1.778831E+00	2.398423E+00	4.179416E-01
3.007965E-01	3.321474E+01	4.245238E+00	1.812237E+00	2.243644E+00	4.011835E-01
3.146244E-01	3.210791E+01	4.286647E+00	1.845952E+00	2.095305E+00	3.858033E-01
3.287631E-01	3.103568E+01	4.286398E+00	1.879978E+00	1.952857E+00	3.717614E-01
3.432125E-01	2.999879E+01	4.242117E+00	1.914314E+00	1.816682E+00	3.590125E-01
3.579727E-01	2.899762E+01	4.154255E+00	1.948962E+00	1.685785E+00	3.475103E-01
3.730436E-01	2.803198E+01	4.025695E+00	1.983920E+00	1.561019E+00	3.372054E-01
3.884252E-01	2.710135E+01	3.861240E+00	2.019189E+00	1.441862E+00	3.280484E-01
4.041176E-01	2.620473E+01	3.667089E+00	2.054770E+00	1.327810E+00	3.199912E-01
4.201207E-01	2.534073E+01	3.450284E+00	2.090659E+00	1.220189E+00	3.129842E-01
4.364346E-01	2.450777E+01	3.218238E+00	2.126861E+00	1.117577E+00	3.069829E-01
4.530592E-01	2.370403E+01	2.978293E+00	2.163373E+00	1.020411E+00	3.019423E-01
4.699945E-01	2.292766E+01	2.737366E+00	2.200195E+00	9.291125E-01	2.978208E-01
4.872406E-01	2.217683E+01	2.501689E+00	2.237329E+00	8.432044E-01	2.945804E-01
5.047974E-01	2.144987E+01	2.276603E+00	2.274774E+00	7.622170E-01	2.921863E-01
5.226650E-01	2.074521E+01	2.066451E+00	2.312528E+00	6.875155E-01	2.906052E-01
5.408433E-01	2.006160E+01	1.874526E+00	2.350594E+00	6.177323E-01	2.898101E-01
5.593323E-01	1.939809E+01	1.703089E+00	2.388970E+00	5.533286E-01	2.897761E-01
5.781321E-01	1.875389E+01	1.553420E+00	2.427657E+00	4.947943E-01	2.904819E-01
5.972427E-01	1.812848E+01	1.425909E+00	2.466655E+00	4.416912E-01	2.919104E-01
6.166639E-01	1.752153E+01	1.320170E+00	2.505964E+00	3.936101E-01	2.940489E-01
6.363959E-01	1.693259E+01	1.235181E+00	2.545584E+00	3.515154E-01	2.968862E-01

Rhodium

Atomic Number	45
Lattice	fcc
Lattice constant a_o	7.24 a.u.
Pressure at a_o	-7.65 Kbar
Total energy (solid)	-9366.870 Ry
Total energy (atom)	-9366.418 Ry
Zero point energy	0.003 Ry
Cohesive energy	0.449 Ry
Bulk modulus	2.61 Mbar
Fermi energy	0.643 Ry
Interstitial electrons (per atom)	0.911
χ/χ_o	1.79
$N(E_f)$	1.35 states/eV-atom
ρ_o	6.362×10^4 electrons/Bohr^3
$(d\rho(r,E)/dE)_{E=E_f,r=0}$	4.5546 electrons/Bohr^3-Ry

Symmetry point	Band energies in Rydbergs (relative to muffin tin zero)					
Γ	0.138	0.430	0.430	0.430	0.568	0.568
X	0.188	0.231	0.674	0.712	0.712	
L	0.234	0.418	0.418	0.680	0.680	0.720
W	0.278	0.386	0.386	0.556	0.712	
K	0.251	0.311	0.510	0.589	0.670	

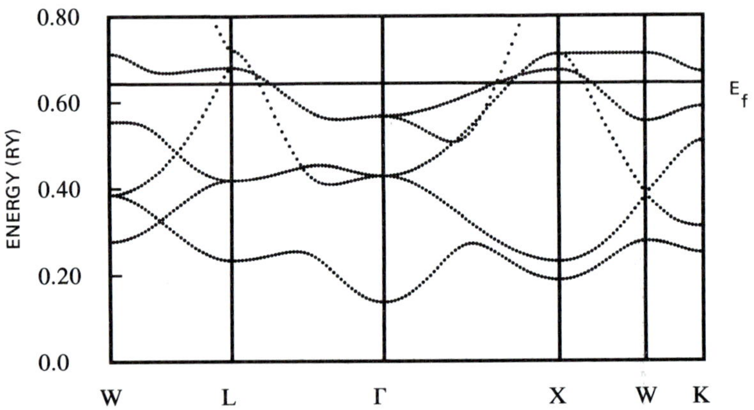

Z=45

Moruzzi
Janak
Williams

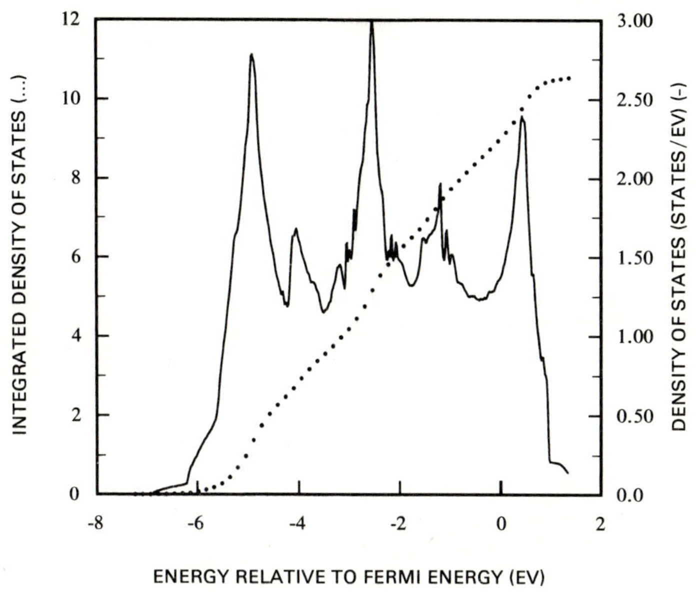

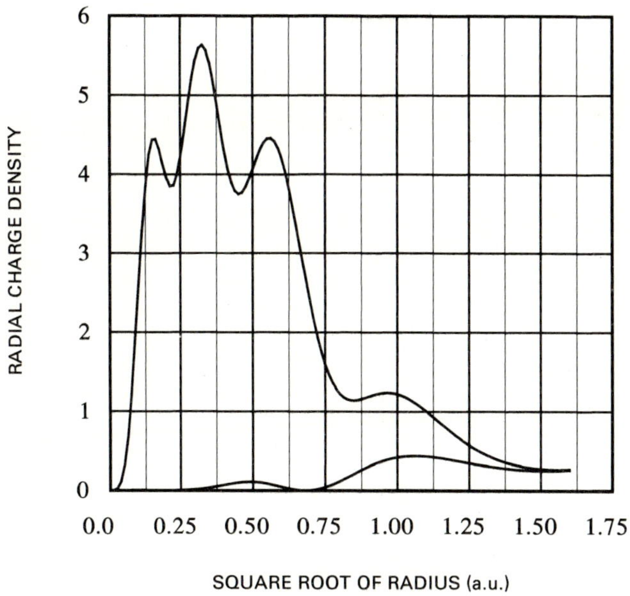

Moruzzi
Janak
Williams

Z=45

Rhodium

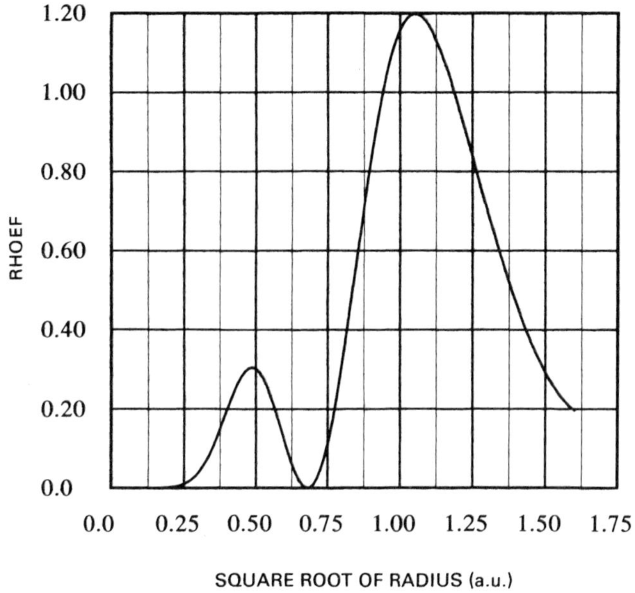

$RHOEF=r^2\partial\rho(E,r)/\partial E$ (at $E=E_f$)

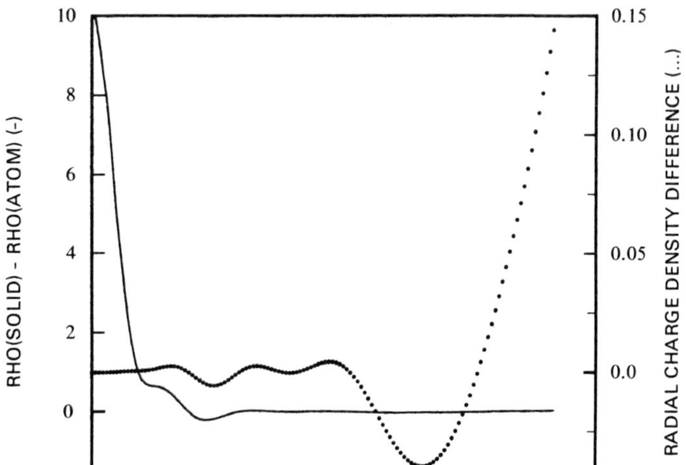

Moruzzi
Janak
Williams

Z=45

Rhodium

r	2z(r)	$r^2\rho$	r	2z(r)	$r^2\rho$
1.562333E - 04	8.993465E+01	1.531162E - 03	6.600855E - 01	1.648479E+01	1.181516E+00
6.249330E - 04	8.973796E+01	2.348673E - 02	6.805521E - 01	1.591438E+01	1.155431E+00
1.406099E - 03	8.940865E+01	1.108310E - 01	7.013311E - 01	1.536008E+01	1.142543E+00
2.499732E - 03	8.894510E+01	3.174654E - 01	7.224226E - 01	1.482082E+01	1.140179E+00
3.905831E - 03	8.834750E+01	6.830437E - 01	7.438265E - 01	1.429570E+01	1.145754E+00
5.624395E - 03	8.761844E+01	1.213862E+00	7.655429E - 01	1.378371E+01	1.156848E+00
7.655427E - 03	8.676379E+01	1.874728E+00	7.875718E - 01	1.328412E+01	1.171258E+00
9.998929E - 03	8.579236E+01	2.594485E+00	8.099132E - 01	1.279636E+01	1.187038E+00
1.265489E - 02	8.471530E+01	3.283052E+00	8.325670E - 01	1.232002E+01	1.202522E+00
1.562332E - 02	8.354474E+01	3.854109E+00	8.555333E - 01	1.185481E+01	1.216330E+00
1.890422E - 02	8.229259E+01	4.246176E+00	8.788121E - 01	1.140057E+01	1.227365E+00
2.249759E - 02	8.096964E+01	4.436093E+00	9.024033E - 01	1.095728E+01	1.234800E+00
2.640342E - 02	7.958478E+01	4.441914E+00	9.263070E - 01	1.052495E+01	1.238062E+00
3.062172E - 02	7.814597E+01	4.315517E+00	9.505231E - 01	1.010362E+01	1.236800E+00
3.515248E - 02	7.666046E+01	4.128235E+00	9.750518E - 01	9.693369E+00	1.230863E+00
3.999572E - 02	7.513559E+01	3.953821E+00	9.998928E - 01	9.294279E+00	1.220267E+00
4.515141E - 02	7.357889E+01	3.853205E+00	1.025046E+00	8.906397E+00	1.205170E+00
5.061958E - 02	7.199753E+01	3.864085E+00	1.050512E+00	8.529748E+00	1.185839E+00
5.640021E - 02	7.039746E+01	3.996710E+00	1.076290E+00	8.164619E+00	1.162625E+00
6.249330E - 02	6.878299E+01	4.235579E+00	1.102382E+00	7.810028E+00	1.135944E+00
6.889886E - 02	6.715764E+01	4.545609E+00	1.128785E+00	7.466833E+00	1.106244E+00
7.561690E - 02	6.552484E+01	4.880754E+00	1.155500E+00	7.134922E+00	1.073995E+00
8.264738E - 02	6.388860E+01	5.193117E+00	1.182529E+00	6.813197E+00	1.039673E+00
8.999032E - 02	6.225339E+01	5.440854E+00	1.209870E+00	6.502430E+00	1.003743E+00
9.764576E - 02	6.062378E+01	5.593883E+00	1.237523E+00	6.202415E+00	9.666520E - 01
1.056136E - 01	5.900397E+01	5.636922E+00	1.265489E+00	5.912448E+00	9.288210E - 01
1.138940E - 01	5.739737E+01	5.569888E+00	1.293767E+00	5.632286E+00	8.906377E - 01
1.224868E - 01	5.580670E+01	5.406137E+00	1.322358E+00	5.361674E+00	8.524538E - 01
1.313921E - 01	5.423380E+01	5.169144E+00	1.351261E+00	5.100811E+00	8.145827E - 01
1.406099E - 01	5.267999E+01	4.888405E+00	1.380477E+00	4.848959E+00	7.773007E - 01
1.501401E - 01	5.114656E+01	4.595171E+00	1.410005E+00	4.606310E+00	7.408434E - 01
1.599829E - 01	4.963475E+01	4.318617E+00	1.439845E+00	4.372122E+00	7.054114E - 01
1.701380E - 01	4.814622E+01	4.082848E+00	1.469998E+00	4.146124E+00	6.711684E - 01
1.806056E - 01	4.668263E+01	3.904897E+00	1.500463E+00	3.928528E+00	6.382472E - 01
1.913857E - 01	4.524580E+01	3.793844E+00	1.531241E+00	3.718608E+00	6.067486E - 01
2.024783E - 01	4.383690E+01	3.750901E+00	1.562332E+00	3.516122E+00	5.767481E - 01
2.138833E - 01	4.245699E+01	3.770311E+00	1.593735E+00	3.321312E+00	5.482949E - 01
2.256008E - 01	4.110652E+01	3.840814E+00	1.625450E+00	3.133482E+00	5.214196E - 01
2.376308E - 01	3.978590E+01	3.947412E+00	1.657478E+00	2.952434E+00	4.961326E - 01
2.499732E - 01	3.849564E+01	4.073211E+00	1.689818E+00	2.778440E+00	4.724287E - 01
2.626281E - 01	3.723643E+01	4.201148E+00	1.722471E+00	2.610848E+00	4.502904E - 01
2.755954E - 01	3.600935E+01	4.315450E+00	1.755437E+00	2.449498E+00	4.296893E - 01
2.888753E - 01	3.481558E+01	4.402742E+00	1.788714E+00	2.294708E+00	4.105881E - 01
3.024676E - 01	3.365642E+01	4.452770E+00	1.822305E+00	2.145864E+00	3.929432E - 01
3.163723E - 01	3.253299E+01	4.458757E+00	1.856207E+00	2.003309E+00	3.767052E - 01
3.305895E - 01	3.144606E+01	4.417439E+00	1.890422E+00	1.866461E+00	3.618236E - 01
3.451192E - 01	3.039601E+01	4.328805E+00	1.924950E+00	1.735207E+00	3.482438E - 01
3.599614E - 01	2.938277E+01	4.195681E+00	1.959789E+00	1.609926E+00	3.359108E - 01
3.751161E - 01	2.840570E+01	4.023157E+00	1.994942E+00	1.490064E+00	3.247706E - 01
3.905831E - 01	2.746384E+01	3.817988E+00	2.030407E+00	1.375546E+00	3.147698E - 01
4.063627E - 01	2.655566E+01	3.587961E+00	2.066184E+00	1.266781E+00	3.058562E - 01
4.224547E - 01	2.567952E+01	3.341336E+00	2.102274E+00	1.163241E+00	2.979808E - 01
4.388592E - 01	2.483348E+01	3.086311E+00	2.138677E+00	1.064875E+00	2.910968E - 01
4.555762E - 01	2.401558E+01	2.830641E+00	2.175391E+00	9.721161E - 01	2.851605E - 01
4.726056E - 01	2.322392E+01	2.581281E+00	2.212419E+00	8.844637E - 01	2.801321E - 01
4.899475E - 01	2.245670E+01	2.344171E+00	2.249759E+00	8.018947E - 01	2.759749E - 01
5.076019E - 01	2.171242E+01	2.124106E+00	2.287411E+00	7.248725E - 01	2.726560E - 01
5.255687E - 01	2.098985E+01	1.924668E+00	2.325376E+00	6.529166E - 01	2.701473E - 01
5.438480E - 01	2.028801E+01	1.748249E+00	2.363652E+00	5.865064E - 01	2.684231E - 01
5.624397E - 01	1.960628E+01	1.596122E+00	2.402242E+00	5.251859E - 01	2.674633E - 01
5.813439E - 01	1.894421E+01	1.468538E+00	2.441144E+00	4.689755E - 01	2.672514E - 01
6.005606E - 01	1.830145E+01	1.364885E+00	2.480359E+00	4.179099E - 01	2.677742E - 01
6.200898E - 01	1.767760E+01	1.283824E+00	2.519886E+00	3.725078E - 01	2.690225E - 01
6.399314E - 01	1.707224E+01	1.223465E+00	2.559726E+00	3.323523E - 01	2.709934E - 01

Z=45

Palladium

Atomic Number	46
Lattice	fcc
Lattice constant a_o	7.42 a.u.
Pressure at a_o	9.81 Kbar
Total energy (solid)	-9870.692 Ry
Total energy (atom)	-9870.419 Ry
Zero point energy	0.002 Ry
Cohesive energy	0.271 Ry
Bulk modulus	1.70 Mbar
Fermi energy	0.501 Ry
Interstitial electrons (per atom)	0.768
χ/χ_o	4.46
$N(E_f)$	2.31 states/eV-atom
ρ_o	6.803×10^4 electrons/Bohr^3
$(d\rho(r,E)/dE)_{E=E_f,r=0}$	7.9059 electrons/Bohr^3-Ry

Symmetry point	Band energies in Rydbergs (relative to muffin tin zero)					
Γ	0.050	0.304	0.304	0.304	0.411	0.411
X	0.116	0.148	0.492	0.521	0.521	
L	0.147	0.297	0.297	0.497	0.497	0.598
W	0.189	0.272	0.272	0.406	0.521	
K	0.167	0.212	0.372	0.430	0.490	

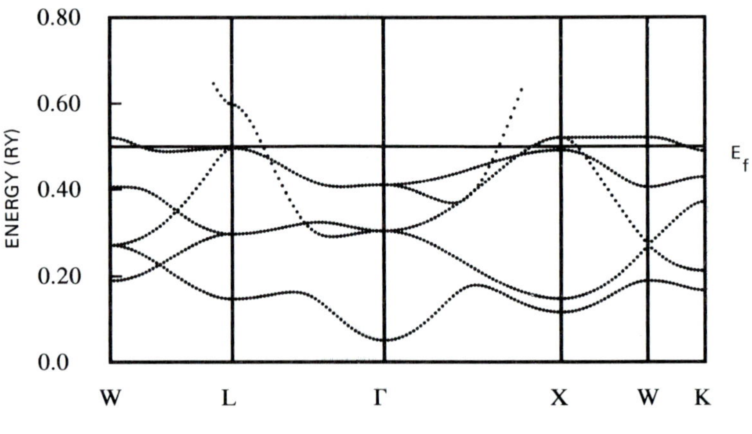

Z=46

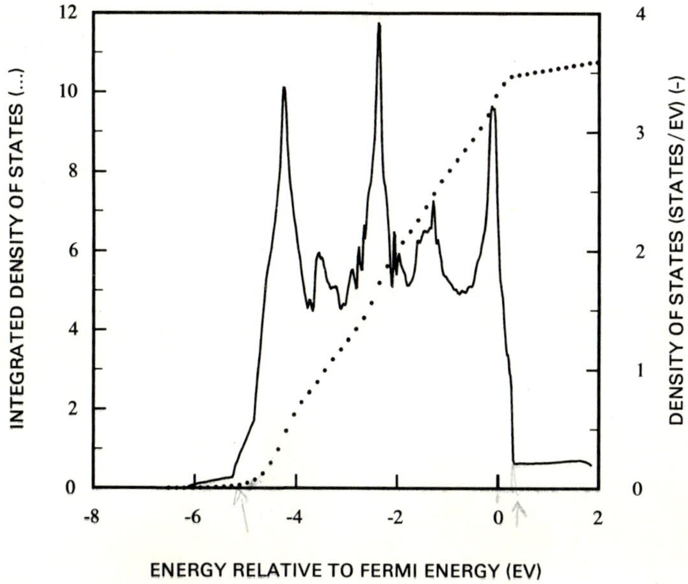

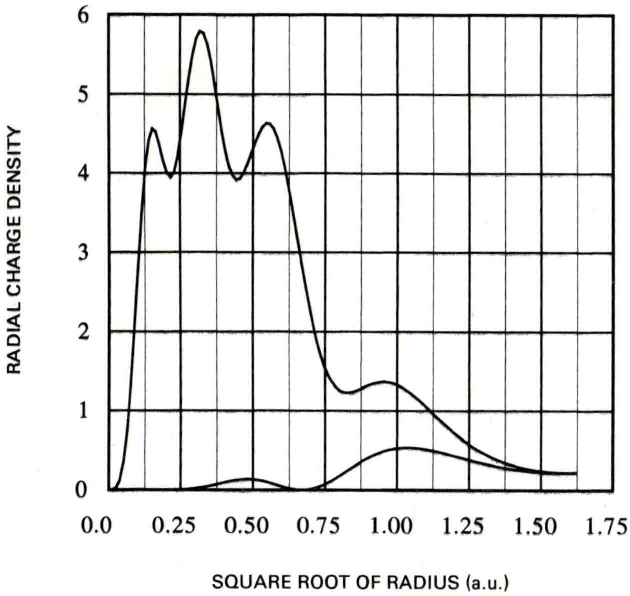

Moruzzi
Janak
Williams

Z=46

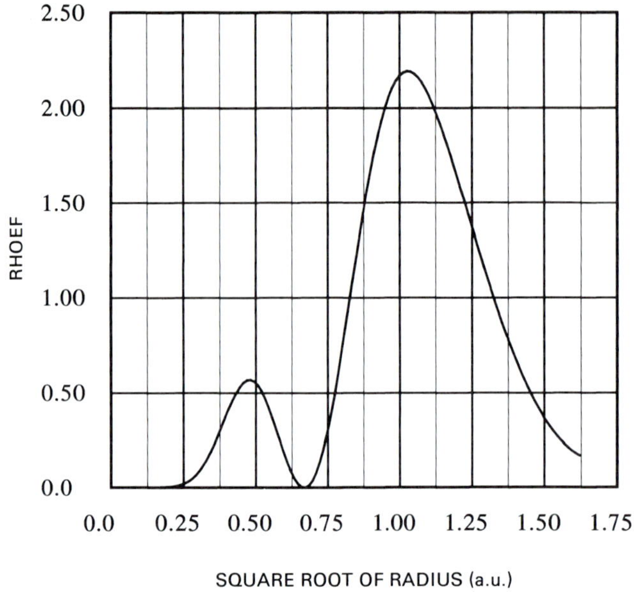

RHOEF=$r^2 \partial\rho(E,r)/\partial E$ (at $E=E_f$)

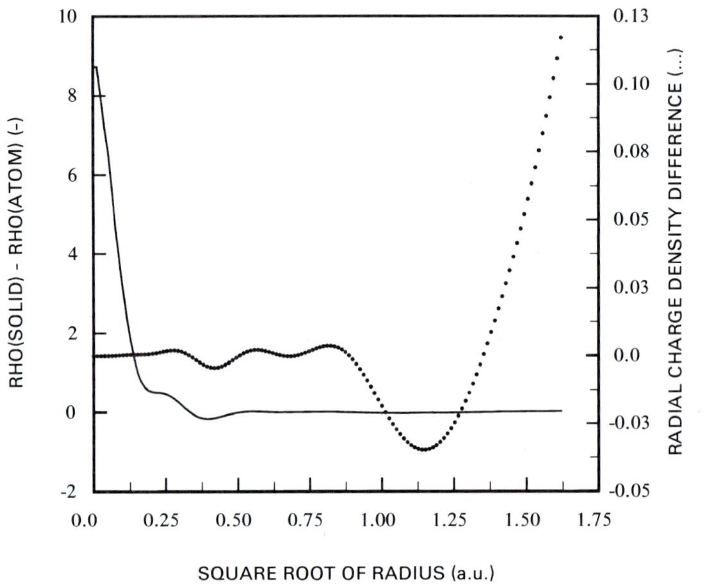

Moruzzi
Janak
Williams

Z=46

Palladium

r	2z(r)	$r^2\rho$	r	2z(r)	$r^2\rho$
1.601175E - 04	9.193085E+01	1.718658E - 03	6.764965E - 01	1.620532E+01	1.227979E+00
6.404701E - 04	9.172275E+01	2.630985E - 02	6.974719E - 01	1.562178E+01	1.229867E+00
1.441058E - 03	9.137424E+01	1.237381E - 01	7.187675E - 01	1.505342E+01	1.240576E+00
2.561880E - 03	9.088370E+01	3.527811E - 01	7.403834E - 01	1.449935E+01	1.257249E+00
4.002936E - 03	9.025151E+01	7.544798E - 01	7.623194E - 01	1.395882E+01	1.277292E+00
5.764227E - 03	8.948079E+01	1.331040E+00	7.845758E - 01	1.343119E+01	1.298429E+00
7.845756E - 03	8.857832E+01	2.038130E+00	8.071524E - 01	1.291616E+01	1.318723E+00
1.024752E - 02	8.755405E+01	2.793160E+00	8.300492E - 01	1.241353E+01	1.336585E+00
1.296952E - 02	8.642014E+01	3.496329E+00	8.532662E - 01	1.192318E+01	1.350775E+00
1.601175E - 02	8.518968E+01	4.056892E+00	8.768035E - 01	1.144520E+01	1.360377E+00
1.937422E - 02	8.387521E+01	4.416052E+00	9.006610E - 01	1.097968E+01	1.364779E+00
2.305692E - 02	8.248813E+01	4.559889E+00	9.248387E - 01	1.052673E+01	1.363635E+00
2.705986E - 02	8.103735E+01	4.519540E+00	9.493367E - 01	1.008647E+01	1.356835E+00
3.138303E - 02	7.953130E+01	4.360130E+00	9.741549E - 01	9.659022E+00	1.344462E+00
3.602644E - 02	7.797772E+01	4.162760E+00	9.992934E - 01	9.244481E+00	1.326761E+00
4.099008E - 02	7.638466E+01	4.005134E+00	1.024752E+00	8.843076E+00	1.304096E+00
4.627396E - 02	7.476022E+01	3.945607E+00	1.050530E+00	8.454464E+00	1.276930E+00
5.187807E - 02	7.311168E+01	4.013768E+00	1.076630E+00	8.078616E+00	1.245780E+00
5.780242E - 02	7.144469E+01	4.208374E+00	1.103049E+00	7.715427E+00	1.211204E+00
6.404698E - 02	6.976347E+01	4.501708E+00	1.129788E+00	7.365230E+00	1.173770E+00
7.061177E - 02	6.807166E+01	4.848242E+00	1.156849E+00	7.026853E+00	1.134048E+00
7.749683E - 02	6.637332E+01	5.195115E+00	1.184229E+00	6.701035E+00	1.092587E+00
8.470213E - 02	6.467305E+01	5.492168E+00	1.211929E+00	6.387024E+00	1.049906E+00
9.222764E - 02	6.297591E+01	5.699882E+00	1.239949E+00	6.085010E+00	1.006487E+00
1.000734E - 01	6.128670E+01	5.794322E+00	1.268291E+00	5.793698E+00	9.627730E - 01
1.082394E - 01	5.960944E+01	5.768878E+00	1.296951E+00	5.513732E+00	9.191548E - 01
1.167256E - 01	5.794751E+01	5.633150E+00	1.325933E+00	5.244276E+00	8.759808E - 01
1.255321E - 01	5.630322E+01	5.409684E+00	1.355234E+00	4.984979E+00	8.335516E - 01
1.346588E - 01	5.467827E+01	5.129434E+00	1.384856E+00	4.735485E+00	7.921210E - 01
1.441057E - 01	5.307417E+01	4.826775E+00	1.414798E+00	4.495928E+00	7.518997E - 01
1.538729E - 01	5.149226E+01	4.534843E+00	1.445060E+00	4.265470E+00	7.130578E - 01
1.639603E - 01	4.993436E+01	4.281745E+00	1.475642E+00	4.043767E+00	6.757287E - 01
1.743680E - 01	4.840234E+01	4.087933E+00	1.506545E+00	3.830489E+00	6.400117E - 01
1.850958E - 01	4.689818E+01	3.964910E+00	1.537768E+00	3.625314E+00	6.059765E - 01
1.961439E - 01	4.542329E+01	3.915127E+00	1.569311E+00	3.428420E+00	5.736651E - 01
2.075123E - 01	4.397882E+01	3.932916E+00	1.601174E+00	3.239025E+00	5.430983E - 01
2.192008E - 01	4.256541E+01	4.006126E+00	1.633358E+00	3.056854E+00	5.142771E - 01
2.312096E - 01	4.118356E+01	4.118189E+00	1.665862E+00	2.881639E+00	4.871858E - 01
2.435387E - 01	3.983376E+01	4.250294E+00	1.698687E+00	2.713135E+00	4.617957E - 01
2.561880E - 01	3.851695E+01	4.383409E+00	1.731831E+00	2.551598E+00	4.380667E - 01
2.691575E - 01	3.723419E+01	4.500025E+00	1.765295E+00	2.396327E+00	4.159515E - 01
2.824473E - 01	3.598685E+01	4.585448E+00	1.799080E+00	2.247120E+00	3.953956E - 01
2.960573E - 01	3.477625E+01	4.628650E+00	1.833185E+00	2.103798E+00	3.763410E - 01
3.099875E - 01	3.360350E+01	4.622639E+00	1.867610E+00	1.966676E+00	3.587255E - 01
3.242379E - 01	3.246935E+01	4.564467E+00	1.902356E+00	1.834627E+00	3.424877E - 01
3.388087E - 01	3.137404E+01	4.454907E+00	1.937422E+00	1.708488E+00	3.275633E - 01
3.536996E - 01	3.031729E+01	4.297888E+00	1.972808E+00	1.587647E+00	3.138909E - 01
3.689107E - 01	2.929828E+01	4.099825E+00	2.008513E+00	1.472481E+00	3.014093E - 01
3.844422E - 01	2.831567E+01	3.868861E+00	2.044540E+00	1.361910E+00	2.900612E - 01
4.002938E - 01	2.736780E+01	3.614154E+00	2.080887E+00	1.256830E+00	2.797896E - 01
4.164656E - 01	2.645265E+01	3.345181E+00	2.117554E+00	1.156678E+00	2.705426E - 01
4.329577E - 01	2.556813E+01	3.071166E+00	2.154541E+00	1.061391E+00	2.622714E - 01
4.497701E - 01	2.471210E+01	2.800626E+00	2.191848E+00	9.714064E - 01	2.549300E - 01
4.669027E - 01	2.388261E+01	2.540995E+00	2.229476E+00	8.857087E - 01	2.484782E - 01
4.843554E - 01	2.307796E+01	2.298425E+00	2.267424E+00	8.052458E - 01	2.428774E - 01
5.021285E - 01	2.229668E+01	2.077646E+00	2.305692E+00	7.295090E - 01	2.380953E - 01
5.202218E - 01	2.153770E+01	1.881958E+00	2.344280E+00	6.584905E - 01	2.341024E - 01
5.386353E - 01	2.080029E+01	1.713273E+00	2.383188E+00	5.926812E - 01	2.308738E - 01
5.573691E - 01	2.008400E+01	1.572244E+00	2.422418E+00	5.311145E - 01	2.283893E - 01
5.764230E - 01	1.938846E+01	1.458411E+00	2.461967E+00	4.747906E - 01	2.266317E - 01
5.957972E - 01	1.871338E+01	1.370409E+00	2.501836E+00	4.232486E - 01	2.255889E - 01
6.154917E - 01	1.805829E+01	1.306136E+00	2.542026E+00	3.765320E - 01	2.252524E - 01
6.355064E - 01	1.742252E+01	1.262974E+00	2.582535E+00	3.351783E - 01	2.256181E - 01
6.558414E - 01	1.680519E+01	1.237964E+00	2.623365E+00	2.982740E - 01	2.266864E - 01

Z=46

Silver

Atomic Number	47
Lattice	fcc
Lattice constant a_o	7.79 a.u.
Pressure at a_o	1.85 Kbar
Total energy (solid)	-10389.961 Ry
Total energy (atom)	-10389.748 Ry
Zero point energy	0.001 Ry
Cohesive energy	0.212 Ry
Bulk modulus	1.02 Mbar
Fermi energy	0.481 Ry
Interstitial electrons (per atom)	0.661
χ/χ_o	1.12
$N(E_f)$	0.27 states/eV-atom
ρ_o	7.265×10^4 electrons/Bohr^3
$(d\rho(r,E)/dE)_{E=E_f,r=0}$	12.101 electrons/Bohr^3-Ry

Symmetry point	Band energies in Rydbergs (relative to muffin tin zero)					
Γ	-0.052	0.103	0.103	0.103	0.171	0.171
X	-0.014	0.001	0.224	0.243	0.243	0.606
L	0.005	0.099	0.099	0.227	0.227	0.446
W	0.037	0.088	0.088	0.173	0.243	
K	0.019	0.050	0.157	0.187	0.223	

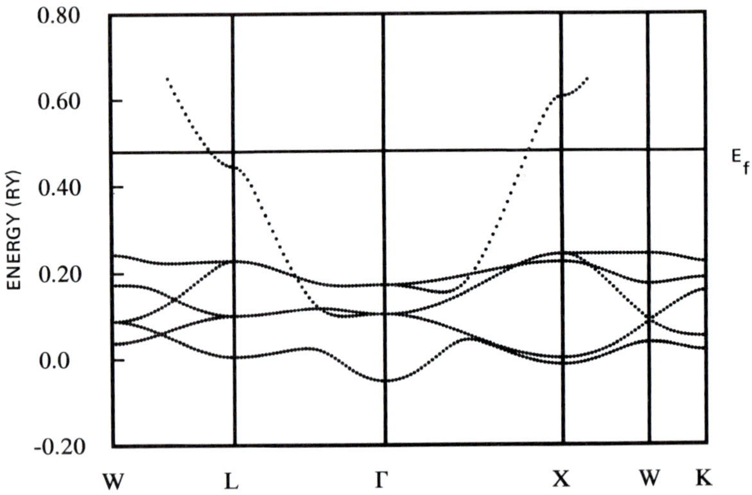

Z=47

Moruzzi
Janak
Williams

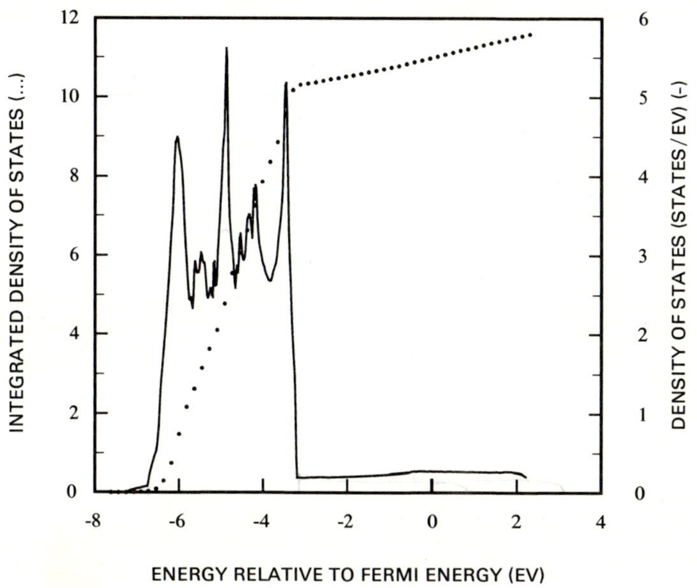
12
10
8
6
4
2
0
6
5
4
3
2
1
0
INTEGRATED DENSITY OF STATES (...)
DENSITY OF STATES (STATES/EV) (-)
-8
-6
-4
-2
0
2
4
ENERGY RELATIVE TO FERMI ENERGY (EV)

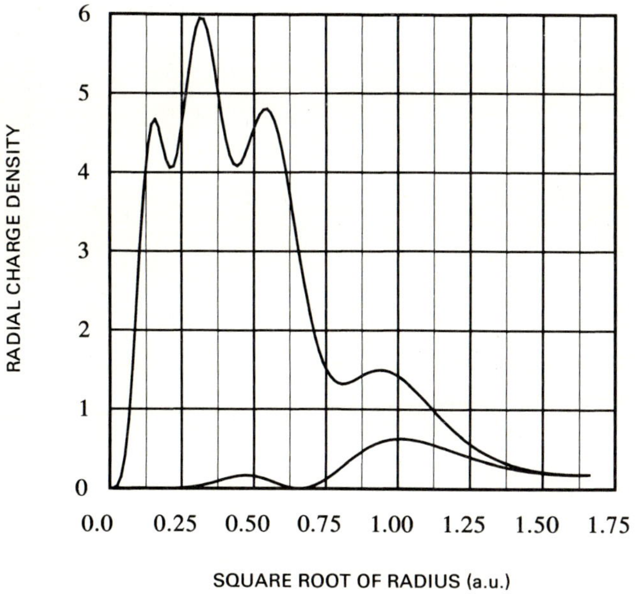
6
5
4
3
2
1
0
RADIAL CHARGE DENSITY
0.0
0.25
0.50
0.75
1.00
1.25
1.50
1.75
SQUARE ROOT OF RADIUS (a.u.)

Z=47

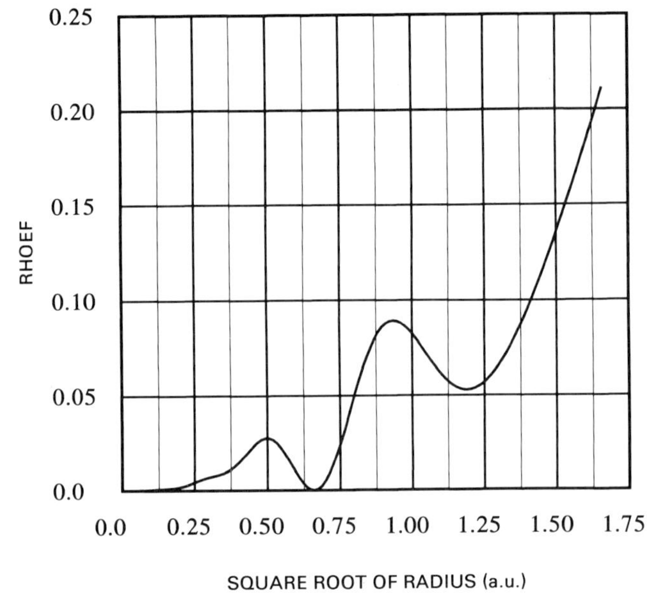

RHOEF=$r^2 \partial \rho(E,r)/\partial E$ (at $E=E_f$)

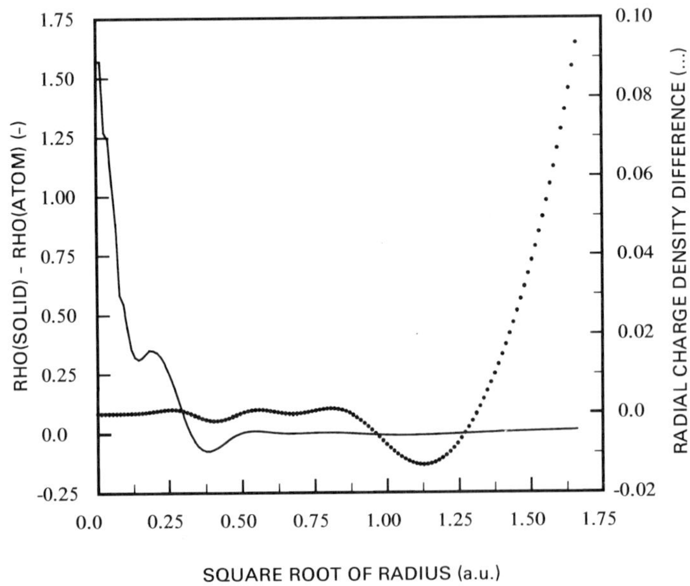

r	2z(r)	$r^2\rho$	r	2z(r)	$r^2\rho$
1.681018E - 04	9.392511E+01	2.020681E - 03	7.102301E - 01	1.545949E+01	1.356776E+00
6.724072E - 04	9.369969E+01	3.083413E - 02	7.322514E - 01	1.486686E+01	1.382295E+00
1.512916E - 03	9.332213E+01	1.442425E - 01	7.546090E - 01	1.428885E+01	1.409210E+00
2.689629E - 03	9.279076E+01	4.081737E - 01	7.773027E - 01	1.372517E+01	1.435081E+00
4.202545E - 03	9.210628E+01	8.646057E - 01	8.003327E - 01	1.317556E+01	1.457911E+00
6.051663E - 03	9.127287E+01	1.507605E+00	8.236988E - 01	1.264007E+01	1.476153E+00
8.236986E - 03	9.029875E+01	2.277086E+00	8.474011E - 01	1.211880E+01	1.488677E+00
1.075852E - 02	8.919569E+01	3.072407E+00	8.714397E - 01	1.161190E+01	1.494744E+00
1.361625E - 02	8.797774E+01	3.780281E+00	8.958145E - 01	1.111956E+01	1.493961E+00
1.681018E - 02	8.665935E+01	4.306556E+00	9.205254E - 01	1.064200E+01	1.486229E+00
2.034032E - 02	8.525404E+01	4.600986E+00	9.455726E - 01	1.017935E+01	1.471698E+00
2.420666E - 02	8.377364E+01	4.667796E+00	9.709560E - 01	9.731737E+00	1.450715E+00
2.840920E - 02	8.222763E+01	4.560360E+00	9.966756E - 01	9.299185E+00	1.423779E+00
3.294795E - 02	8.062503E+01	4.363482E+00	1.022731E+00	8.881953E+00	1.391489E+00
3.782290E - 02	7.897464E+01	4.169934E+00	1.049123E+00	8.479424E+00	1.354528E+00
4.303406E - 02	7.728554E+01	4.058277E+00	1.075851E+00	8.091471E+00	1.313603E+00
4.858142E - 02	7.556644E+01	4.077254E+00	1.102916E+00	7.718384E+00	1.269433E+00
5.446498E - 02	7.382439E+01	4.239304E+00	1.130316E+00	7.359895E+00	1.222722E+00
6.068475E - 02	7.206464E+01	4.522956E+00	1.158052E+00	7.015182E+00	1.174146E+00
6.724072E - 02	7.029143E+01	4.881783E+00	1.186126E+00	6.683892E+00	1.124335E+00
7.413286E - 02	6.850911E+01	5.256805E+00	1.214535E+00	6.366126E+00	1.073866E+00
8.136123E - 02	6.672273E+01	5.589165E+00	1.243280E+00	6.060971E+00	1.023260E+00
8.892584E - 02	6.493793E+01	5.830667E+00	1.272362E+00	5.768488E+00	9.729735E - 01
9.682661E - 02	6.316016E+01	5.950687E+00	1.301780E+00	5.487724E+00	9.234024E - 01
1.050636E - 01	6.139426E+01	5.939011E+00	1.331534E+00	5.218225E+00	8.748851E - 01
1.136368E - 01	5.964400E+01	5.804924E+00	1.361624E+00	4.960536E+00	8.276969E - 01
1.225462E - 01	5.791217E+01	5.573436E+00	1.392051E+00	4.712698E+00	7.820666E - 01
1.317918E - 01	5.620085E+01	5.279757E+00	1.422813E+00	4.475268E+00	7.381678E - 01
1.413736E - 01	5.451169E+01	4.963153E+00	1.453912E+00	4.247804E+00	6.961327E - 01
1.512916E - 01	5.284653E+01	4.661216E+00	1.485347E+00	4.029387E+00	6.560538E - 01
1.615458E - 01	5.120734E+01	4.405245E+00	1.517118E+00	3.819607E+00	6.179883E - 01
1.721362E - 01	4.959628E+01	4.217158E+00	1.549226E+00	3.618570E+00	5.819625E - 01
1.830629E - 01	4.801559E+01	4.108090E+00	1.581670E+00	3.425413E+00	5.479798E - 01
1.943256E - 01	4.646695E+01	4.078512E+00	1.614450E+00	3.240269E+00	5.160199E - 01
2.059247E - 01	4.495125E+01	4.119603E+00	1.647565E+00	3.062815E+00	4.860471E - 01
2.178599E - 01	4.346931E+01	4.215467E+00	1.681018E+00	2.891729E+00	4.580116E - 01
2.301313E - 01	4.202173E+01	4.345777E+00	1.714806E+00	2.728222E+00	4.318506E - 01
2.427390E - 01	4.060927E+01	4.488449E+00	1.748931E+00	2.570515E+00	4.074975E - 01
2.556828E - 01	3.923303E+01	4.622063E+00	1.783392E+00	2.419353E+00	3.848763E - 01
2.689629E - 01	3.789447E+01	4.727818E+00	1.818189E+00	2.274502E+00	3.639086E - 01
2.825791E - 01	3.659505E+01	4.790899E+00	1.853322E+00	2.135242E+00	3.445149E - 01
2.965316E - 01	3.533609E+01	4.801238E+00	1.888791E+00	2.001872E+00	3.266146E - 01
3.108202E - 01	3.411856E+01	4.753725E+00	1.924597E+00	1.873708E+00	3.101285E - 01
3.254451E - 01	3.294289E+01	4.647945E+00	1.960739E+00	1.750581E+00	2.949784E - 01
3.404061E - 01	3.180885E+01	4.487583E+00	1.997217E+00	1.632844E+00	2.810887E - 01
3.557034E - 01	3.071561E+01	4.279591E+00	2.034031E+00	1.520360E+00	2.683871E - 01
3.713369E - 01	2.966176E+01	4.033266E+00	2.071182E+00	1.412003E+00	2.568058E - 01
3.873065E - 01	2.864542E+01	3.759298E+00	2.108668E+00	1.309169E+00	2.462790E - 01
4.036124E - 01	2.766435E+01	3.468883E+00	2.146491E+00	1.210767E+00	2.367469E - 01
4.202545E - 01	2.671619E+01	3.172983E+00	2.184650E+00	1.116708E+00	2.281530E - 01
4.372327E - 01	2.579860E+01	2.881684E+00	2.223145E+00	1.027426E+00	2.204453E - 01
4.545472E - 01	2.490941E+01	2.603756E+00	2.261977E+00	9.423622E - 01	2.135767E - 01
4.721979E - 01	2.404677E+01	2.346351E+00	2.301146E+00	8.614728E - 01	2.075040E - 01
4.901848E - 01	2.320924E+01	2.114849E+00	2.340649E+00	7.857205E - 01	2.021881E - 01
5.085079E - 01	2.239580E+01	1.912851E+00	2.380489E+00	7.135778E - 01	1.975956E - 01
5.271672E - 01	2.160574E+01	1.742249E+00	2.420666E+00	6.460301E - 01	1.936956E - 01
5.461627E - 01	2.083868E+01	1.603389E+00	2.461178E+00	5.830777E - 01	1.904622E - 01
5.654944E - 01	2.009435E+01	1.495305E+00	2.502027E+00	5.242187E - 01	1.878737E - 01
5.851623E - 01	1.937231E+01	1.415956E+00	2.543212E+00	4.694649E - 01	1.859120E - 01
6.051664E - 01	1.867203E+01	1.362483E+00	2.584733E+00	4.193426E - 01	1.845633E - 01
6.255068E - 01	1.799254E+01	1.331476E+00	2.626590E+00	3.738860E - 01	1.838172E - 01
6.461833E - 01	1.733275E+01	1.319201E+00	2.668784E+00	3.321274E - 01	1.836677E - 01
6.671960E - 01	1.669147E+01	1.321812E+00	2.711313E+00	2.956221E - 01	1.841115E - 01
6.885450E - 01	1.606740E+01	1.335528E+00	2.754180E+00	2.629248E - 01	1.851507E - 01

Z=47

Cadmium

Atomic Number	48
Lattice	fcc
Lattice constant a_o	8.40 a.u.
Pressure at a_o	-3.6 Kbar
Total energy (solid)	-10924.544 Ry
Total energy (atom)	-10924.440 Ry
Zero point energy	0.001 Ry
Cohesive energy	0.103 Ry
Bulk modulus	0.76 Mbar
Fermi energy	0.461 Ry
Interstitial electrons (per atom)	0.630
χ/χ_o	1.18
$N(E_f)$	0.36 states/eV-atom
ρ_o	7.747×10^4 electrons/Bohr^3
$(d\rho(r,E)/dE)_{E=E_f,r=0}$	26.931 electrons/Bohr^3-Ry

Symmetry point	Band energies in Rydbergs (relative to muffin tin zero)						
Γ	-0.266	-0.266	-0.266	-0.236	-0.236	-0.169	
X	-0.310	-0.306	-0.212	-0.202	-0.202	0.396	0.580
L	-0.295	-0.267	-0.267	-0.210	-0.210	0.274	0.380
W	-0.283	-0.265	-0.265	-0.231	-0.202	0.602	0.602
K	-0.297	-0.279	-0.232	-0.226	-0.211	0.508	0.557

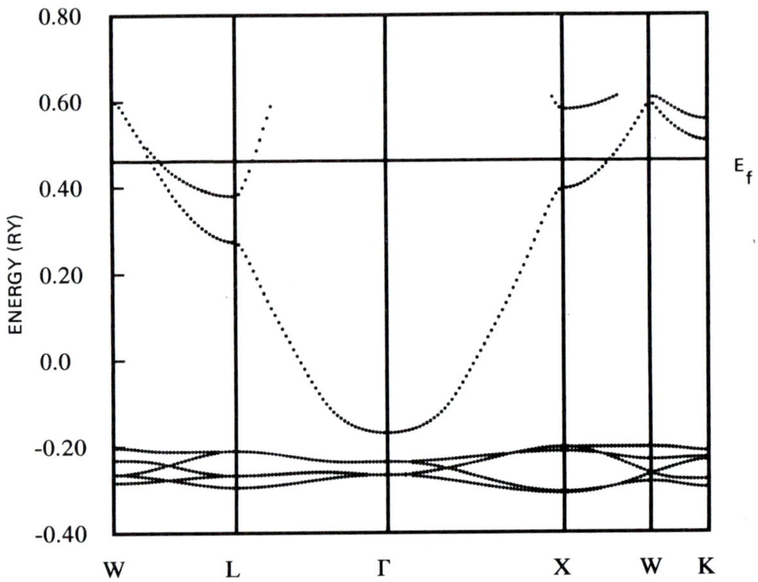

Z=48

Moruzzi
Janak
Williams

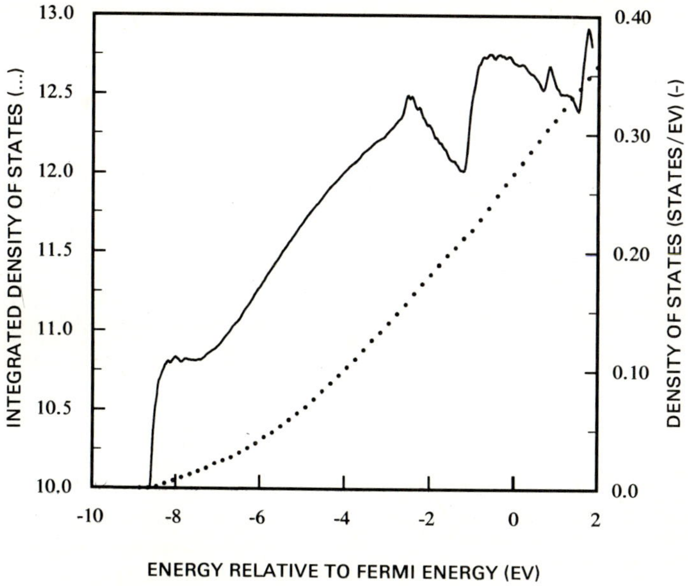

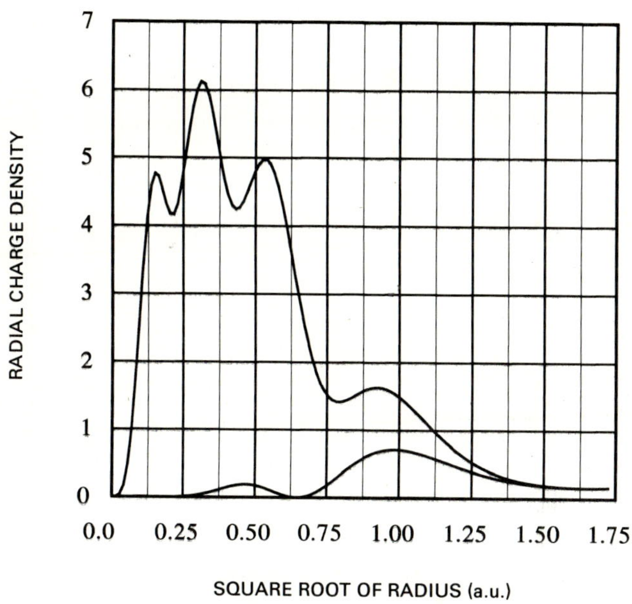

Moruzzi
Janak
Williams

Z=48

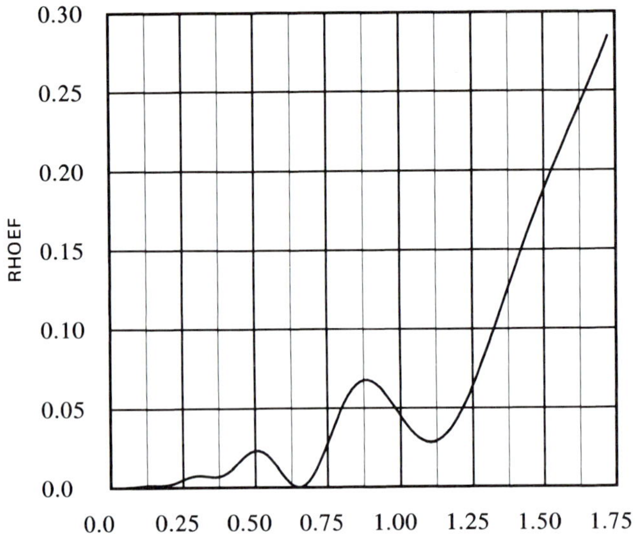

SQUARE ROOT OF RADIUS (a.u.)

$RHOEF=r^2\partial\rho(E,r)/\partial E$ (at $E=E_f$)

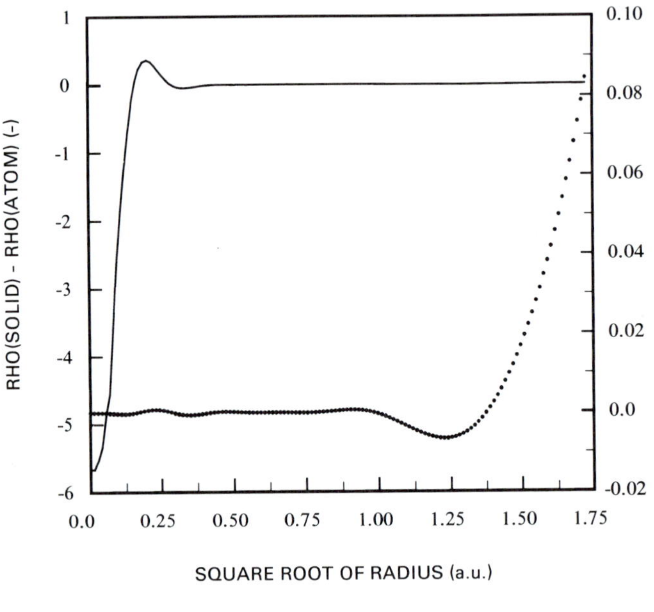

SQUARE ROOT OF RADIUS (a.u.)

Z=48

Cadmium

r	2z(r)	$r^2\rho$	r	2z(r)	$r^2\rho$
1.812651E - 04	9.591679E+01	2.501555E - 03	7.658451E - 01	1.426704E+01	1.573793E+00
7.250605E - 04	9.566624E+01	3.798919E - 02	7.895908E - 01	1.367561E+01	1.597384E+00
1.631386E - 03	9.524649E+01	1.762987E - 01	8.136991E - 01	1.310082E+01	1.613947E+00
2.900242E - 03	9.465579E+01	4.933405E - 01	8.381699E - 01	1.254287E+01	1.622450E+00
4.531626E - 03	9.389563E+01	1.030131E+00	8.630032E - 01	1.200202E+01	1.622345E+00
6.525543E - 03	9.297186E+01	1.765201E+00	8.881990E - 01	1.147854E+01	1.613500E+00
8.881990E - 03	9.189525E+01	2.612325E+00	9.137574E - 01	1.097260E+01	1.596127E+00
1.160097E - 02	9.068036E+01	3.444194E+00	9.396783E - 01	1.048438E+01	1.570705E+00
1.468247E - 02	8.934392E+01	4.131665E+00	9.659618E - 01	1.001387E+01	1.537924E+00
1.812651E - 02	8.790231E+01	4.582955E+00	9.926078E - 01	9.561003E+00	1.498606E+00
2.193308E - 02	8.637051E+01	4.768735E+00	1.019616E+00	9.125766E+00	1.453664E+00
2.610217E - 02	8.476042E+01	4.725989E+00	1.046987E+00	8.708056E+00	1.404053E+00
3.063380E - 02	8.308289E+01	4.541930E+00	1.074720E+00	8.306879E+00	1.350726E+00
3.552796E - 02	8.134808E+01	4.325678E+00	1.102817E+00	7.922362E+00	1.294605E+00
4.078465E - 02	7.956650E+01	4.177858E+00	1.131275E+00	7.554058E+00	1.236563E+00
4.640387E - 02	7.774846E+01	4.166985E+00	1.160096E+00	7.201469E+00	1.177402E+00
5.238562E - 02	7.590269E+01	4.317554E+00	1.189280E+00	6.864056E+00	1.117847E+00
5.872989E - 02	7.403568E+01	4.610662E+00	1.218826E+00	6.540742E+00	1.058539E+00
6.543666E - 02	7.215247E+01	4.994354E+00	1.248734E+00	6.231970E+00	1.000026E+00
7.250601E - 02	7.025807E+01	5.399459E+00	1.279006E+00	5.936121E+00	9.427781E - 01
7.993788E - 02	6.835852E+01	5.756197E+00	1.309640E+00	5.653111E+00	8.871780E - 01
8.773232E - 02	6.646028E+01	6.008053E+00	1.340636E+00	5.382356E+00	8.335333E - 01
9.588921E - 02	6.456984E+01	6.120689E+00	1.371995E+00	5.123272E+00	7.820823E - 01
1.044087E - 01	6.269257E+01	6.085278E+00	1.403717E+00	4.874784E+00	7.329988E - 01
1.132907E - 01	6.083282E+01	5.916762E+00	1.435801E+00	4.637380E+00	6.863993E - 01
1.225352E - 01	5.899371E+01	5.648373E+00	1.468247E+00	4.409508E+00	6.423558E - 01
1.321422E - 01	5.717755E+01	5.324057E+00	1.501056E+00	4.191700E+00	6.008915E - 01
1.421118E - 01	5.538638E+01	4.990465E+00	1.534227E+00	3.982452E+00	5.620014E - 01
1.524439E - 01	5.362260E+01	4.689781E+00	1.567761E+00	3.781843E+00	5.256478E - 01
1.631386E - 01	5.188866E+01	4.454398E+00	1.601658E+00	3.589446E+00	4.917706E - 01
1.741958E - 01	5.018716E+01	4.303827E+00	1.635918E+00	3.404359E+00	4.602920E - 01
1.856155E - 01	4.852045E+01	4.243910E+00	1.670539E+00	3.227248E+00	4.311184E - 01
1.973977E - 01	4.689006E+01	4.267922E+00	1.705523E+00	3.057261E+00	4.041484E - 01
2.095425E - 01	4.529688E+01	4.359163E+00	1.740870E+00	2.893566E+00	3.792741E - 01
2.220497E - 01	4.374170E+01	4.494317E+00	1.776579E+00	2.736385E+00	3.563815E - 01
2.349195E - 01	4.222542E+01	4.647057E+00	1.812651E+00	2.585452E+00	3.353570E - 01
2.481519E - 01	4.074933E+01	4.791379E+00	1.849085E+00	2.440519E+00	3.160868E - 01
2.617468E - 01	3.931502E+01	4.904325E+00	1.885881E+00	2.301358E+00	2.984591E - 01
2.757042E - 01	3.792424E+01	4.967908E+00	1.923041E+00	2.167241E+00	2.823659E - 01
2.900242E - 01	3.657845E+01	4.970196E+00	1.960563E+00	2.039011E+00	2.677014E - 01
3.047066E - 01	3.527867E+01	4.905579E+00	1.998447E+00	1.915462E+00	2.543674E - 01
3.197516E - 01	3.402522E+01	4.774415E+00	2.036695E+00	1.796941E+00	2.422686E - 01
3.351592E - 01	3.281764E+01	4.582153E+00	2.075304E+00	1.683304E+00	2.313165E - 01
3.509293E - 01	3.165474E+01	4.338212E+00	2.114276E+00	1.574417E+00	2.214282E - 01
3.670619E - 01	3.053461E+01	4.054675E+00	2.153610E+00	1.470152E+00	2.125264E - 01
3.835570E - 01	2.945486E+01	3.745041E+00	2.193308E+00	1.369886E+00	2.045404E - 01
4.004146E - 01	2.841283E+01	3.423052E+00	2.233367E+00	1.274549E+00	1.974038E - 01
4.176348E - 01	2.740570E+01	3.101732E+00	2.273789E+00	1.183020E+00	1.910575E - 01
4.352175E - 01	2.643094E+01	2.792651E+00	2.314574E+00	1.095733E+00	1.854465E - 01
4.531628E - 01	2.548628E+01	2.505390E+00	2.355721E+00	1.013129E+00	1.805215E - 01
4.714705E - 01	2.457004E+01	2.247271E+00	2.397231E+00	9.336092E - 01	1.762387E - 01
4.901409E - 01	2.368091E+01	2.023254E+00	2.439103E+00	8.586585E - 01	1.725577E - 01
5.091737E - 01	2.281822E+01	1.836025E+00	2.481338E+00	7.877214E - 01	1.694441E - 01
5.285690E - 01	2.198154E+01	1.686185E+00	2.523935E+00	7.202508E - 01	1.668673E - 01
5.483270E - 01	2.117052E+01	1.572568E+00	2.566895E+00	6.572493E - 01	1.648000E - 01
5.684474E - 01	2.038481E+01	1.492559E+00	2.610217E+00	5.976734E - 01	1.632203E - 01
5.889303E - 01	1.962358E+01	1.442480E+00	2.653902E+00	5.420306E - 01	1.621089E - 01
6.097758E - 01	1.888571E+01	1.417935E+00	2.697949E+00	4.903228E - 01	1.614504E - 01
6.309838E - 01	1.816988E+01	1.414150E+00	2.742359E+00	4.425673E - 01	1.612329E - 01
6.525544E - 01	1.747455E+01	1.426251E+00	2.787132E+00	3.982618E - 01	1.614481E - 01
6.744875E - 01	1.679836E+01	1.449512E+00	2.832267E+00	3.584827E - 01	1.620898E - 01
6.967831E - 01	1.614020E+01	1.479538E+00	2.877765E+00	3.222266E - 01	1.631565E - 01
7.194412E - 01	1.549920E+01	1.512406E+00	2.923625E+00	2.900696E - 01	1.646484E - 01
7.424619E - 01	1.487488E+01	1.544747E+00	2.969848E+00	2.620704E - 01	1.665695E - 01

Z=48

Indium

Atomic Number	49
Lattice	fcc
Lattice constant a_o	8.95 a.u.
Pressure at a_o	-14.6 Kbar
Total energy (solid)	-11474.491 Ry
Total energy (atom)	-11474.277 Ry
Zero point energy	0.001 Ry
Cohesive energy	0.213 Ry
Bulk modulus	0.35 Mbar
Fermi energy	0.405 Ry
Interstitial electrons (per atom)	0.651
χ/χ_o	1.26
$N(E_f)$	0.50 states/eV-atom
ρ_o	8.251×10^4 electrons/Bohr^3
$(d\rho(r,E)/dE)_{E=E_f,r=0}$	36.811 electrons/Bohr^3-Ry

Symmetry point	Band energies in Rydbergs (relative to muffin tin zero)								
Γ	-0.750	-0.750	-0.750	-0.739	-0.739	-0.275			
X	-0.763	-0.758	-0.729	-0.726	-0.726	0.238	0.279		
L	-0.753	-0.749	-0.749	-0.729	-0.729	0.116	0.144		
W	-0.751	-0.746	-0.746	-0.736	-0.726	0.333	0.388	0.388	0.478
K	-0.757	-0.749	-0.734	-0.734	-0.729	0.287	0.316	0.384	

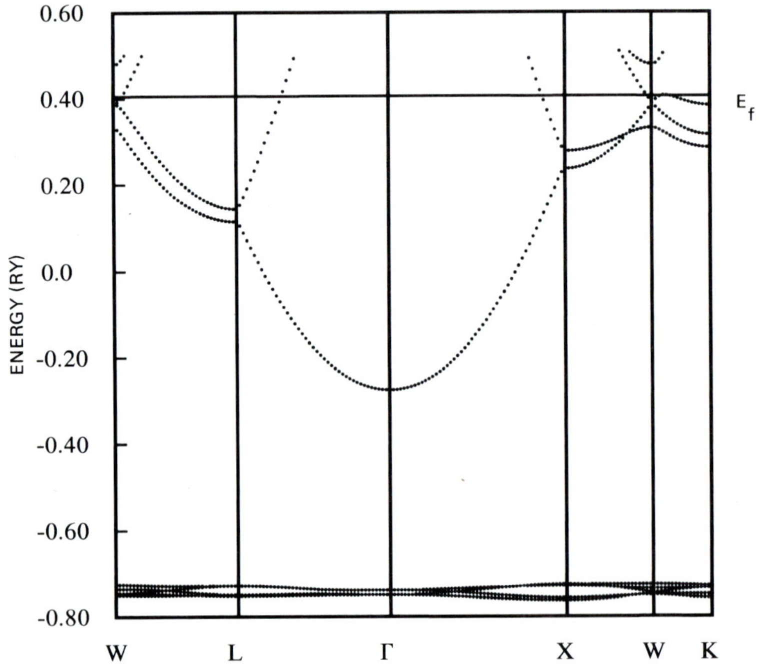

Z=49

Moruzzi
Janak
Williams

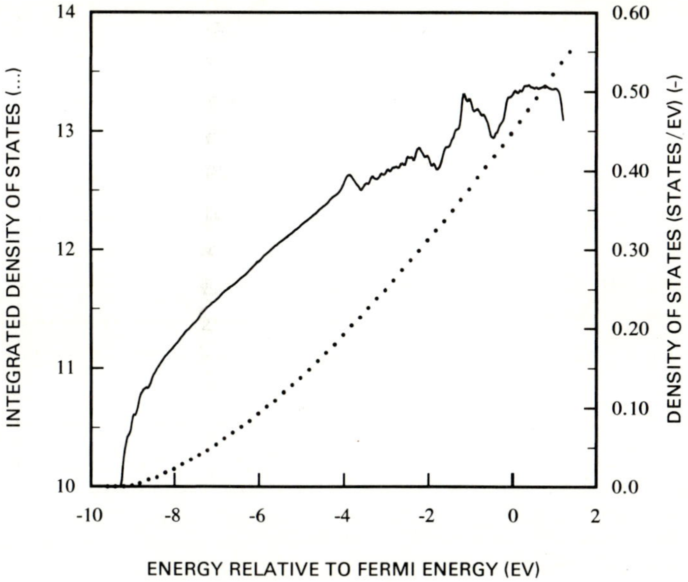

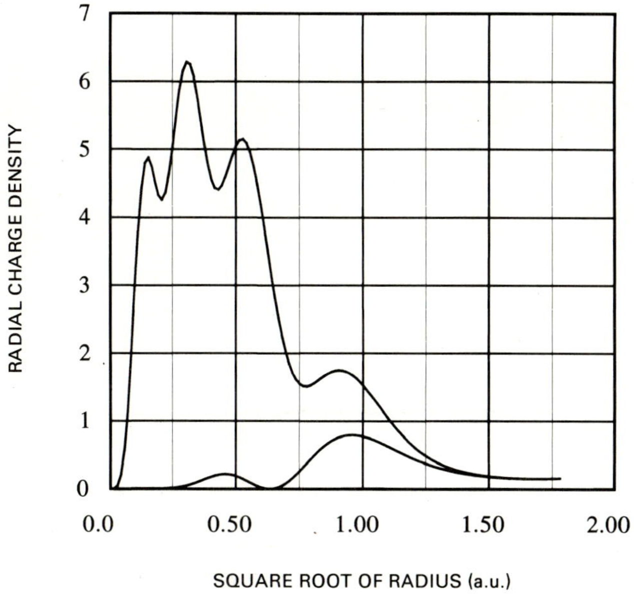

Moruzzi
Janak
Williams

Z=49

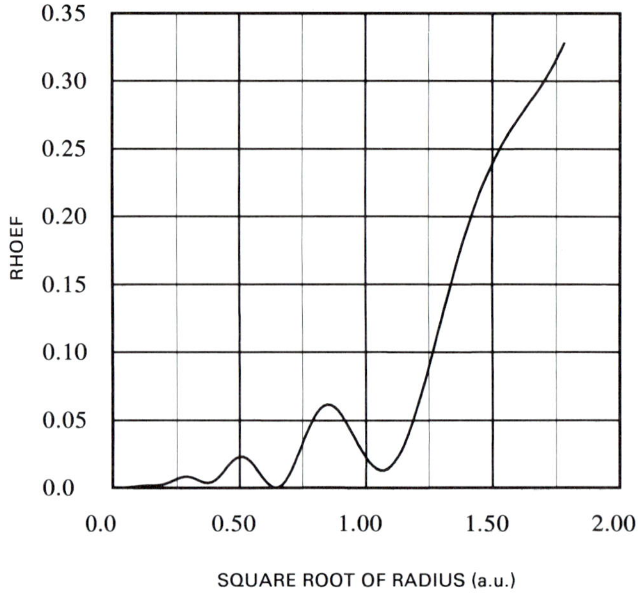

RHOEF=$r^2\partial\rho(E,r)/\partial E$ (at $E=E_f$)

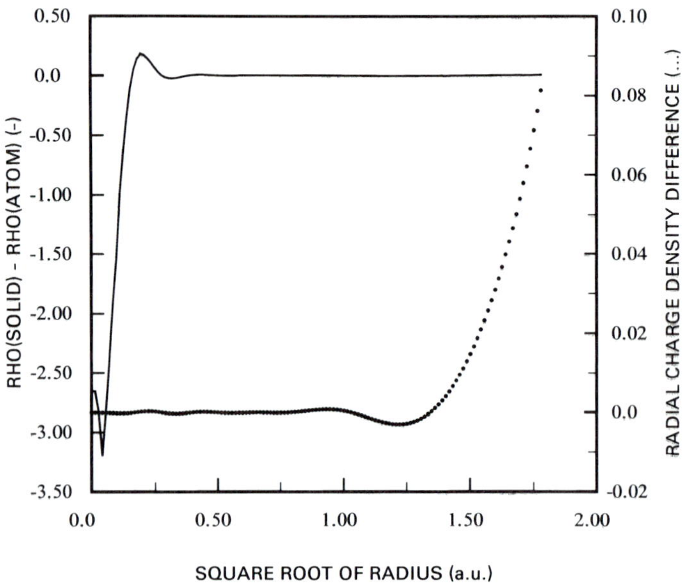

Moruzzi
Janak
Williams

Z=49

Indium

r	2z(r)	$r^2\rho$	r	2z(r)	$r^2\rho$
1.931337E - 04	9.790872E+01	3.019977E - 03	8.159897E - 01	1.334867E+01	1.749853E+00
7.725346E - 04	9.763374E+01	4.565269E - 02	8.412902E - 01	1.276769E+01	1.745787E+00
1.738203E - 03	9.717310E+01	2.102547E - 01	8.669770E - 01	1.220673E+01	1.731013E+00
3.090139E - 03	9.652492E+01	5.821226E - 01	8.930500E - 01	1.166604E+01	1.705981E+00
4.828341E - 03	9.569165E+01	1.199024E+00	9.195094E - 01	1.114567E+01	1.671463E+00
6.952811E - 03	9.468102E+01	2.020805E+00	9.463549E - 01	1.064563E+01	1.628469E+00
9.463549E - 03	9.350636E+01	2.933231E+00	9.735867E - 01	1.016574E+01	1.578155E+00
1.236055E - 02	9.218530E+01	3.783762E+00	1.001204E+00	9.705925E+00	1.521754E+00
1.564382E - 02	9.073680E+01	4.432726E+00	1.029209E+00	9.265249E+00	1.460526E+00
1.931337E - 02	8.917900E+01	4.798355E+00	1.057599E+00	8.844132E+00	1.395686E+00
2.336917E - 02	8.752757E+01	4.879039E+00	1.086376E+00	8.440964E+00	1.328403E+00
2.781124E - 02	8.579549E+01	4.747240E+00	1.115540E+00	8.055639E+00	1.259734E+00
3.263959E - 02	8.399449E+01	4.520913E+00	1.145089E+00	7.687494E+00	1.190641E+00
3.785419E - 02	8.213637E+01	4.325054E+00	1.175025E+00	7.335820E+00	1.121963E+00
4.345507E - 02	8.023314E+01	4.256918E+00	1.205347E+00	6.999891E+00	1.054419E+00
4.944222E - 02	7.829561E+01	4.364462E+00	1.236055E+00	6.678973E+00	9.886096E - 01
5.581563E - 02	7.633199E+01	4.641340E+00	1.267150E+00	6.372336E+00	9.250178E - 01
6.257528E - 02	7.434836E+01	5.036437E+00	1.298631E+00	6.079257E+00	8.640252E - 01
6.972122E - 02	7.235048E+01	5.472447E+00	1.330498E+00	5.799042E+00	8.059121E - 01
7.725346E - 02	7.034494E+01	5.867098E+00	1.362751E+00	5.531010E+00	7.508757E - 01
8.517194E - 02	6.833926E+01	6.151664E+00	1.395391E+00	5.274519E+00	6.990347E - 01
9.347665E - 02	6.634082E+01	6.283337E+00	1.428416E+00	5.028953E+00	6.504453E - 01
1.021677E - 01	6.435602E+01	6.250206E+00	1.461828E+00	4.793739E+00	6.051069E - 01
1.112450E - 01	6.238977E+01	6.069428E+00	1.495626E+00	4.568326E+00	5.629752E - 01
1.207085E - 01	6.044582E+01	5.780320E+00	1.529811E+00	4.352210E+00	5.239694E - 01
1.305583E - 01	5.852687E+01	5.434695E+00	1.564383E+00	4.144383E+00	4.879810E - 01
1.407944E - 01	5.663547E+01	5.086658E+00	1.599339E+00	3.945465E+00	4.548770E - 01
1.514168E - 01	5.477428E+01	4.783705E+00	1.634683E+00	3.754517E+00	4.245142E - 01
1.624254E - 01	5.294650E+01	4.560388E+00	1.670413E+00	3.570625E+00	3.967363E - 01
1.738203E - 01	5.115500E+01	4.435002E+00	1.706529E+00	3.394496E+00	3.713817E - 01
1.856014E - 01	4.940210E+01	4.409245E+00	1.743031E+00	3.225271E+00	3.482874E - 01
1.977689E - 01	4.768956E+01	4.470348E+00	1.779920E+00	3.062114E+00	3.272927E - 01
2.103225E - 01	4.601820E+01	4.594815E+00	1.817194E+00	2.905795E+00	3.082379E - 01
2.232625E - 01	4.438898E+01	4.752988E+00	1.854856E+00	2.754987E+00	2.909712E - 01
2.365887E - 01	4.280325E+01	4.913591E+00	1.892902E+00	2.610501E+00	2.753456E - 01
2.503012E - 01	4.126276E+01	5.047696E+00	1.931336E+00	2.471044E+00	2.612236E - 01
2.643999E - 01	3.976935E+01	5.131701E+00	1.970156E+00	2.337445E+00	2.484740E - 01
2.788849E - 01	3.832484E+01	5.149153E+00	2.009362E+00	2.208447E+00	2.369770E - 01
2.937562E - 01	3.693051E+01	5.091485E+00	2.048955E+00	2.084379E+00	2.266198E - 01
3.090138E - 01	3.558682E+01	4.957747E+00	2.088933E+00	1.965591E+00	2.172986E - 01
3.246577E - 01	3.429340E+01	4.753627E+00	2.129298E+00	1.850852E+00	2.089197E - 01
3.406878E - 01	3.304884E+01	4.490013E+00	2.170050E+00	1.740528E+00	2.013966E - 01
3.571041E - 01	3.185098E+01	4.181314E+00	2.211186E+00	1.635012E+00	1.946517E - 01
3.739067E - 01	3.069710E+01	3.843777E+00	2.252710E+00	1.533106E+00	1.886153E - 01
3.910956E - 01	2.958406E+01	3.493961E+00	2.294621E+00	1.435215E+00	1.832249E - 01
4.086708E - 01	2.850882E+01	3.147437E+00	2.336917E+00	1.341236E+00	1.784248E - 01
4.266322E - 01	2.746849E+01	2.817822E+00	2.379600E+00	1.251069E+00	1.741664E - 01
4.449799E - 01	2.646075E+01	2.516132E+00	2.422668E+00	1.164630E+00	1.704063E - 01
4.637139E - 01	2.548386E+01	2.250435E+00	2.466124E+00	1.081851E+00	1.671078E - 01
4.828341E - 01	2.453673E+01	2.025787E+00	2.509965E+00	1.002669E+00	1.642388E - 01
5.023406E - 01	2.361880E+01	1.844388E+00	2.554193E+00	9.270464E - 01	1.617720E - 01
5.222334E - 01	2.272978E+01	1.705913E+00	2.598806E+00	8.549430E - 01	1.596851E - 01
5.425124E - 01	2.186925E+01	1.607932E+00	2.643806E+00	7.863463E - 01	1.579598E - 01
5.631777E - 01	2.103651E+01	1.546404E+00	2.689193E+00	7.212453E - 01	1.565815E - 01
5.842293E - 01	2.023036E+01	1.516155E+00	2.734965E+00	6.596450E - 01	1.555394E - 01
6.056671E - 01	1.944933E+01	1.511374E+00	2.781124E+00	6.015586E - 01	1.548260E - 01
6.274912E - 01	1.869160E+01	1.526008E+00	2.827669E+00	5.470194E - 01	1.544370E - 01
6.497016E - 01	1.795569E+01	1.554125E+00	2.874601E+00	4.955348E - 01	1.543713E - 01
6.722983E - 01	1.724016E+01	1.590186E+00	2.921919E+00	4.482195E - 01	1.546299E - 01
6.952811E - 01	1.654416E+01	1.629251E+00	2.969623E+00	4.046000E - 01	1.552173E - 01
7.186503E - 01	1.586715E+01	1.667106E+00	3.017713E+00	3.647485E - 01	1.561403E - 01
7.424058E - 01	1.520895E+01	1.700329E+00	3.066190E+00	3.282195E - 01	1.574084E - 01
7.665474E - 01	1.456965E+01	1.726304E+00	3.115052E+00	2.961642E - 01	1.590328E - 01
7.910755E - 01	1.394942E+01	1.743190E+00	3.164302E+00	2.676216E - 01	1.610282E - 01

Z=49

Moruzzi
Janak
Williams

V. EFFECTS OF SPIN POLARIZATION

In this chapter, we describe the results of spin-polarized energy-band calculations for Fe, Co, and Ni. The magnetic susceptibility calculations described in Chapter III correctly indicate that, of all the metals with $Z \leq 49$, only Fe, Co, and Ni exhibit the possibility of spontaneous ferromagnetic order. Thus, even without performing spin-polarized calculations, we can say that the local-density theory correctly predicts the presence or absence of ferromagnetism in the metallic elements up through In. Naturally, the spin-polarized calculations verify the presence of ferromagnetism in Fe, Co, and Ni, in the sense that the total energy of the ferromagnetic state is found to be lower. In addition (and probably independently of the local-density approximation), our calculations[16] show that the presence of ferromagnetism has large effects on the mechanical properties (equilibrium lattice constant and bulk modulus) which account for much of the discrepancy between theory and experiment for Fe through Ni seen in Figure 1.1. Similar results for Fe were obtained independently by Poulsen et al.[34]

After first describing the computational procedure, we consider the general implications of the results, and how they affect the trends discussed in Chapter III. Finally, we give the detailed results of the calculations in a format similar to that used in Chapter IV.

1. SPIN-POLARIZED CALCULATIONS

Self-consistent spin-polarized energy-band calculations are a fairly straightforward generalization of the nonpolarized calculations described in Chapter II. For each self-consistency iteration, two complete energy-band calculations are performed, one for each spin, and the spin-polarized charge densities $\rho\uparrow(r)$ and $\rho\downarrow(r)$ are calculated. From these, the total charge density $\rho(r)=\rho\uparrow(r)+\rho\downarrow(r)$ and the magnetization density $m(r)=\rho\uparrow(r)-\rho\downarrow(r)$ are constructed. Knowing the charge and magnetization densities, new spin-polarized potentials for the next iteration are constructed via eq. 2.1 (the potentials for spin up and spin down are different because the exchange-correlation potential μ_{xc} has an explicit spin dependence). The total magnetic moment per atom (which could be obtained by integrating m(r) over the unit cell) is given by the difference in the spin-up and spin-down integrated densities of states at the Fermi energy. Thus, the magnetic moment is completely determined by the exchange-correlation functional described in eqs. 2.5-2.9; no adjustable parameters have been introduced in order to fit the magnetic moment to experiment. Calculations using a similar approximation for exchange and correlation have been performed for Ni by Wang and Callaway.[41]

Our spin-polarized calculations were performed using the muffin-tin approximation; this amounts to assuming that $\rho\uparrow$ and $\rho\downarrow$ are both constant (with different values) in the interstitial region. The fact that the interstitial charge density is different for the two spins implies that no single reference energy exists for which the interstitial potential vanishes for both spins. For the presentation of our results, we have chosen the interstitial potential seen by the spin-up electrons as the energy reference. The spin-down interstitial potential is therefore nonzero; its value is given at the end of the tabulation of the spin-down potential. Note that the band energies for spin-up and spin-down electrons can be calculated independently with respect to different references, and later presented relative to a common reference. All the energies, both spin up and spin down, given in the tables and figures below are referred to the same energy zero.

2. GENERAL DESCRIPTION OF RESULTS AND TRENDS

In Figure 5.1, the calculated spin magnetic moment (Bohr magnetons/atom) and its pressure derivative, $\partial \ln M/\partial P$, are compared to the experimental results for Fe, Co, and Ni. In the real materials, there is a small orbital moment (the g-factor is not quite two), and the experimental spin moment M_S has been obtained from the experimental total moment M and gyromagnetic ratio[42] (magnetomechanical factor) g' according to

$$M_S = \frac{2(g'-1)}{g'} M \, . \qquad 5.1$$

Taking the derivative of this expression with respect to pressure, we find

$$\frac{\partial \ln M}{\partial P} = \frac{\partial \ln M_S}{\partial P} - \frac{1}{g'-1}\frac{\partial \ln g'}{\partial P} \, . \qquad 5.2$$

However, there do not appear to be any published values for $\partial \ln g'/\partial P$, so, in the second part of Figure 5.1, the calculated values of the pressure derivative of the spin moment are com-

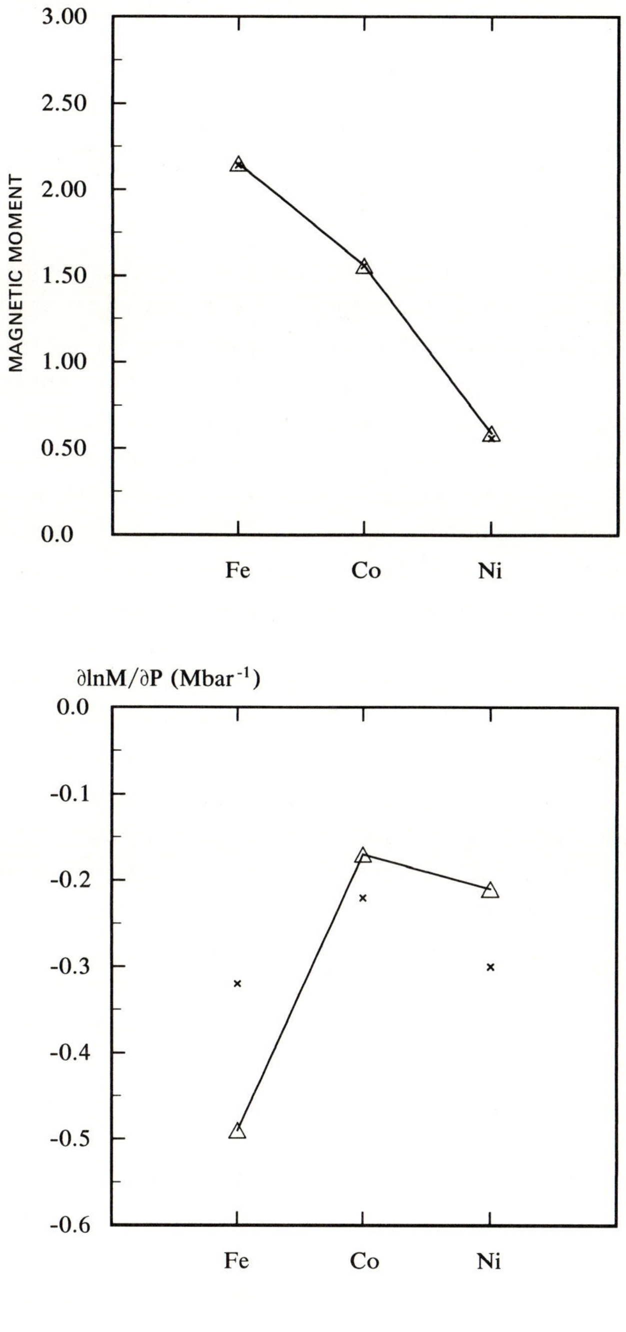

igure 5.1 Calculated (△) and experimental (×) spin magnetic moment M (Bohr magnetons) and pressure erivative of the logarithm of M (Mbar^{-1}) for Fe, Co and Ni.

pared directly to experimental values[43,44] of the pressure derivative of the total moment. The agreement of the spin moments, as shown in Figure 5.1, is rather good; the error in the calculated moment is less than 1% for Fe and Co, and is about 5% for Ni. The differences between measured and calculated pressure derivatives of the moment are probably due to a weak dependence of the gyromagnetic ratio on pressure: only a 1% change in g' per 100 kilobars is required to explain the discrepancies, although the results suggest that the pressure derivative of g' changes sign between Fe and Co.

The changes in the equilibrium atomic volume (expressed in terms of the Wigner-Seitz radius) and the bulk modulus due to the presence of ferromagnetism, as obtained from self-consistent spin-polarized calculations at several lattice constants, are shown in Figure 5.2. In this figure, the crosses are experimental values[15,45-47], the circles connected by lines are the results of the nonmagnetic calculations (the same data as appeared in Figure 1.1), and the triangles connected by lines are the results of the spin-polarized calculations for Fe, fcc Co, and Ni; the triangles for Mn and Cr represent estimates, obtained in the manner described below. It is clear that the presence of ferromagnetism can lead to large increases in the atomic volume, and large decreases in the bulk modulus. There is a fairly simple explanation for these effects:[16,34,48] the character of the d-orbitals changes gradually from bonding to antibonding as the energy increases through a d-band. The spin splitting of the bands (raising of the spin-down bands with respect to the spin-up bands) causes some of the spin-down electrons to enter previously unoccupied spin-up states, in order to keep the chemical potentials for the two spin systems equal. But these newly-occupied spin-up states are higher in the d-band than were the spin-down states which were emptied; they correspond to less-bonding d-orbitals than the orbitals which were emptied. The filling of less-bonding orbitals at the expense of more-bonding orbitals reduces the net interatomic attraction, allowing the lattice to expand. The magnetically induced increase in the atomic volume reduces the compression to which the s-electrons are subjected, thereby sharply reducing the bulk modulus.[16,36,37]

It is not necessary to go through all the labor of self-consistent spin-polarized calculations in order to establish the size of these effects. An estimate can be obtained from the Stoner model,[16,34] if it is assumed that the density of states at the Fermi energy scales with volume inversely as does the bandwidth.[49] Estimates obtained in this way for Fe, Co, and Ni[16] are in remarkably good agreement with the results of the spin-polarized self-consistent calculations presented in Figure 5.2.

It is much harder to obtain estimates of the size of the repulsion in antiferromagnetically ordered systems, because of the lower translational symmetry, but we can speculate that similar repulsive forces also exist in antiferromagnets. All that is required is the existence of a magnetic moment over a distance scale sufficient to establish a band structure (a few atoms). This local moment will be sufficient to induce a spin splitting of the local bands, and thus a local repulsion between the atoms. It does not matter if, on a larger scale, the moments are

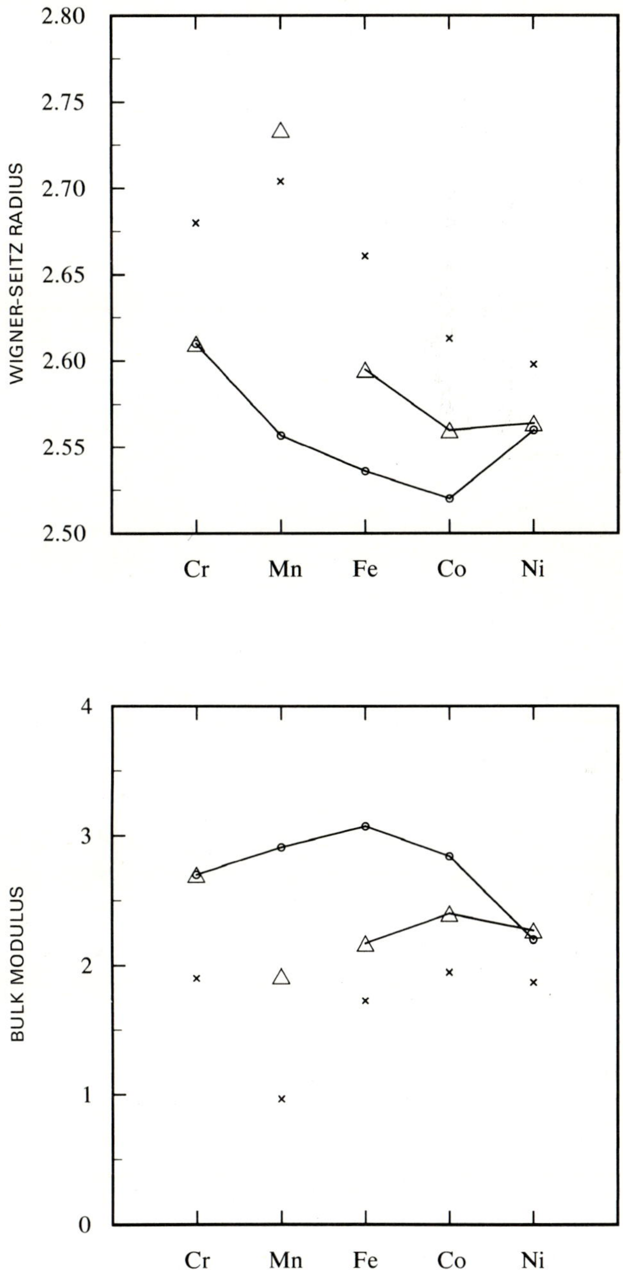

Figure 5.2 Wigner-Seitz radii (a.u.) and bulk moduli (Mbar) for Cr, Mn, Fe, Co, and Ni. (×) = experimental values. (○) = nonmagnetic calculations. (△) = spin-polarized calculations (values for Cr and Mn are estimates based on the Stoner model).

disordered or antiferromagnetically ordered instead of being in perfect ferromagnetic alignment, because the volume expansion is independent of the spin direction. The estimates of the volume increase and bulk modulus decrease for Cr and Mn shown in Figure 5.2 were obtained from the Stoner model as though these materials were ferromagnetic.[16] Even though they are overestimates (because we should be computing the volume dependence of the susceptibility at the wavevector q of the antiferromagnetic ordering, rather than the susceptibility at $q=0$, and the former must be larger than the latter if the antiferromagnetic state is the most stable), they suggest that the repulsive effect is quite small in Cr. This is consistent with the small volume changes observed in Cr at the Néel temperature.[50] If this argument also applies to Mn, then its anomalously small bulk modulus (in comparison to Tc or Re), and possibly even its polymorphism, are consequences of its magnetism.

The changes in atomic volume and bulk modulus of Fe, Co, and, to a lesser extent, Ni as the temperature is raised through the Curie temperature T_c are much smaller than might be guessed from Figure 5.2. Fe, Co, and Ni are all known to have short-range magnetic order above T_c, either from neutron diffraction data,[51,52] or from the persistence of the spin splitting seen in photoemission.[53,54] Our nonmagnetic calculations, on the other hand, are performed for a hypothetical state of the material in which the microscopic magnetization density m(r) is identically zero at every point in the crystal. The experimental fact that local moments exist above T_c means that this hypothetical state is not the state of lowest free energy above T_c; thus our nonmagnetic calculations are probably not appropriate for these systems at any reasonable temperature, and certainly should not be associated with the paramagnetic state just above T_c. According to the arguments given in the previous paragraph, we would expect disordered local moments to lead to repulsive forces similar in magnitude to those obtained in our perfectly-ordered ferromagnetic calculations. However, even though some progress has been made in describing the disorder in the spin system[55-57], it would still be difficult to perform reliable calculations, because of the loss of translational symmetry. Thus we have no simple way of actually calculating the change in mechanical properties at T_c.

The theory of the disordered magnetic state above T_c has not yet progressed to the point where one can reliably predict those materials in which disordered states occur. If any materials exist in which there are *no* local moments above T_c, so that our nonmagnetic calculations *can* be associated with the paramagnetic state, then such materials should show large volume contractions and large bulk-modulus increases as the temperature is raised through T_c. The volume contraction is basically the same as the "spontaneous volume magnetostriction" introduced in thermodynamic descriptions of magnetoelasticity,[58] although it can be as large as several percent. The bulk modulus increase, on the other hand, is mostly due to the dependence of the nonmagnetic bulk modulus on atomic volume, as was discussed in connection with Figure 3.2. This is a different effect from the changes in bulk modulus usually considered in discussions of magnetoelastic effects, coming from second derivatives of the magnetic part of the free energy,[58] and can be much larger. This is of possible importance in the behavior of invar alloys.[48,59]

3. DETAILED CALCULATIONS

The format of the presentation of the detailed results for spin-polarized Fe, Co, and Ni closely parallels the format used in Chapter IV (see p. 30). On the first page for each element, we give a table of results similar to the table given for the nonmagnetic calculations. The susceptibility enhancement was not calculated for the ferromagnetic state; ferromagnetic quantities appearing in this table include calculated values for the lowering of the total energy due to spin polarization, for the total spin magnetic moment per atom and its pressure derivative, and for the interstitial magnetic moment (which in all three cases is antiparallel to the total moment). The remainder of the first page contains a table of the eigenvalues at symmetry points for both spin-up and spin-down bands.

On the second page for each element is given a plot along symmetry axes of the energy bands for both spin directions, and a plot of the density of states and the integrated density of states for each spin direction is given on the third page (note carefully that the majority- and minority-spin densities of states are plotted on different scales). On the fourth page for each element are given plots of r^2 times the total and valence charge densities (electrons/Bohr; the total charge density at the nucleus is given in the table on the first page), and the magnetization density m(r) in Bohr magnetons/Bohr^3 (*not* multiplied by r^2), as functions of $\sqrt{r}$. These magnetization profiles are close to those observed by neutron diffraction,[60] though they may appear unfamiliar because the square-root scale emphasizes the region near the nucleus, and the resolution of the neutron-diffraction experiment is not high enough to see the magnetization reversal there. That the magnetization density is negative near the nucleus is observable through nuclear magnetic resonance,[61] and is a consequence of exchange polarization;[62] the presence of spin-polarized orbitals with different occupations means that the exchange forces on spin-up and spin-down electrons are different. In other words, the presence of spin polarization in the 3d-orbitals can induce a spin polarization in the s core orbitals and thus a net magnetization density near the nucleus, even though the 3d-wavefunctions vanish there.

Finally, on the last two pages are given tables of the spin-up and spin-down charge densities and potentials as functions of radius r. As pointed out earlier in this Chapter, both potentials cannot simultaneously be zero in the interstitial region. We have arbitrarily chosen the energy reference in order to make the spin-up interstitial potential zero, so that the spin-down interstitial potential takes the constant value given below the table.

Iron

Atomic Number	26
Lattice	bcc
Lattice constant a_o	5.27 a.u.
Pressure at a_o	0.1 Kbar
Total energy (solid)	-2522.833 Ry
Total energy (atom)	-2522.639 Ry
Zero point energy	0.004 Ry
Cohesive energy	0.460 Ry
Energy of spin polarization (per atom)	0.021 Ry
Bulk modulus	2.17 Mbar
Fermi energy	0.756 Ry
Spin magnetic moment	2.15 μ_B/atom
dln M/dP	-4.9×10^{-4} $Kbar^{-1}$
Interstitial magnetic moment	-0.0315 μ_B/atom
Interstitial electrons (per atom)	1.009
$N(E_f)$	1.00 states/eV-atom
ρ_o	1.186×10^4 electrons/$Bohr^3$

Symmetry point	Majority-spin band energies in Rydbergs					
Γ	0.123	0.574	0.574	0.574	0.685	0.685
H	0.376	0.376	0.763	0.763	0.763	
N	0.371	0.488	0.687	0.703	0.785	0.785
P	0.499	0.499	0.499	0.706	0.706	

Symmetry point	Minority-spin band energies in Rydbergs					
Γ	0.151	0.723	0.723	0.723	0.843	0.843
H	0.483	0.483				
N	0.468	0.622	0.814	0.846	0.866	
P	0.613	0.613	0.613	0.870	0.870	

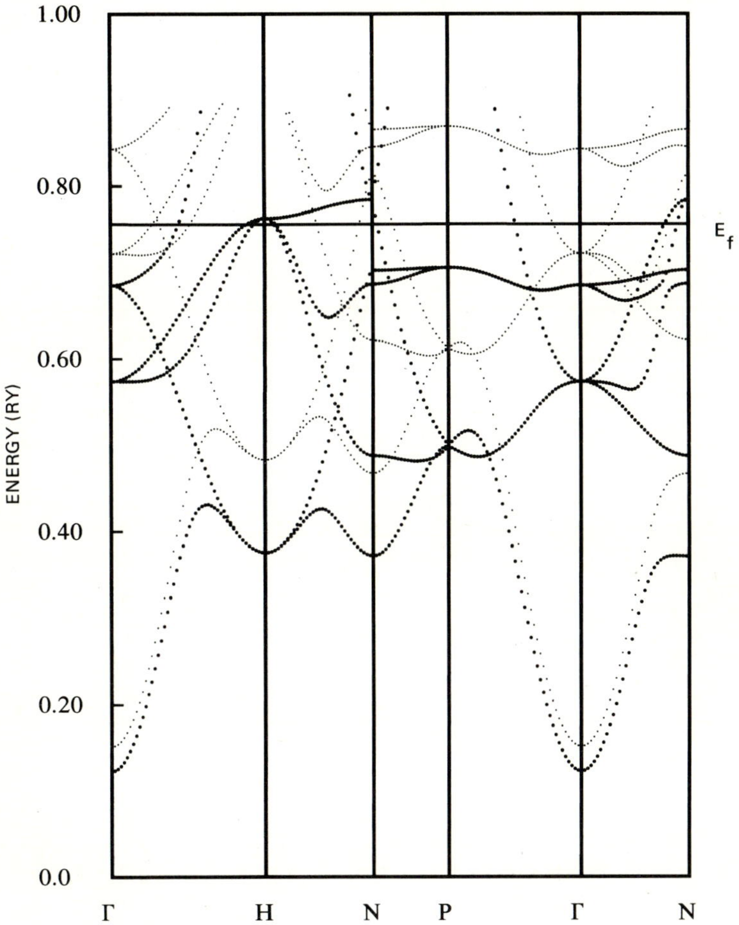

Dark points for majority spin, light points for minority spin.

Z=26

Moruzzi
Janak
Williams

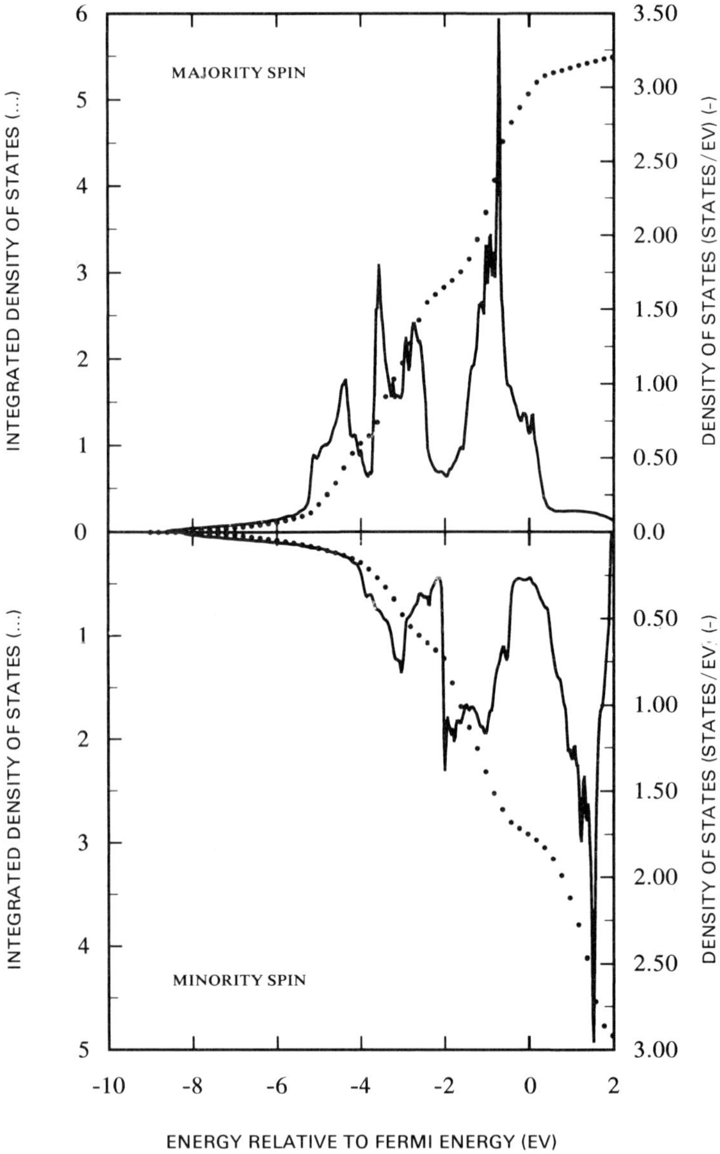

Moruzzi
Janak
Williams

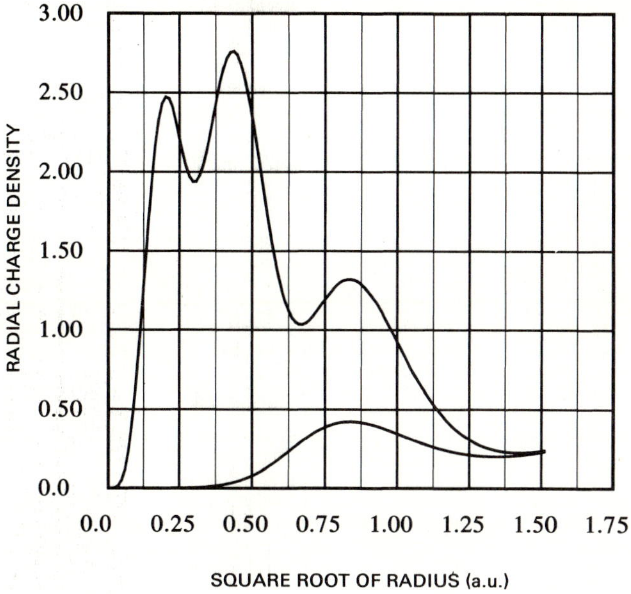

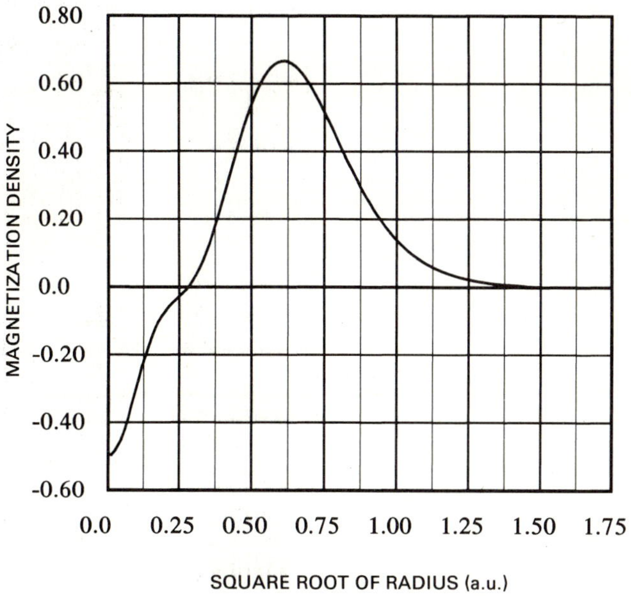

Moruzzi
Janak
Williams

Iron

r	2z(r)	$r^2\rho$	r	2z(r)	$r^2\rho$
1.392808E-04	5.197391E+01	1.142117E-04	5.884613E-01	1.293660E+01	7.026405E-01
5.571232E-04	5.189546E+01	1.788113E-03	6.067072E-01	1.249539E+01	7.168094E-01
1.253527E-03	5.176433E+01	8.730423E-03	6.252315E-01	1.206275E+01	7.287029E-01
2.228493E-03	5.157974E+01	2.622891E-02	6.440344E-01	1.163897E+01	7.379858E-01
3.482020E-03	5.134122E+01	5.999699E-02	6.631159E-01	1.122433E+01	7.444264E-01
5.014107E-03	5.104857E+01	1.148929E-01	6.824759E-01	1.081908E+01	7.478889E-01
6.824758E-03	5.070207E+01	1.937595E-01	7.021145E-01	1.042350E+01	7.483237E-01
8.913971E-03	5.030286E+01	2.966049E-01	7.220317E-01	1.003782E+01	7.457575E-01
1.128174E-02	4.985278E+01	4.202744E-01	7.422274E-01	9.662195E+00	7.402821E-01
1.392808E-02	4.935471E+01	5.586643E-01	7.627016E-01	9.296782E+00	7.320419E-01
1.685297E-02	4.881186E+01	7.034323E-01	7.834545E-01	8.941651E+00	7.212254E-01
2.005643E-02	4.822832E+01	8.450580E-01	8.044859E-01	8.596836E+00	7.080525E-01
2.353846E-02	4.760832E+01	9.740755E-01	8.257959E-01	8.262340E+00	6.927665E-01
2.729904E-02	4.695619E+01	1.082270E+00	8.473844E-01	7.938098E+00	6.756240E-01
3.133818E-02	4.627603E+01	1.163658E+00	8.692515E-01	7.623990E+00	6.568897E-01
3.565589E-02	4.557159E+01	1.215109E+00	8.913971E-01	7.319887E+00	6.368273E-01
4.025215E-02	4.484615E+01	1.236569E+00	9.138213E-01	7.025591E+00	6.156957E-01
4.512698E-02	4.410252E+01	1.230884E+00	9.365241E-01	6.740906E+00	5.937443E-01
5.028037E-02	4.334317E+01	1.203262E+00	9.595054E-01	6.465580E+00	5.712088E-01
5.571232E-02	4.257028E+01	1.160542E+00	9.827653E-01	6.199370E+00	5.483096E-01
6.142283E-02	4.178596E+01	1.110322E+00	1.006304E+00	5.941986E+00	5.252489E-01
6.741190E-02	4.099239E+01	1.060108E+00	1.030120E+00	5.693365E+00	5.022107E-01
7.367951E-02	4.019186E+01	1.016571E+00	1.054216E+00	5.452581E+00	4.793607E-01
8.022571E-02	3.938672E+01	9.849861E-01	1.078590E+00	5.219999E+00	4.568416E-01
8.705050E-02	3.857922E+01	9.688749E-01	1.103243E+00	4.995106E+00	4.347810E-01
9.415382E-02	3.777139E+01	9.698753E-01	1.128174E+00	4.777608E+00	4.132860E-01
1.015357E-01	3.696474E+01	9.878125E-01	1.153384E+00	4.566999E+00	3.924463E-01
1.091961E-01	3.616035E+01	1.020910E+00	1.178872E+00	4.363447E+00	3.723353E-01
1.171351E-01	3.535904E+01	1.066139E+00	1.204639E+00	4.166245E+00	3.530115E-01
1.253527E-01	3.456155E+01	1.119598E+00	1.230684E+00	3.975581E+00	3.345186E-01
1.338488E-01	3.376866E+01	1.176930E+00	1.257009E+00	3.790776E+00	3.168885E-01
1.426235E-01	3.298120E+01	1.233714E+00	1.283611E+00	3.611821E+00	3.001412E-01
1.516768E-01	3.220033E+01	1.285797E+00	1.310493E+00	3.438514E+00	2.842873E-01
1.610086E-01	3.142723E+01	1.329567E+00	1.337652E+00	3.270657E+00	2.693277E-01
1.706190E-01	3.066316E+01	1.362144E+00	1.365090E+00	3.108072E+00	2.552568E-01
1.805079E-01	2.990926E+01	1.381494E+00	1.392808E+00	2.950377E+00	2.420624E-01
1.906754E-01	2.916653E+01	1.386469E+00	1.420803E+00	2.797866E+00	2.297263E-01
2.011214E-01	2.843578E+01	1.376793E+00	1.449077E+00	2.650182E+00	2.182267E-01
2.118461E-01	2.771754E+01	1.352977E+00	1.477630E+00	2.506989E+00	2.075384E-01
2.228492E-01	2.701210E+01	1.316218E+00	1.506461E+00	2.368399E+00	1.976334E-01
2.341310E-01	2.631949E+01	1.268256E+00	1.535570E+00	2.234535E+00	1.884811E-01
2.456913E-01	2.563953E+01	1.211235E+00	1.564959E+00	2.104877E+00	1.800507E-01
2.575302E-01	2.497188E+01	1.147539E+00	1.594625E+00	1.979571E+00	1.723100E-01
2.696476E-01	2.431609E+01	1.079657E+00	1.624571E+00	1.858549E+00	1.652270E-01
2.820436E-01	2.367171E+01	1.010053E+00	1.654795E+00	1.741755E+00	1.587694E-01
2.947181E-01	2.303827E+01	9.410452E-01	1.685297E+00	1.629143E+00	1.529056E-01
3.076712E-01	2.241541E+01	8.747280E-01	1.716079E+00	1.520442E+00	1.476054E-01
3.209029E-01	2.180287E+01	8.128934E-01	1.747138E+00	1.416059E+00	1.428385E-01
3.344132E-01	2.120059E+01	7.569942E-01	1.778476E+00	1.315740E+00	1.385769E-01
3.482020E-01	2.060860E+01	7.081153E-01	1.810093E+00	1.219233E+00	1.347939E-01
3.622693E-01	2.002705E+01	6.669763E-01	1.841989E+00	1.126735E+00	1.314639E-01
3.766153E-01	1.945613E+01	6.339427E-01	1.874162E+00	1.038448E+00	1.285629E-01
3.912398E-01	1.889601E+01	6.090517E-01	1.906614E+00	9.539136E-01	1.260694E-01
4.061428E-01	1.834677E+01	5.920531E-01	1.939345E+00	8.733404E-01	1.239627E-01
4.213244E-01	1.780827E+01	5.824519E-01	1.972355E+00	7.967220E-01	1.222244E-01
4.367846E-01	1.728033E+01	5.795557E-01	2.005643E+00	7.240517E-01	1.208377E-01
4.525233E-01	1.676251E+01	5.825265E-01	2.039209E+00	6.553401E-01	1.197879E-01
4.685406E-01	1.625435E+01	5.904266E-01	2.073055E+00	5.903688E-01	1.190614E-01
4.848365E-01	1.575538E+01	6.022634E-01	2.107179E+00	5.295991E-01	1.186466E-01
5.014108E-01	1.526514E+01	6.170316E-01	2.141582E+00	4.726058E-01	1.185341E-01
5.182638E-01	1.478327E+01	6.337476E-01	2.176262E+00	4.198615E-01	1.187152E-01
5.353954E-01	1.430952E+01	6.514786E-01	2.211222E+00	3.709596E-01	1.191839E-01
5.528055E-01	1.384381E+01	6.693653E-01	2.246460E+00	3.261608E-01	1.199350E-01
5.704942E-01	1.338613E+01	6.866408E-01	2.281977E+00	2.855158E-01	1.209651E-01

Z=26; majority spin; $V_{interstitial} = 0$

r	2z(r)	$r^2\rho$	r	2z(r)	$r^2\rho$
1.392808E - 04	5.197391E+01	1.142213E - 04	5.884613E - 01	1.279294E+01	5.348995E - 01
5.571232E - 04	5.189546E+01	1.788263E - 03	6.067072E - 01	1.235015E+01	5.464405E - 01
1.253527E - 03	5.176431E+01	8.731153E - 03	6.252315E - 01	1.191610E+01	5.562511E - 01
2.228493E - 03	5.157974E+01	2.623112E - 02	6.440344E - 01	1.149103E+01	5.639946E - 01
3.482020E - 03	5.134122E+01	6.000207E - 02	6.631159E - 01	1.107517E+01	5.694323E - 01
5.014107E - 03	5.104855E+01	1.149026E - 01	6.824759E - 01	1.066878E+01	5.724158E - 01
6.824758E - 03	5.070206E+01	1.937757E - 01	7.021145E - 01	1.027208E+01	5.728785E - 01
8.913971E - 03	5.030287E+01	2.966298E - 01	7.220317E - 01	9.885312E+00	5.708263E - 01
1.128174E - 02	4.985281E+01	4.203097E - 01	7.422274E - 01	9.508615E+00	5.663255E - 01
1.392808E - 02	4.935472E+01	5.587103E - 01	7.627016E - 01	9.142137E+00	5.594944E - 01
1.685297E - 02	4.881187E+01	7.034916E - 01	7.834545E - 01	8.785946E+00	5.504917E - 01
2.005643E - 02	4.822833E+01	8.451298E - 01	8.044859E - 01	8.440075E+00	5.395066E - 01
2.353846E - 02	4.760834E+01	9.741578E - 01	8.257959E - 01	8.104526E+00	5.267509E - 01
2.729904E - 02	4.695622E+01	1.082362E+00	8.473844E - 01	7.779238E+00	5.124497E - 01
3.133818E - 02	4.627606E+01	1.163759E+00	8.692515E - 01	7.464095E+00	4.968353E - 01
3.565589E - 02	4.557161E+01	1.215219E+00	8.913971E - 01	7.158971E+00	4.801406E - 01
4.025215E - 02	4.484616E+01	1.236687E+00	9.138213E - 01	6.863676E+00	4.625939E - 01
4.512698E - 02	4.410255E+01	1.231004E+00	9.365241E - 01	6.578019E+00	4.444151E - 01
5.028037E - 02	4.334320E+01	1.203384E+00	9.595054E - 01	6.301760E+00	4.258120E - 01
5.571232E - 02	4.257031E+01	1.160664E+00	9.827653E - 01	6.034660E+00	4.069783E - 01
6.142283E - 02	4.178600E+01	1.110433E+00	1.006304E+00	5.776445E+00	3.880913E - 01
6.741190E - 02	4.099242E+01	1.060198E+00	1.030120E+00	5.527063E+00	3.693122E - 01
7.367951E - 02	4.019188E+01	1.016623E+00	1.054216E+00	5.285596E+00	3.507828E - 01
8.022571E - 02	3.938669E+01	9.849707E - 01	1.078590E+00	5.052430E+00	3.326287E - 01
8.705050E - 02	3.857916E+01	9.687479E - 01	1.103243E+00	4.827063E+00	3.149577E - 01
9.415382E - 02	3.777127E+01	9.695825E - 01	1.128174E+00	4.609223E+00	2.978606E - 01
1.015357E - 01	3.696452E+01	9.872749E - 01	1.153384E+00	4.398420E+00	2.814119E - 01
1.091961E - 01	3.616003E+01	1.020028E+00	1.178872E+00	4.194840E+00	2.656720E - 01
1.171351E - 01	3.535855E+01	1.064784E+00	1.204639E+00	3.997797E+00	2.506871E - 01
1.253527E - 01	3.456082E+01	1.117617E+00	1.230684E+00	3.807505E+00	2.364905E - 01
1.338488E - 01	3.376767E+01	1.174139E+00	1.257009E+00	3.623302E+00	2.231045E - 01
1.426235E - 01	3.297984E+01	1.229895E+00	1.283611E+00	3.445210E+00	2.105413E - 01
1.516768E - 01	3.219853E+01	1.280704E+00	1.310493E+00	3.273040E+00	1.988041E - 01
1.610086E - 01	3.142487E+01	1.322919E+00	1.337652E+00	3.106624E+00	1.878886E - 01
1.706190E - 01	3.066014E+01	1.353633E+00	1.365090E+00	2.945806E+00	1.777837E - 01
1.805079E - 01	2.990543E+01	1.370783E+00	1.392808E+00	2.790221E+00	1.684732E - 01
1.906754E - 01	2.916171E+01	1.373200E+00	1.420803E+00	2.640185E+00	1.599356E - 01
2.011214E - 01	2.842975E+01	1.360586E+00	1.449077E+00	2.495357E+00	1.521469E - 01
2.118461E - 01	2.771008E+01	1.333439E+00	1.477630E+00	2.355413E+00	1.450793E - 01
2.228492E - 01	2.700288E+01	1.292941E+00	1.506461E+00	2.220472E+00	1.387035E - 01
2.341310E - 01	2.630818E+01	1.240838E+00	1.535570E+00	2.090664E+00	1.329885E - 01
2.456913E - 01	2.562570E+01	1.179268E+00	1.564959E+00	1.965464E+00	1.279027E - 01
2.575302E - 01	2.495509E+01	1.110629E+00	1.594625E+00	1.845010E+00	1.234137E - 01
2.696476E - 01	2.429578E+01	1.037425E+00	1.624571E+00	1.729224E+00	1.194899E - 01
2.820436E - 01	2.364726E+01	9.621413E - 01	1.654795E+00	1.618031E+00	1.160994E - 01
2.947181E - 01	2.300897E+01	8.871268E - 01	1.685297E+00	1.511350E+00	1.132113E - 01
3.076712E - 01	2.238051E+01	8.145071E - 01	1.716079E+00	1.408891E+00	1.107961E - 01
3.209029E - 01	2.176161E+01	7.461159E - 01	1.747138E+00	1.311018E+00	1.088246E - 01
3.344132E - 01	2.115216E+01	6.834481E - 01	1.778476E+00	1.217435E+00	1.072696E - 01
3.482020E - 01	2.055228E+01	6.276364E - 01	1.810093E+00	1.127854E+00	1.061054E - 01
3.622693E - 01	1.996225E+01	5.794541E - 01	1.841989E+00	1.042421E+00	1.053072E - 01
3.766153E - 01	1.938243E+01	5.393177E - 01	1.874162E+00	9.612894E - 01	1.048520E - 01
3.912398E - 01	1.881328E+01	5.073207E - 01	1.906614E+00	8.839548E - 01	1.047189E - 01
4.061428E - 01	1.825513E+01	4.832701E - 01	1.939345E+00	8.105736E - 01	1.048879E - 01
4.213244E - 01	1.770815E+01	4.667278E - 01	1.972355E+00	7.410973E - 01	1.053408E - 01
4.367846E - 01	1.717238E+01	4.570589E - 01	2.005643E+00	6.754738E - 01	1.060611E - 01
4.525233E - 01	1.664755E+01	4.534808E - 01	2.039209E+00	6.136621E - 01	1.070336E - 01
4.685406E - 01	1.613330E+01	4.551096E - 01	2.073055E+00	5.554149E - 01	1.082447E - 01
4.848365E - 01	1.562914E+01	4.610039E - 01	2.107179E+00	5.011435E - 01	1.096818E - 01
5.014108E - 01	1.513454E+01	4.702053E - 01	2.141582E+00	4.503957E - 01	1.113344E - 01
5.182638E - 01	1.464904E+01	4.817726E - 01	2.176262E+00	4.036074E - 01	1.131921E - 01
5.353954E - 01	1.417229E+01	4.948099E - 01	2.211222E+00	3.603426E - 01	1.152470E - 01
5.528055E - 01	1.370408E+01	5.084891E - 01	2.246460E+00	3.208399E - 01	1.174914E - 01
5.704942E - 01	1.324427E+01	5.220679E - 01	2.281977E+00	2.851183E - 01	1.199189E - 01

Z=26; minority spin; $V_{interstitial}$ = -0.010991 Ry.

Cobalt

Atomic Number	27
Lattice	fcc
Lattice constant a_o	6.54 a.u.
Pressure at a_o	12.0 Kbar
Total energy (solid)	-2760.768 Ry
Total energy (atom)	-2760.288 Ry
Zero point energy	0.004 Ry
Cohesive energy	0.476 Ry
Energy of spin polarization (per atom)	0.008 Ry
Bulk modulus	2.40 Mbar
Fermi energy	0.727 Ry
Spin magnetic moment	1.56 μ_B/atom
dln M/dP	-1.7×10^{-4} $Kbar^{-1}$
Interstitial magnetic moment	-0.0332 μ_B/atom
Interstitial electrons (per atom)	0.774
$N(E_f)$	1.24 states/eV-atom
ρ_o	1.332×10^4 electrons/$Bohr^3$

Symmetry point	Majority-spin band energies in Rydbergs					
Γ	0.065	0.532	0.532	0.532	0.622	0.622
X	0.322	0.379	0.685	0.706	0.706	
L	0.342	0.524	0.524	0.688	0.688	0.704
W	0.396	0.481	0.481	0.616	0.706	
K	0.382	0.421	0.566	0.635	0.683	

Symmetry point	Minority-spin band energies in Rydbergs					
Γ	0.079	0.643	0.643	0.643	0.743	0.743
X	0.395	0.471	0.812	0.835	0.835	
L	0.414	0.633	0.633	0.717	0.815	0.815
W	0.475	0.573	0.573	0.731	0.835	
K	0.463	0.509	0.659	0.754	0.810	

Z=27

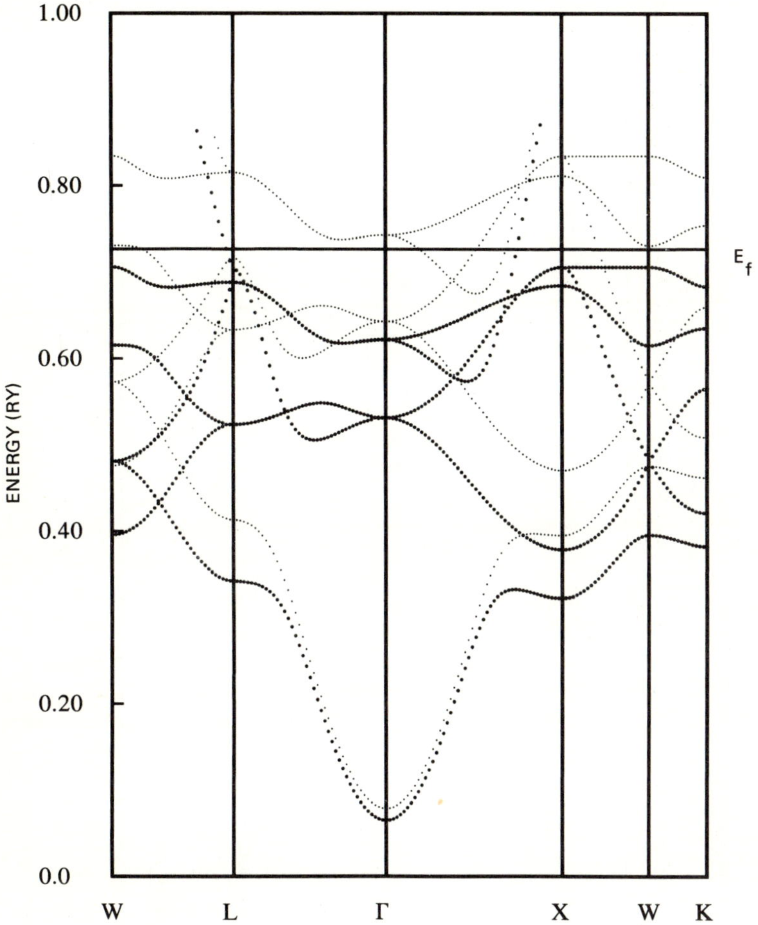

Dark points for majority spin, light points for minority spin.

Z=27

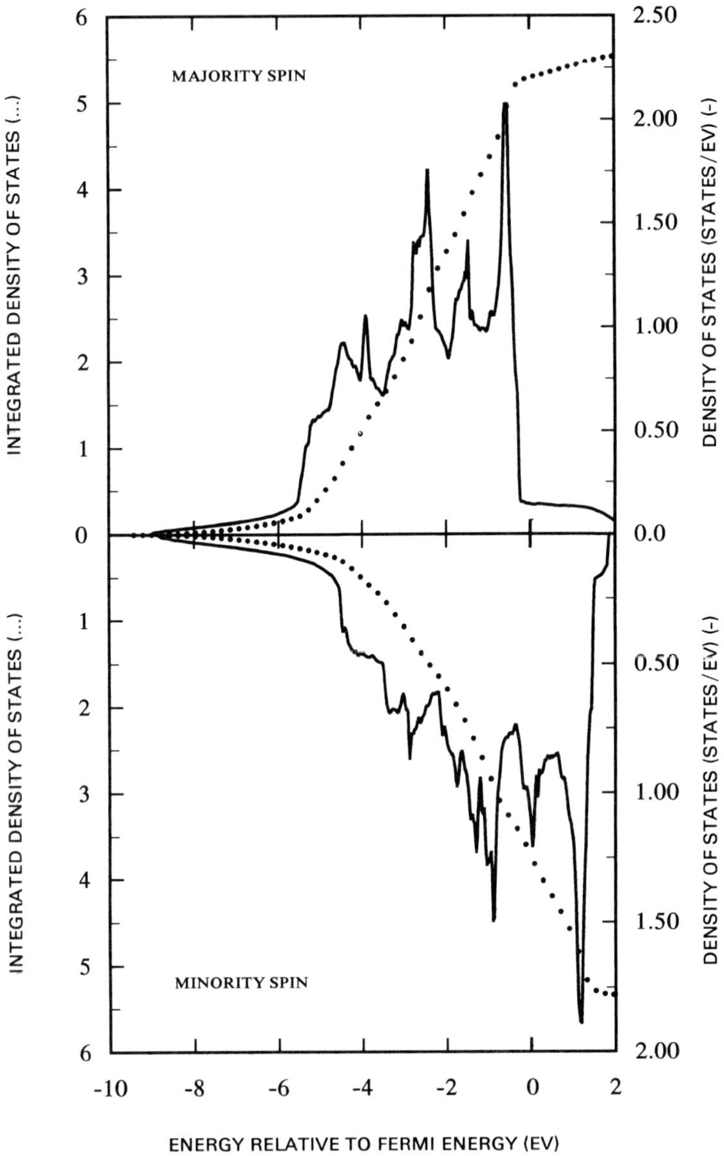

Moruzzi
Janak
Williams

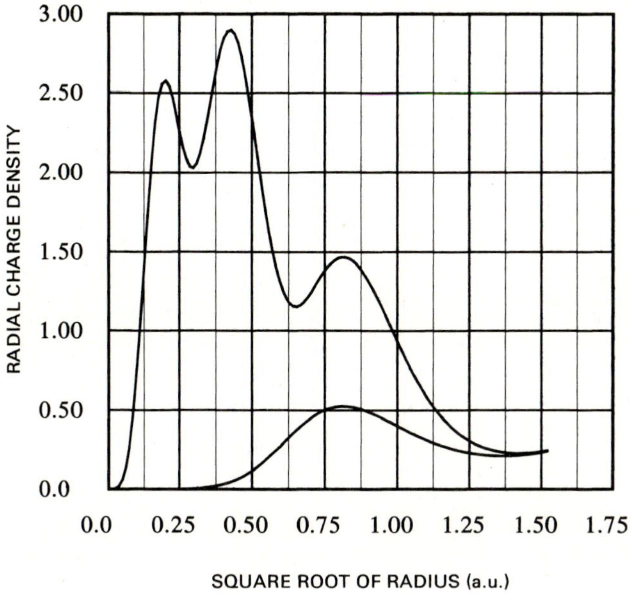
3.00
2.50
2.00
1.50
1.00
0.50
0.0
RADIAL CHARGE DENSITY
0.0 0.25 0.50 0.75 1.00 1.25 1.50 1.75
SQUARE ROOT OF RADIUS (a.u.)

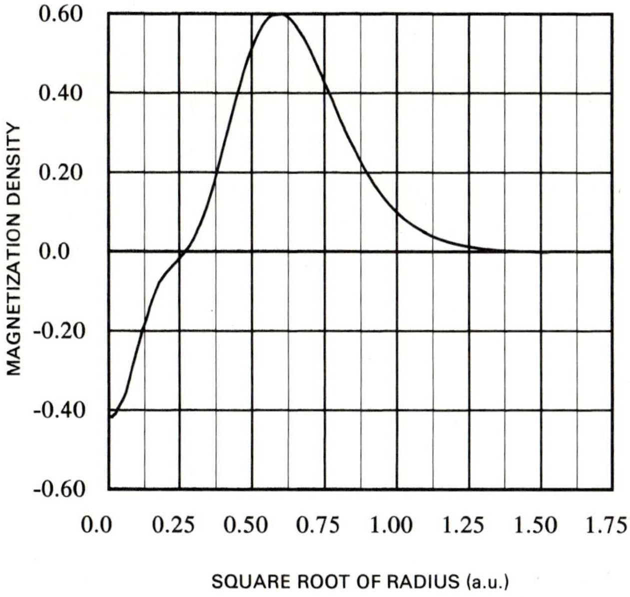
0.60
0.40
0.20
0.0
-0.20
-0.40
-0.60
MAGNETIZATION DENSITY
0.0 0.25 0.50 0.75 1.00 1.25 1.50 1.75
SQUARE ROOT OF RADIUS (a.u.)

Cobalt

r	2z(r)	$r^2\rho$	r	2z(r)	$r^2\rho$
1.411278E - 04	5.397198E+01	1.315903E - 04	5.962651E - 01	1.271349E+01	7.843740E - 01
5.645114E - 04	5.388783E+01	2.057857E - 03	6.147529E - 01	1.225203E+01	7.940590E - 01
1.270151E - 03	5.374712E+01	1.002848E - 02	6.335229E - 01	1.180108E+01	8.005088E - 01
2.258046E - 03	5.354904E+01	3.004906E - 02	6.525751E - 01	1.136099E+01	8.035777E - 01
3.528196E - 03	5.329309E+01	6.850189E - 02	6.719097E - 01	1.093199E+01	8.032221E - 01
5.080599E - 03	5.297910E+01	1.306363E - 01	6.915264E - 01	1.051442E+01	7.994870E - 01
6.915264E - 03	5.260754E+01	2.192331E - 01	7.114255E - 01	1.010839E+01	7.924939E - 01
9.032182E - 03	5.217969E+01	3.337128E - 01	7.316067E - 01	9.714101E+00	7.824249E - 01
1.143136E - 02	5.169785E+01	4.698542E - 01	7.520703E - 01	9.331579E+00	7.695104E - 01
1.411278E - 02	5.116522E+01	6.201679E - 01	7.728161E - 01	8.960897E+00	7.540160E - 01
1.707647E - 02	5.058549E+01	7.748427E - 01	7.938441E - 01	8.601986E+00	7.362311E - 01
2.032241E - 02	4.996317E+01	9.230744E - 01	8.151544E - 01	8.254753E+00	7.164592E - 01
2.385060E - 02	4.930292E+01	1.054523E+00	8.367470E - 01	7.919070E+00	6.950086E - 01
2.766106E - 02	4.860933E+01	1.160688E+00	8.586218E - 01	7.594733E+00	6.721851E - 01
3.175376E - 02	4.788675E+01	1.235925E+00	8.807789E - 01	7.281527E+00	6.482850E - 01
3.612873E - 02	4.713907E+01	1.278028E+00	9.032182E - 01	6.979188E+00	6.235909E - 01
4.078595E - 02	4.636964E+01	1.288320E+00	9.259398E - 01	6.687411E+00	5.983689E - 01
4.572542E - 02	4.558138E+01	1.271241E+00	9.489436E - 01	6.405900E+00	5.728631E - 01
5.094715E - 02	4.477676E+01	1.233612E+00	9.722297E - 01	6.134306E+00	5.472971E - 01
5.645114E - 02	4.395818E+01	1.183662E+00	9.957981E - 01	5.872304E+00	5.218707E - 01
6.223738E - 02	4.312793E+01	1.129998E+00	1.019649E+00	5.619560E+00	4.967595E - 01
6.830585E - 02	4.228850E+01	1.080640E+00	1.043781E+00	5.375710E+00	4.721184E - 01
7.465661E - 02	4.144243E+01	1.042251E+00	1.068196E+00	5.140389E+00	4.480782E - 01
8.128959E - 02	4.059225E+01	1.019604E+00	1.092894E+00	4.913259E+00	4.247487E - 01
8.820486E - 02	3.974028E+01	1.015323E+00	1.117873E+00	4.693979E+00	4.022190E - 01
9.540242E - 02	3.888834E+01	1.029866E+00	1.143135E+00	4.482230E+00	3.805599E - 01
1.028821E - 01	3.803783E+01	1.061744E+00	1.168679E+00	4.277699E+00	3.598263E - 01
1.106442E - 01	3.718961E+01	1.107871E+00	1.194506E+00	4.080091E+00	3.400570E - 01
1.186885E - 01	3.634462E+01	1.164031E+00	1.220614E+00	3.889126E+00	3.212756E - 01
1.270151E - 01	3.550356E+01	1.225364E+00	1.247005E+00	3.704539E+00	3.034963E - 01
1.356238E - 01	3.466754E+01	1.286838E+00	1.273679E+00	3.526086E+00	2.867213E - 01
1.445149E - 01	3.383772E+01	1.343674E+00	1.300633E+00	3.353768E+00	2.709424E - 01
1.536882E - 01	3.301538E+01	1.391686E+00	1.327871E+00	3.186911E+00	2.561468E - 01
1.631438E - 01	3.220187E+01	1.427522E+00	1.355392E+00	3.025551E+00	2.423134E - 01
1.728816E - 01	3.139851E+01	1.448810E+00	1.383194E+00	2.869512E+00	2.294161E - 01
1.829017E - 01	3.060641E+01	1.454227E+00	1.411278E+00	2.718868E+00	2.174248E - 01
1.932040E - 01	2.982645E+01	1.443463E+00	1.439645E+00	2.573001E+00	2.063071E - 01
2.037886E - 01	2.905922E+01	1.417140E+00	1.468294E+00	2.432017E+00	1.960278E - 01
2.146555E - 01	2.830505E+01	1.376676E+00	1.497225E+00	2.296027E+00	1.865495E - 01
2.258045E - 01	2.756403E+01	1.324118E+00	1.526439E+00	2.164462E+00	1.778355E - 01
2.372359E - 01	2.683592E+01	1.261963E+00	1.555934E+00	2.037690E+00	1.698478E - 01
2.489495E - 01	2.612044E+01	1.192973E+00	1.585712E+00	1.915162E+00	1.625493E - 01
2.609454E - 01	2.541711E+01	1.119998E+00	1.615772E+00	1.797268E+00	1.559027E - 01
2.732235E - 01	2.472552E+01	1.045833E+00	1.646114E+00	1.683702E+00	1.498728E - 01
2.857839E - 01	2.404523E+01	9.730672E - 01	1.676740E+00	1.574182E+00	1.444253E - 01
2.986265E - 01	2.337599E+01	9.039962E - 01	1.707646E+00	1.469107E+00	1.395268E - 01
3.117514E - 01	2.271762E+01	8.405436E - 01	1.738835E+00	1.368203E+00	1.351461E - 01
3.251585E - 01	2.207013E+01	7.842121E - 01	1.770308E+00	1.271199E+00	1.312543E - 01
3.388479E - 01	2.143369E+01	7.360692E - 01	1.802061E+00	1.178520E+00	1.278232E - 01
3.528196E - 01	2.080849E+01	6.967508E - 01	1.834097E+00	1.089902E+00	1.248274E - 01
3.670735E - 01	2.019479E+01	6.664830E - 01	1.866415E+00	1.005327E+00	1.222433E - 01
3.816097E - 01	1.959273E+01	6.451269E - 01	1.899015E+00	9.247668E - 01	1.200495E - 01
3.964281E - 01	1.900229E+01	6.322263E - 01	1.931899E+00	8.479766E - 01	1.182263E - 01
4.115288E - 01	1.842329E+01	6.270666E - 01	1.965064E+00	7.754111E - 01	1.167557E - 01
4.269117E - 01	1.785542E+01	6.287365E - 01	1.998511E+00	7.068374E - 01	1.156222E - 01
4.425769E - 01	1.729823E+01	6.361842E - 01	2.032241E+00	6.422567E - 01	1.148120E - 01
4.585243E - 01	1.675124E+01	6.482783E - 01	2.066253E+00	5.816822E - 01	1.143132E - 01
4.747540E - 01	1.621400E+01	6.638582E - 01	2.100547E+00	5.251296E - 01	1.141157E - 01
4.912660E - 01	1.568618E+01	6.817795E - 01	2.135122E+00	4.728519E - 01	1.142108E - 01
5.080602E - 01	1.516756E+01	7.009512E - 01	2.169981E+00	4.244233E - 01	1.145924E - 01
5.251367E - 01	1.465803E+01	7.203678E - 01	2.205122E+00	3.801082E - 01	1.152552E - 01
5.424954E - 01	1.415768E+01	7.391300E - 01	2.240545E+00	3.399600E - 01	1.161960E - 01
5.601364E - 01	1.366664E+01	7.564606E - 01	2.276251E+00	3.040326E - 01	1.174128E - 01
5.780597E - 01	1.318512E+01	7.717134E - 01	2.312239E+00	2.723931E - 01	1.189055E - 01

Z=27; majority spin; $V_{interstitial} = 0$

r	2z(r)	$r^2\rho$	r	2z(r)	$r^2\rho$
1.411278E - 04	5.397198E+01	1.315986E - 04	5.962651E - 01	1.260521E+01	6.460119E - 01
5.645114E - 04	5.388783E+01	2.057987E - 03	6.147529E - 01	1.214290E+01	6.546717E - 01
1.270151E - 03	5.374712E+01	1.002911E - 02	6.335229E - 01	1.169117E+01	6.605594E - 01
2.258046E - 03	5.354904E+01	3.005097E - 02	6.525751E - 01	1.125036E+01	6.635158E - 01
3.528196E - 03	5.329308E+01	6.850630E - 02	6.719097E - 01	1.082069E+01	6.634795E - 01
5.080599E - 03	5.297910E+01	1.306446E - 01	6.915264E - 01	1.040247E+01	6.604751E - 01
6.915264E - 03	5.260753E+01	2.192469E - 01	7.114255E - 01	9.995817E+00	6.546002E - 01
9.032182E - 03	5.217967E+01	3.337338E - 01	7.316067E - 01	9.600917E+00	6.460103E - 01
1.143136E - 02	5.169785E+01	4.698838E - 01	7.520703E - 01	9.217810E+00	6.349085E - 01
1.411278E - 02	5.116521E+01	6.202066E - 01	7.728161E - 01	8.846554E+00	6.215321E - 01
1.707647E - 02	5.058549E+01	7.748924E - 01	7.938441E - 01	8.487086E+00	6.061416E - 01
2.032241E - 02	4.996318E+01	9.231328E - 01	8.151544E - 01	8.139318E+00	5.890113E - 01
2.385060E - 02	4.930292E+01	1.054591E+00	8.367470E - 01	7.803122E+00	5.704211E - 01
2.766106E - 02	4.860934E+01	1.160764E+00	8.586218E - 01	7.478298E+00	5.506484E - 01
3.175376E - 02	4.788676E+01	1.236004E+00	8.807789E - 01	7.164635E+00	5.299631E - 01
3.612873E - 02	4.713908E+01	1.278114E+00	9.032182E - 01	6.861880E+00	5.086223E - 01
4.078595E - 02	4.636966E+01	1.288409E+00	9.259398E - 01	6.569731E+00	4.868672E - 01
4.572542E - 02	4.558139E+01	1.271333E+00	9.489436E - 01	6.287898E+00	4.649197E - 01
5.094715E - 02	4.477678E+01	1.233701E+00	9.722297E - 01	6.016044E+00	4.429821E - 01
5.645114E - 02	4.395819E+01	1.183745E+00	9.957981E - 01	5.753850E+00	4.212344E - 01
6.223738E - 02	4.312793E+01	1.130063E+00	1.019649E+00	5.500996E+00	3.998350E - 01
6.830585E - 02	4.228848E+01	1.080675E+00	1.043781E+00	5.257124E+00	3.789216E - 01
7.465661E - 02	4.144240E+01	1.042232E+00	1.068196E+00	5.021887E+00	3.586106E - 01
8.128959E - 02	4.059221E+01	1.019499E+00	1.092894E+00	4.794953E+00	3.389981E - 01
8.820486E - 02	3.974019E+01	1.015084E+00	1.117873E+00	4.575999E+00	3.201624E - 01
9.540242E - 02	3.888818E+01	1.029428E+00	1.143135E+00	4.364718E+00	3.021632E - 01
1.028821E - 01	3.803757E+01	1.061024E+00	1.168679E+00	4.160812E+00	2.850462E - 01
1.106442E - 01	3.718921E+01	1.106764E+00	1.194506E+00	3.964003E+00	2.688424E - 01
1.186885E - 01	3.634404E+01	1.162404E+00	1.220614E+00	3.774023E+00	2.535690E - 01
1.270151E - 01	3.550273E+01	1.223056E+00	1.247005E+00	3.590623E+00	2.392336E - 01
1.356238E - 01	3.466646E+01	1.283663E+00	1.273679E+00	3.413574E+00	2.258340E - 01
1.445149E - 01	3.383626E+01	1.339419E+00	1.300633E+00	3.242894E+00	2.133583E - 01
1.536882E - 01	3.301347E+01	1.386106E+00	1.327871E+00	3.077919E+00	2.017900E - 01
1.631438E - 01	3.219943E+01	1.420346E+00	1.355392E+00	2.918700E+00	1.911056E - 01
1.728816E - 01	3.139539E+01	1.439746E+00	1.383194E+00	2.765069E+00	1.812772E - 01
1.829017E - 01	3.060248E+01	1.442959E+00	1.411278E+00	2.617105E+00	1.722736E - 01
1.932040E - 01	2.982152E+01	1.429657E+00	1.439645E+00	2.474206E+00	1.640612E - 01
2.037886E - 01	2.905310E+01	1.400451E+00	1.468294E+00	2.336468E+00	1.566043E - 01
2.146555E - 01	2.829749E+01	1.356751E+00	1.497225E+00	2.204010E+00	1.498662E - 01
2.258045E - 01	2.755472E+01	1.300607E+00	1.526439E+00	2.076254E+00	1.438097E - 01
2.372359E - 01	2.682451E+01	1.234514E+00	1.555934E+00	1.953559E+00	1.383977E - 01
2.489495E - 01	2.610652E+01	1.161249E+00	1.585712E+00	1.835365E+00	1.335939E - 01
2.609454E - 01	2.540025E+01	1.083681E+00	1.615772E+00	1.722043E+00	1.293621E - 01
2.732235E - 01	2.470520E+01	1.004622E+00	1.646114E+00	1.613274E+00	1.256680E - 01
2.857839E - 01	2.402089E+01	9.266941E - 01	1.676740E+00	1.508735E+00	1.224785E - 01
2.986265E - 01	2.334702E+01	8.522273E - 01	1.707646E+00	1.408818E+00	1.197615E - 01
3.117514E - 01	2.268338E+01	7.831820E - 01	1.738835E+00	1.313210E+00	1.174874E - 01
3.251585E - 01	2.203009E+01	7.211049E - 01	1.770308E+00	1.221605E+00	1.156282E - 01
3.388479E - 01	2.138731E+01	6.671077E - 01	1.802061E+00	1.134398E+00	1.141572E - 01
3.528196E - 01	2.075546E+01	6.218768E - 01	1.834097E+00	1.051293E+00	1.130502E - 01
3.670735E - 01	2.013492E+01	5.856877E - 01	1.866415E+00	9.722331E - 01	1.122847E - 01
3.816097E - 01	1.952608E+01	5.584531E - 01	1.899015E+00	8.971578E - 01	1.118402E - 01
3.964281E - 01	1.892915E+01	5.397699E - 01	1.931899E+00	8.257923E - 01	1.116981E - 01
4.115288E - 01	1.834415E+01	5.289748E - 01	1.965064E+00	7.585534E - 01	1.118410E - 01
4.269117E - 01	1.777089E+01	5.252063E - 01	1.998511E+00	6.951799E - 01	1.122540E - 01
4.425769E - 01	1.720901E+01	5.274622E - 01	2.032241E+00	6.356470E - 01	1.129236E - 01
4.585243E - 01	1.665802E+01	5.346563E - 01	2.066253E+00	5.799335E - 01	1.138382E - 01
4.747540E - 01	1.611746E+01	5.456699E - 01	2.100547E+00	5.280373E - 01	1.149873E - 01
4.912660E - 01	1.558689E+01	5.593961E - 01	2.135122E+00	4.801895E - 01	1.163621E - 01
5.080602E - 01	1.506601E+01	5.747764E - 01	2.169981E+00	4.359388E - 01	1.179557E - 01
5.251367E - 01	1.455464E+01	5.908315E - 01	2.205122E+00	3.955346E - 01	1.197619E - 01
5.424954E - 01	1.405274E+01	6.066830E - 01	2.240545E+00	3.590146E - 01	1.217763E - 01
5.601364E - 01	1.356041E+01	6.215679E - 01	2.276251E+00	3.264169E - 01	1.239954E - 01
5.780597E - 01	1.307780E+01	6.348484E - 01	2.312239E+00	2.977930E - 01	1.264170E - 01

Z=27; minority spin; $V_{interstitial}$ = -0.015074 Ry.

Nickel

Atomic Number	28
Lattice	fcc
Lattice constant a_o	6.55 a.u.
Pressure at a_o	10.1 Kbar
Total energy (solid)	-3011.655 Ry
Total energy (atom)	-3011.233 Ry
Zero point energy	0.003 Ry
Cohesive energy	0.419 Ry
Energy of spin polarization (per atom)	0.006 Ry
Bulk modulus	2.27 Mbar
Fermi energy	0.683 Ry
Spin magnetic moment	0.59 μ_B/atom
dln M/dP	-2.1×10^{-4} $Kbar^{-1}$
Interstitial magnetic moment	-0.0184 μ_B/atom
Interstitial electrons (per atom)	0.718
$N(E_f)$	1.69 states/eV-atom
ρ_o	1.488×10^4 electrons/$Bohr^3$

Symmetry point	Majority-spin band energies in Rydbergs					
Γ	0.017	0.504	0.504	0.504	0.584	0.584
X	0.311	0.366	0.640	0.659	0.659	
L	0.323	0.497	0.497	0.644	0.644	0.655
W	0.380	0.456	0.456	0.579	0.659	
K	0.368	0.402	0.531	0.596	0.639	

Symmetry point	Minority-spin band energies in Rydbergs					
Γ	0.019	0.548	0.548	0.548	0.632	0.632
X	0.340	0.403	0.690	0.710	0.710	
L	0.350	0.541	0.541	0.656	0.693	0.693
W	0.412	0.492	0.492	0.625	0.710	
K	0.399	0.436	0.568	0.643	0.689	

Z=28

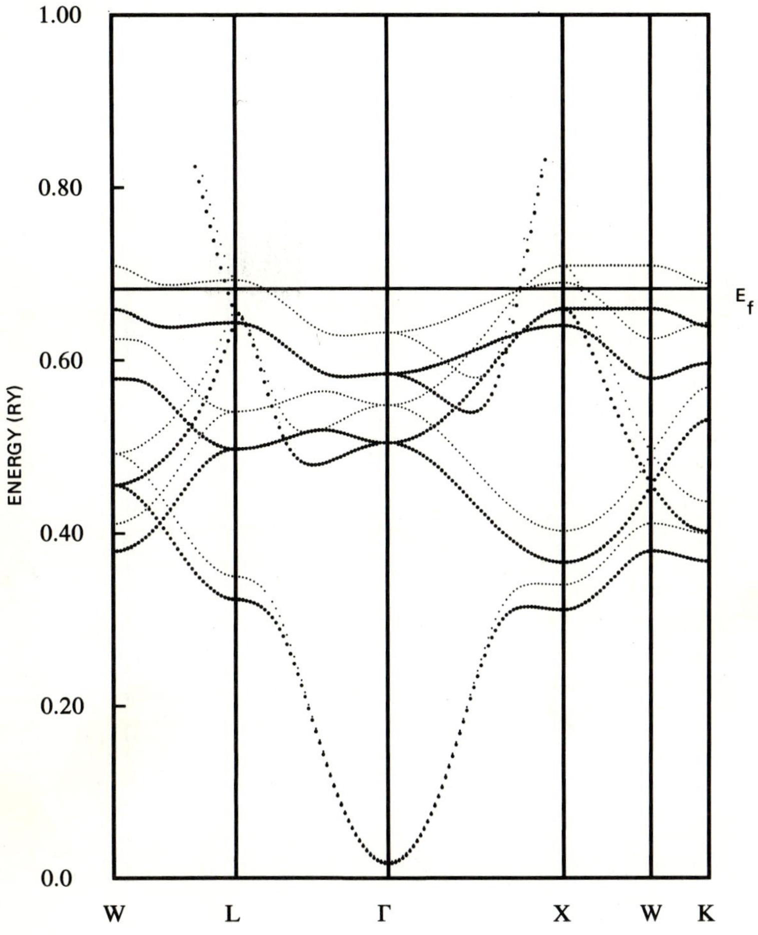

Dark points for majority spin, light points for minority spin.

Z=28

Moruzzi
Janak
Williams

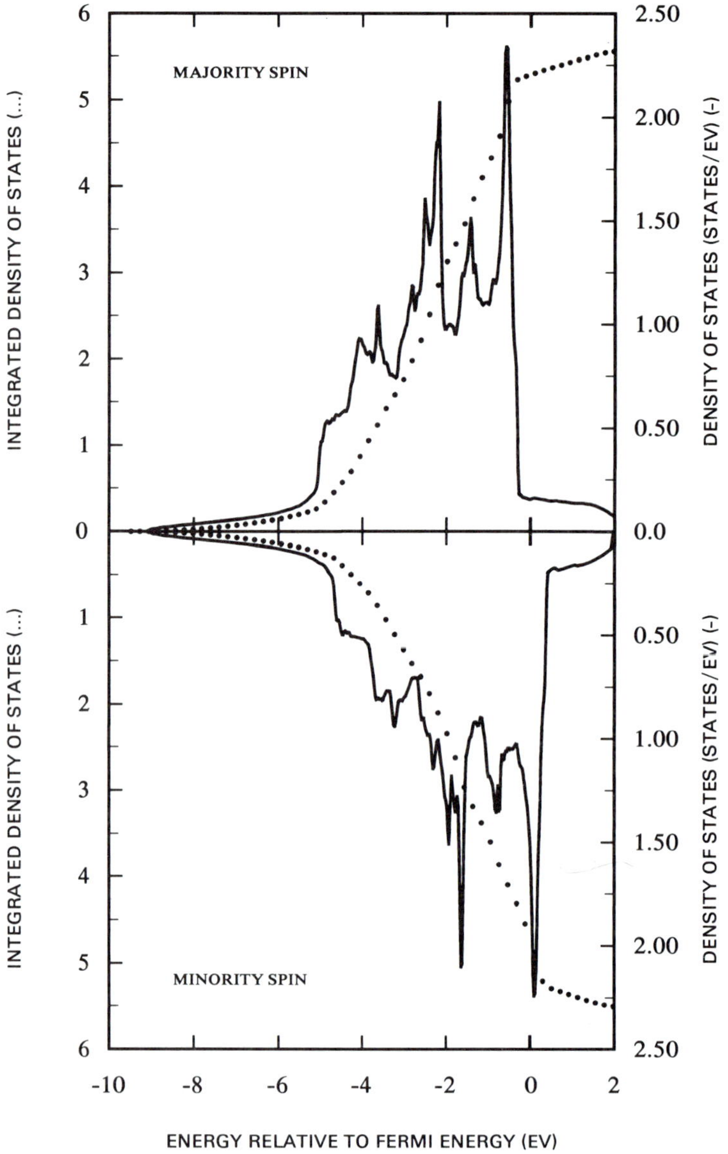

Moruzzi
Janak
Williams

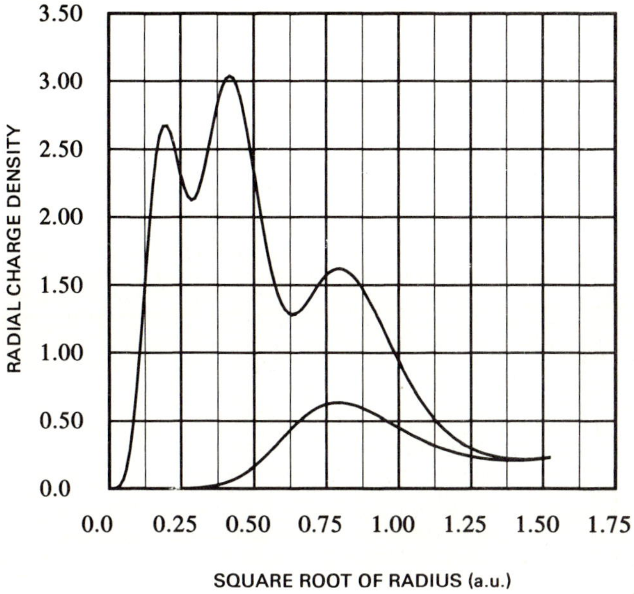

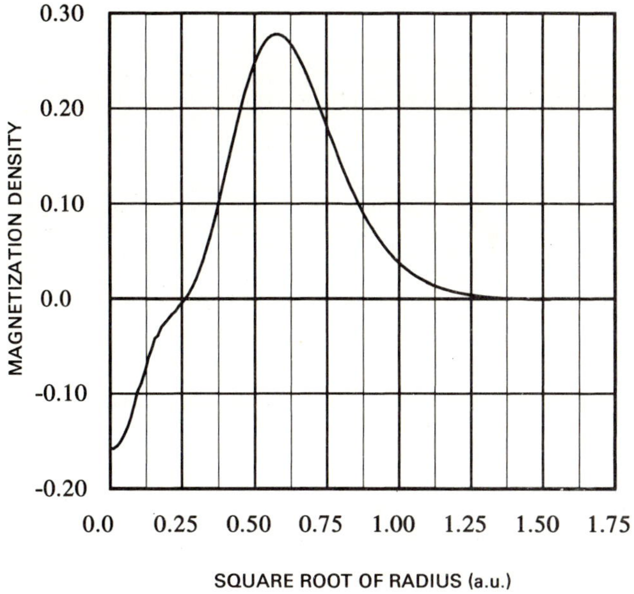

Moruzzi
Janak
Williams

Nickel

r	2z(r)	$r^2\rho$	r	2z(r)	$r^2\rho$
1.413436E - 04	5.597035E+01	1.474994E - 04	5.971768E - 01	1.259430E+01	8.344430E - 01
5.653745E - 04	5.588126E+01	2.304609E - 03	6.156928E - 01	1.211472E+01	8.390115E - 01
1.272093E - 03	5.573230E+01	1.121446E - 02	6.344915E - 01	1.164756E+01	8.398550E - 01
2.261498E - 03	5.552258E+01	3.353351E - 02	6.535729E - 01	1.119313E+01	8.369868E - 01
3.533591E - 03	5.525163E+01	7.624352E - 02	6.729370E - 01	1.075169E+01	8.305119E - 01
5.088370E - 03	5.491930E+01	1.449309E - 01	6.925837E - 01	1.032337E+01	8.206109E - 01
6.925836E - 03	5.452615E+01	2.422968E - 01	7.125132E - 01	9.908319E+00	8.075264E - 01
9.045992E - 03	5.407375E+01	3.672061E - 01	7.327253E - 01	9.506559E+00	7.915446E - 01
1.144883E - 02	5.356459E+01	5.144576E - 01	7.532201E - 01	9.118055E+00	7.729828E - 01
1.413436E - 02	5.300223E+01	6.753194E - 01	7.739977E - 01	8.742716E+00	7.521777E - 01
1.710258E - 02	5.239073E+01	8.386908E - 01	7.950579E - 01	8.380385E+00	7.294734E - 01
2.035348E - 02	5.173497E+01	9.926661E - 01	8.164008E - 01	8.030877E+00	7.052119E - 01
2.388707E - 02	5.103986E+01	1.126216E+00	8.380263E - 01	7.693917E+00	6.797272E - 01
2.770335E - 02	5.031027E+01	1.230667E+00	8.599346E - 01	7.369226E+00	6.533366E - 01
3.180232E - 02	4.955070E+01	1.300778E+00	8.821256E - 01	7.056446E+00	6.263403E - 01
3.618397E - 02	4.876512E+01	1.335241E+00	9.045992E - 01	6.755249E+00	5.990124E - 01
4.084831E - 02	4.795695E+01	1.336618E+00	9.273555E - 01	6.465247E+00	5.716048E - 01
4.579533E - 02	4.712918E+01	1.310757E+00	9.503945E - 01	6.186060E+00	5.443403E - 01
5.102505E - 02	4.628436E+01	1.265825E+00	9.737162E - 01	5.917274E+00	5.174169E - 01
5.653745E - 02	4.542505E+01	1.211168E+00	9.973206E - 01	5.658486E+00	4.910054E - 01
6.233254E - 02	4.455376E+01	1.156126E+00	1.021207E+00	5.409506E+00	4.652503E - 01
6.841028E - 02	4.367317E+01	1.109009E+00	1.045377E+00	5.169477E+00	4.402747E - 01
7.477075E - 02	4.278607E+01	1.076310E+00	1.069829E+00	4.938358E+00	4.161735E - 01
8.141392E - 02	4.189505E+01	1.062231E+00	1.094564E+00	4.715527E+00	3.930265E - 01
8.833975E - 02	4.100238E+01	1.068501E+00	1.119582E+00	4.500863E+00	3.708898E - 01
9.554827E - 02	4.010974E+01	1.094508E+00	1.144883E+00	4.293770E+00	3.498037E - 01
1.030395E - 01	3.921835E+01	1.137610E+00	1.170466E+00	4.094159E+00	3.297939E - 01
1.108134E - 01	3.832916E+01	1.193641E+00	1.196332E+00	3.901717E+00	3.108711E - 01
1.188700E - 01	3.744301E+01	1.257450E+00	1.222481E+00	3.715900E+00	2.930362E - 01
1.272092E - 01	3.656094E+01	1.323472E+00	1.248912E+00	3.536674E+00	2.762780E - 01
1.358312E - 01	3.568423E+01	1.386249E+00	1.275626E+00	3.363531E+00	2.605792E - 01
1.447359E - 01	3.481416E+01	1.440852E+00	1.302623E+00	3.196480E+00	2.459138E - 01
1.539232E - 01	3.395221E+01	1.483223E+00	1.329902E+00	3.035296E+00	2.322512E - 01
1.633932E - 01	3.309981E+01	1.510377E+00	1.357464E+00	2.879529E+00	2.195570E - 01
1.731459E - 01	3.225822E+01	1.520506E+00	1.385308E+00	2.729234E+00	2.077923E - 01
1.831813E - 01	3.142850E+01	1.512978E+00	1.413436E+00	2.583994E+00	1.969174E - 01
1.934994E - 01	3.061136E+01	1.488255E+00	1.441846E+00	2.443895E+00	1.868902E - 01
2.041002E - 01	2.980725E+01	1.447747E+00	1.470538E+00	2.308791E+00	1.776680E - 01
2.149836E - 01	2.901634E+01	1.393629E+00	1.499514E+00	2.178311E+00	1.692092E - 01
2.261498E - 01	2.823851E+01	1.328625E+00	1.528772E+00	2.052331E+00	1.614714E - 01
2.375986E - 01	2.747346E+01	1.255792E+00	1.558313E+00	1.930988E+00	1.544133E - 01
2.493301E - 01	2.672083E+01	1.178320E+00	1.588137E+00	1.814180E+00	1.479956E - 01
2.613443E - 01	2.598015E+01	1.099339E+00	1.618242E+00	1.701822E+00	1.421794E - 01
2.736412E - 01	2.525108E+01	1.021770E+00	1.648632E+00	1.593342E+00	1.369290E - 01
2.862208E - 01	2.453339E+01	9.481921E - 01	1.679303E+00	1.489398E+00	1.322091E - 01
2.990831E - 01	2.382695E+01	8.807573E - 01	1.710258E+00	1.389436E+00	1.279876E - 01
3.122280E - 01	2.313187E+01	8.211389E - 01	1.741494E+00	1.293641E+00	1.242336E - 01
3.256557E - 01	2.244836E+01	7.705054E - 01	1.773014E+00	1.201710E+00	1.209193E - 01
3.393660E - 01	2.177673E+01	7.295337E - 01	1.804816E+00	1.113846E+00	1.180181E - 01
3.533590E - 01	2.111728E+01	6.984323E - 01	1.836902E+00	1.029762E+00	1.155061E - 01
3.676347E - 01	2.047015E+01	6.769934E - 01	1.869269E+00	9.496665E - 01	1.133615E - 01
3.821931E - 01	1.983539E+01	6.646569E - 01	1.901919E+00	8.735328E - 01	1.115644E - 01
3.970342E - 01	1.921283E+01	6.605756E - 01	1.934853E+00	8.010920E - 01	1.100972E - 01
4.121580E - 01	1.860214E+01	6.636893E - 01	1.968068E+00	7.325724E - 01	1.089439E - 01
4.275644E - 01	1.800291E+01	6.727930E - 01	2.001567E+00	6.677201E - 01	1.080909E - 01
4.432536E - 01	1.741469E+01	6.866055E - 01	2.035348E+00	6.067801E - 01	1.075261E - 01
4.592254E - 01	1.683710E+01	7.038267E - 01	2.069411E+00	5.497596E - 01	1.072391E - 01
4.754799E - 01	1.626988E+01	7.231894E - 01	2.103758E+00	4.964220E - 01	1.072220E - 01
4.920171E - 01	1.571289E+01	7.435045E - 01	2.138388E+00	4.467909E - 01	1.074677E - 01
5.088370E - 01	1.516615E+01	7.636862E - 01	2.173299E+00	4.013826E - 01	1.079709E - 01
5.259396E - 01	1.462981E+01	7.827827E - 01	2.208493E+00	3.597416E - 01	1.087281E - 01
5.433249E - 01	1.410410E+01	7.999867E - 01	2.243971E+00	3.219177E - 01	1.097375E - 01
5.609928E - 01	1.358938E+01	8.146424E - 01	2.279731E+00	2.882028E - 01	1.109982E - 01
5.789435E - 01	1.308598E+01	8.262466E - 01	2.315774E+00	2.584172E - 01	1.125110E - 01

Z=28; majority spin; $V_{interstitial} = 0$

r	2z(r)	$r^2\rho$	r	2z(r)	$r^2\rho$
1.413436E - 04	5.597035E+01	1.475024E - 04	5.971768E - 01	1.255239E+01	7.765000E - 01
5.653745E - 04	5.588126E+01	2.304660E - 03	6.156928E - 01	1.207257E+01	7.810183E - 01
1.272093E - 03	5.573228E+01	1.121471E - 02	6.344915E - 01	1.160518E+01	7.820031E - 01
2.261498E - 03	5.552258E+01	3.353421E - 02	6.535729E - 01	1.115055E+01	7.794583E - 01
3.533591E - 03	5.525163E+01	7.624525E - 02	6.729370E - 01	1.070890E+01	7.734779E - 01
5.088370E - 03	5.491928E+01	1.449341E - 01	6.925837E - 01	1.028040E+01	7.642310E - 01
6.925836E - 03	5.452614E+01	2.423022E - 01	7.125132E - 01	9.865174E+00	7.519471E - 01
9.045992E - 03	5.407376E+01	3.672141E - 01	7.327253E - 01	9.463248E+00	7.368993E - 01
1.144883E - 02	5.356461E+01	5.144693E - 01	7.532201E - 01	9.074590E+00	7.193919E - 01
1.413436E - 02	5.300223E+01	6.753349E - 01	7.739977E - 01	8.699109E+00	6.997482E - 01
1.710258E - 02	5.239073E+01	8.387092E - 01	7.950579E - 01	8.336648E+00	6.782994E - 01
2.035348E - 02	5.173495E+01	9.926882E - 01	8.164008E - 01	7.987021E+00	6.553742E - 01
2.388707E - 02	5.103986E+01	1.126240E+00	8.380263E - 01	7.649959E+00	6.312946E - 01
2.770335E - 02	5.031027E+01	1.230697E+00	8.599346E - 01	7.325182E+00	6.063668E - 01
3.180232E - 02	4.955069E+01	1.300809E+00	8.821256E - 01	7.012336E+00	5.808794E - 01
3.618397E - 02	4.876511E+01	1.335275E+00	9.045992E - 01	6.711097E+00	5.550969E - 01
4.084831E - 02	4.795697E+01	1.336654E+00	9.273555E - 01	6.421075E+00	5.292611E - 01
4.579533E - 02	4.712918E+01	1.310791E+00	9.503945E - 01	6.141897E+00	5.035866E - 01
5.102505E - 02	4.628436E+01	1.265860E+00	9.737162E - 01	5.873154E+00	4.782633E - 01
5.653745E - 02	4.542503E+01	1.211195E+00	9.973206E - 01	5.614443E+00	4.534552E - 01
6.233254E - 02	4.455374E+01	1.156143E+00	1.021207E+00	5.365581E+00	4.293003E - 01
6.841028E - 02	4.367317E+01	1.109008E+00	1.045377E+00	5.125712E+00	4.059151E - 01
7.477075E - 02	4.278604E+01	1.076278E+00	1.069829E+00	4.894806E+00	3.833912E - 01
8.141392E - 02	4.189502E+01	1.062147E+00	1.094564E+00	4.672241E+00	3.618025E - 01
8.833975E - 02	4.100230E+01	1.068343E+00	1.119582E+00	4.457900E+00	3.412015E - 01
9.554827E - 02	4.010963E+01	1.094238E+00	1.144883E+00	4.251198E+00	3.216262E - 01
1.030395E - 01	3.921819E+01	1.137188E+00	1.170466E+00	4.052046E+00	3.030984E - 01
1.108134E - 01	3.832893E+01	1.193010E+00	1.196332E+00	3.860135E+00	2.856274E - 01
1.188700E - 01	3.744272E+01	1.256545E+00	1.222481E+00	3.674930E+00	2.692115E - 01
1.272092E - 01	3.656052E+01	1.322215E+00	1.248912E+00	3.496399E+00	2.538386E - 01
1.358312E - 01	3.568364E+01	1.384543E+00	1.275626E+00	3.324042E+00	2.394899E - 01
1.447359E - 01	3.481343E+01	1.438598E+00	1.302623E+00	3.157869E+00	2.261384E - 01
1.539232E - 01	3.395126E+01	1.480300E+00	1.329902E+00	2.997663E+00	2.137533E - 01
1.633932E - 01	3.309857E+01	1.506658E+00	1.357464E+00	2.842971E+00	2.022997E - 01
1.731459E - 01	3.225668E+01	1.515852E+00	1.385308E+00	2.693853E+00	1.917385E - 01
1.831813E - 01	3.142656E+01	1.507242E+00	1.413436E+00	2.549895E+00	1.820301E - 01
1.934994E - 01	3.060892E+01	1.481287E+00	1.441846E+00	2.411180E+00	1.731327E - 01
2.041002E - 01	2.980424E+01	1.439393E+00	1.470538E+00	2.277565E+00	1.650037E - 01
2.149836E - 01	2.901262E+01	1.383731E+00	1.499514E+00	2.148671E+00	1.576021E - 01
2.261498E - 01	2.823393E+01	1.317029E+00	1.528772E+00	2.024374E+00	1.508861E - 01
2.375986E - 01	2.746790E+01	1.242350E+00	1.558313E+00	1.904808E+00	1.448150E - 01
2.493301E - 01	2.671404E+01	1.162890E+00	1.588137E+00	1.789864E+00	1.393498E - 01
2.613443E - 01	2.597195E+01	1.081795E+00	1.618242E+00	1.679445E+00	1.344529E - 01
2.736412E - 01	2.524126E+01	1.001992E+00	1.648632E+00	1.572977E+00	1.300892E - 01
2.862208E - 01	2.452168E+01	9.260821E - 01	1.679303E+00	1.471105E+00	1.262239E - 01
2.990831E - 01	2.381314E+01	8.562319E - 01	1.710258E+00	1.373263E+00	1.228259E - 01
3.122280E - 01	2.311572E+01	7.941377E - 01	1.741494E+00	1.279628E+00	1.198651E - 01
3.256557E - 01	2.242970E+01	7.409872E - 01	1.773014E+00	1.189886E+00	1.173142E - 01
3.393660E - 01	2.175545E+01	6.974812E - 01	1.804816E+00	1.104223E+00	1.151476E - 01
3.533590E - 01	2.109332E+01	6.638508E - 01	1.836902E+00	1.022344E+00	1.133420E - 01
3.676347E - 01	2.044360E+01	6.399128E - 01	1.869269E+00	9.444415E - 01	1.118762E - 01
3.821931E - 01	1.980638E+01	6.251309E - 01	1.901919E+00	8.704756E - 01	1.107312E - 01
3.970342E - 01	1.918156E+01	6.186808E - 01	1.934853E+00	8.001702E - 01	1.098895E - 01
4.121580E - 01	1.856891E+01	6.195258E - 01	1.968068E+00	7.337409E - 01	1.093360E - 01
4.275644E - 01	1.796796E+01	6.264821E - 01	2.001567E+00	6.709223E - 01	1.090572E - 01
4.432536E - 01	1.737831E+01	6.382878E - 01	2.035348E+00	6.119502E - 01	1.090416E - 01
4.592254E - 01	1.679953E+01	6.536609E - 01	2.069411E+00	5.568161E - 01	1.092791E - 01
4.754799E - 01	1.623134E+01	6.713508E - 01	2.103758E+00	5.052809E - 01	1.097617E - 01
4.920171E - 01	1.567356E+01	6.901789E - 01	2.138388E+00	4.573597E - 01	1.104827E - 01
5.088370E - 01	1.512618E+01	7.090722E - 01	2.173299E+00	4.135583E - 01	1.114364E - 01
5.259396E - 01	1.458932E+01	7.270845E - 01	2.208493E+00	3.734182E - 01	1.126197E - 01
5.433249E - 01	1.406316E+01	7.434132E - 01	2.243971E+00	3.369824E - 01	1.140302E - 01
5.609928E - 01	1.354807E+01	7.574046E - 01	2.279731E+00	3.045382E - 01	1.156667E - 01
5.789435E - 01	1.304436E+01	7.685535E - 01	2.315774E+00	2.758987E - 01	1.175297E - 01

Z=28; minority spin; $V_{interstitial}$ = -0.008751 Ry.

REFERENCES

1. F. Herman and S. Skillman, Atomic Structure Calculations (Prentice-Hall, Englewood, N. J. 1963).
2. W. Kohn and L. J. Sham, Phys. Rev. 140, A1133 (1965).
3. L. Hedin and B. I. Lundqvist, J. Phys. C4, 2064 (1971).
4. J. F. Janak, Phys. Rev. B9, 3985 (1974).
5. P. Hohenberg and W. Kohn, Phys. Rev. 136, B864 (1964).
6. O. Gunnarsson and B. I. Lundqvist, Phys. Rev. B13, 4274 (1976).
7. D. R. Hartree, Proc. Camb. Phil. Soc. 24, 89, 111, 426 (1928); D. R. Hartree, The Calculation of Atomic Structures (Wiley, New York 1957).
8. J. C. Slater, The Self-consistent Field for Molecules and Solids, Vol.4, (McGraw-Hill, New York 1963).
9. K. S. Singwi, A. Sjölander, M. P. Tosi, and R. H. Land, Phys. Rev. B1, 1044 (1970).
10. B. Y. Tong, Phys. Rev. A4, 1375 (1971).
11. O. Gunnarsson, J. Harris, and R. O. Jones, Phys. Rev. B15, 3027 (1977).
12. N. D. Lang, Solid State Phys. 28, 225 (1973).
13. J. F. Janak, V. L. Moruzzi, and A. R. Williams, Phys. Rev. B12, 1257 (1975).
14. V. L. Moruzzi, A. R. Williams, and J. F. Janak, Phys. Rev. B15, 2854 (1977).
15. O. L. Andersen, in Physical Acoustics, edited by W. P. Mason (Academic, New York 1965), Vol.III-B, pp. 77-95; K. A. Gschneider, Jr., Solid State Phys. 16, 276 (1964); American Institute of Physics Handbook, 3rd ed. (McGraw-Hill, New York 1970), Table 9a-2; *ibid.*, Table 9d-3; M. Rosen, Phys. Rev. 165, 351 (1968); G. R. Love, C. C. Koch, H. L. Whaley, and Z. R. McNutt, J. Less-Common Metals 20, 73 (1970); S. B. Palmer and E. W. Lee, Phil. Mag. 24, 311 (1971); and also references 45-47.
16. J. F. Janak and A. R. Williams, Phys. Rev. B14, 4199 (1976).
17. J. B. Mann, Los Alamos Scientific Laboratory Report No. LA-3690 (1967) (unpublished).
18. V. L. Moruzzi, A. R. Williams and J. F. Janak, Phys. Rev. B8, 2546 (1973).
19. J. F. Janak, A. R. Williams and V. L. Moruzzi, Phys. Rev. B11, 1522 (1975).
20. L. J. Sham and W. Kohn, Phys. Rev. 145, 561 (1966).
21. J. Korringa, Physica 13, 392 (1946); W. Kohn and N. Rostoker, Phys. Rev. 94, 1111 (1954).
22. A. R. Williams, J. F. Janak, and V. L. Moruzzi, Phys. Rev. B6, 4509 (1972).
23. D. R. Hartree and W. Hartree, Proc. Roy. Soc. (London) A154, 588 (1936).

24. T. M. Hattox, J. B. Conklin, Jr., J. C. Slater, and S. B. Trickey, J. Phys. Chem. Solids 34, 1627 (1973).
25. O. Gunnarsson, B. I. Lundqvist, and J. W. Wilkins, Phys. Rev. B10, 1319 (1974).
26. U. von Barth and L. Hedin, J. Phys. C5, 1629 (1972); see also A. K. Rajagopal and J. Callaway, Phys. Rev. B7, 1912 (1973).
27. J. C. Slater, J. B. Mann, T. M. Wilson, and J. H. Wood, Phys. Rev. 184, 672 (1969).
28. American Institute of Physics Handbook, 3rd ed. (McGraw-Hill, New York 1970), Table 4e-10 and 9d-3.
29. K. A. Gschneider, Jr., Solid State Phys. 16, 276 (1964).
30. G. Gilat and L. J. Raubenheimer, Phys. Rev. 144, 390 (1966).
31. S. H. Vosko and J. P. Perdew, Can. J. Phys. 53, 1385 (1975).
32. O. Gunnarsson, J. Phys. F6, 587 (1976).
33. J. F. Janak, Phys. Rev. B16, 255 (1977).
34. U. K. Poulsen, J. Kollár, and O. K. Andersen, J. Phys. F6, L241 (1976). See also O. K. Andersen, J. Madsen, U. K. Poulsen, O. Jepsen, and J. Kollár, Physica 86-88B, 249 (1977) (Proceedings ICM 1976).
35. J. Friedel, in Physics of Metals, edited by J. M. Ziman (Cambridge University Press 1969), p. 340.
36. D. G. Pettifor, J. Phys. F4, 613 (1977).
37. C. D. Gelatt, Jr., H. Ehrenreich, and R. E. Watson, Phys. Rev. B15, 1613 (1977).
38. A. R. Miedema, R. Boom, and F. R. de Boer, J. Less-Common Metals 41, 283 (1975).
39. G. D. Gaspari and B. L. Gyorffy, Phys. Rev. Letters 29, 801 (1972).
40. See, for example, L. D. Landau and E. M. Lifshitz, Quantum Mechanics (Pergamon, 1958), p. 235.
41. C. S. Wang and J. Callaway, Phys. Rev. B15, 298 (1977).
42. See, e.g., R. S. Tebble and D. J. Craik, Magnetic Materials (Wiley, New York 1969), and references therein.
43. E. I. Kondorskii and V. L. Sedov, Sov. Phys.-JETP 11, 561 (1960).
44. J. S. Kouvel and C. C. Hartelius, J. Appl. Phys. 35, 940 (1964).
45. J. A. Rayne and B. S. Chandrasekhar, Phys. Rev. 122, 1714 (1961).
46. E. S. Fisher and D. Dever, Trans. Metall. Soc. AIME 239, 48 (1967).
47. G. A. Alers, J. R. Neighbours. and H. Sato, J. Phys. Chem. Solids 13, 40 (1960).
48. M. Shiga and Y. Nakamura, J. Phys. Soc. Japan 26, 24 (1969).
49. V. Heine, Phys. Rev. 153, 673 (1967).
50. M. O. Steinitz, L. H. Schwartz, J. A. Marcus, E. Fawcett, and W. A. Reed, Phys. Rev. Letters 23, 979 (1969).
51. H. A. Mook, J. W. Lynn, and R. M. Nicklow, Phys. Rev. Letters 30, 556 (1973).
52. J. W. Lynn, Phys. Rev. B11, 2624 (1975).
53. J. E. Rowe and J. C. Tracy, Phys. Rev. Letters 27, 799 (1971); D. T. Pierce and W. E. Spicer, Phys. Rev. B7, 1787 (1972).
54. L.-G. Petersson, R. Melander, D. P. Spears, and S. B. M. Hagström, Phys. Rev. B14, 4177 (1976).

55. W. E. Evenson, J. R. Schrieffer, and S. Q. Wang, J. Appl. Phys. 41, 1199 (1970).

56. R. E. Prange and V. Korenman, AIP Conf. Proc. 24, 325 (1974).

57. See the following papers presented at the International Conference on Itinerant Electron Magnetism (Oxford, September 1976): D. M. Edwards, Physica 91B, 3 (1977); J. A. Hertz and M. A. Klenin, Physica 91B, 49 (1977); T. Moriya, Physica 91B, 235 (1977).

58. See, e.g., E. P. Wohlfarth, J. Phys. F 6, L59 (1976).

59. M. Shiga, IEEE Trans. Mag. MAG-8, 666 (1972).

60. See, e.g., G. E. Bacon, Neutron Diffraction, (Oxford, 1975), p. 226.

61. A. M. Portis and R. H. Lindquist, in Magnetism, G. T. Rado and H. Suhl, editors (Academic, 1965), p. 357, and references therein.

62. R. E. Watson and A. J. Freeman, Phys. Rev. 123, 2027 (1961).